Scarcity and Growth Revisited

Natural Resources and the Environment in the New Millennium

R. David Simpson

Michael A. Toman

Robert U. Ayres

EDITORS

RESOURCES FOR THE FUTURE
Washington, DC, USA

An RFF Press book
Published by Resources for the Future
1616 P Street NW
Washington, DC 20036–1400
USA
www.rffpress.org

Library of Congress Cataloging-in-Publication Data
Scarcity and growth revisited : natural resources and the environment in the new millennium /
R. David Simpson, Michael A. Toman, and Robert U. Ayres, editors.
   p.   cm.
  Includes bibliographical references and index.
  ISBN 1-933115-10-6 (hardcover : alk. paper) — ISBN 1-933115-11-4 (pbk. : alk. paper)
  1. Natural resources—Management. 2. Sustainable development. 3. Environmental policy.
I. Simpson, Ralph David. II. Toman, Michael A. III. Ayres, Robert U.
  HC85.S33 2005
  333.7--dc22   2004030206

ISBN 1-933115-10-6 (cloth) ISBN 1-933115-11-4 (paper)

About Resources for the Future *and* RFF Press

RESOURCES FOR THE FUTURE (RFF) improves environmental and natural resource policymaking worldwide through independent social science research of the highest caliber. Founded in 1952, RFF pioneered the application of economics as a tool for developing more effective policy about the use and conservation of natural resources. Its scholars continue to employ social science methods to analyze critical issues concerning pollution control, energy policy, land and water use, hazardous waste, climate change, biodiversity, and the environmental challenges of developing countries.

RFF PRESS supports the mission of RFF by publishing book-length works that present a broad range of approaches to the study of natural resources and the environment. Its authors and editors include RFF staff, researchers from the larger academic and policy communities, and journalists. Audiences for publications by RFF Press include all of the participants in the policymaking process—scholars, the media, advocacy groups, NGOs, professionals in business and government, and the public.

Dedication

THIS BOOK IS DEDICATED to the memory of Jeffrey A. Krautkraemer (1954–2004), one of the contributors to the volume. Jeff was an economics professor at Washington State University, a Gilbert White Fellow at RFF in 1989-1990, and an associate editor of the *Journal of Environmental Economics and Management,* among many other professional activities. He also was a community activist who fought for improved air quality in eastern Washington State. His contributions to the environmental and natural resource economics literature on resource scarcity and on interactions between economic growth and the environment, among other topics, were widely respected and influential in the field. Even more importantly, he was our friend and a friend to pretty much everyone in the economics profession who knew him. He exemplified grace and kindness in the way he lived and worked, and the way he faced his untimely death. He will be keenly missed by the many, many people whose lives he touched and enriched.

R. DAVID SIMPSON, ROBERT U. AYRES, AND MICHAEL A. TOMAN
Editors

Contents

About the Contributors

ROBERT AYRES is institute scholar at the International Institute for Applied Systems Analysis and emeritus professor at INSEAD, where he held the Novartis Chair for Environment and Management. His research interests include industrial ecology, environmental policy and technology evaluation, economic growth, and environmental regulation. His recent books include the *Handbook of Industrial Ecology*. Educated as a theoretical physicist, he was previously professor of engineering and public policy at Carnegie Mellon University.

CHRISTIAN AZAR is professor of energy and environment at Chalmers Institute of Technology, Sweden. His research focuses on climate change mitigation strategies. He is on the editorial board of several international scientific journals and has been a lead author of the Intergovernmental Panel on Climate Change. He is an advisor to the Swedish Minister for the environment and has been an advisor to the EU Commissioner on the environment.

JEROEN C.J.M. VAN DEN BERGH holds appointments in both the Faculty of Economics and Business Administration, and the Institute for Environmental Studies in Free University, Amsterdam. His research covers elements of environmental, spatial, and evolutionary economics. His recent books include *Spatial Ecological-Economic Analysis for Wetland Management*, and *Economics of Industrial Ecology*. He was awarded the Royal/Shell Prize in 2002, and is a member of the Energy Council of the Netherlands.

Sir PARTHA DASGUPTA is Frank Ramsey Professor of Economics at the University of Cambridge, where he is also a fellow of St. John's College. He is a fellow of the British Academy, the Royal Society, a foreign associate of the U.S. National Academy of Sciences, a university fellow at RFF, and a member of the Pontifical

Academy of Social Sciences. He was co-winner of the 2002 Volvo Environment Prize. His most recent book is *Human Well-Being and the Natural Environment.*

JOHN H. DEYOUNG, JR. is chief scientist of the U.S. Geological Survey Minerals Information Team. His research at the USGS has included a broad range of mineral-resource topics, such as the cumulative tonnage-grade distribution of mineral resources and the effect that physical attributes of mineral resources have on metal supply.

SYLVIE FAUCHEUX is president as well as professor of economics at the University of Versailles St-Quentin-en-Yvelines. She formerly directed the Centre d'Economie et d'Ethique pour l'Environnement et le Développement of the university, where she led the establishment of a major interdisciplinary program on environmental, economic, and institutional assessment and management. Faucheux is editor of the *International Journal of Sustainable Development.* Her most recent book is *L'économie face aux changements climatiques.*

JEFFREY A. KRAUTKRAEMER, until his untimely death in December 2004, was professor of economics at Washington State University. Krautkraemer also had been a Gilbert White Fellow at RFF and an associate editor of the *Journal of Environmental Economics and Management.* His primary research focuses were natural resource economics and relationships between economic growth and natural resources, and he published widely on these topics.

RAMÓN LÓPEZ is professor in the Department of Agricultural and Resource Economics at the University of Maryland and senior fellow at the Center for Development Research at the University of Bonn. His recent research studies the links among public policies, social equity, and sustainable development. In 2002 he won the Alexander Von Humboldt Research Prize. Book publications include *Rural Poverty in Latin America* and *The Quality of Growth.*

MOLLY K. MACAULEY is a senior fellow at RFF. Her research includes the economics of regulating new technologies, including commercial space transportation and space remote sensing. She serves on numerous policy committees and is president of the Advisory Board of the Thomas Jefferson Program in Public Policy at the College of William and Mary. She has also served as a visiting professor at Johns Hopkins University.

W. DAVID MENZIE is chief of the International Minerals Section of the U.S. Geological Survey and is an adjunct faculty member in the Environmental Sciences and Public Policy Program at Johns Hopkins University. He has conducted research on a variety of topics in economic geology, including mineral deposit modeling and methods of mineral-resource assessment. His current research interests include the role of minerals in economic development.

DAVID PEARCE is emeritus professor of environmental economics at University College, London. He has recently completed a study of the United Kingdom's climate change levy for the Organisation for Economic Cooperation and Development in Paris, and is now completing a major textbook on the environment and economic development, and a detailed study of the U.K.'s environmental policy. He is currently advising the House of Lords on climate change policy.

JOHN C.V. ("Jack") PEZZEY is senior fellow at the Australian National University's Centre for Resource and Environmental Studies in Canberra. His research is on economy-wide sustainability, including a recent co-edited book, *The Economics of Sustainability*, and articles in *Journal of Environmental Economics and Management* and *Scandinavian Journal of Economics*; and on the economics of pollution taxes and tradable permits, including a recent grant from the Hewlett Foundation.

STEPHEN POLASKY holds the Fesler-Lampert Chair in Ecological/Environmental Economics at the University of Minnesota. He served as senior staff economist for environment and resources for the President's Council of Economic Advisers and as associate editor for the *Journal of Environmental Economics and Management*. He is currently serving on the Science Advisory Board of the U.S. Environmental Protection Agency (EPA). His research interests include biodiversity conservation and integrating ecological and economic analysis.

R. DAVID SIMPSON is with the National Center for Environmental Economics of the U.S. Environmental Protection Agency (EPA). He was earlier a senior fellow at RFF. Simpson has published numerous articles on industrial economics, technological innovation, biodiversity, and conservation. His earlier edited volumes include *Productivity in the Natural Resource Industries*.

DONALD A. SINGER is a senior research geologist with the U.S. Geological Survey. His research has focused on developing methods of quantitatively assessing undiscovered mineral resources. His published papers cover resource assessments, deposit models, quantitative methods, and exploration strategies. The Society of Economic Geologists awarded him its 1999 Silver Medal, and the U.S. Department of the Interior awarded him its Distinguished Service Award in 2005.

SJAK A. SMULDERS is associate professor of economics at Tilburg University. His research has focused on economic growth and resource theory, and on induced technological change and environmental policy in particular. He is an associate editor of *Environmental and Resource Economics*. His publications have appeared in *Economic Journal*, *Journal of Public Economics*, *International Economic Review*, and *Journal of Environmental Economics and Management*.

DAVID TILMAN is Regents Professor and McKnight University Presidential Chair in Ecology at the University of Minnesota. He is a member of the National Academy of Sciences and the American Academy of Arts and Sciences. His interests include biodiversity, the controls of ecosystem composition, stability and productivity, and the societal implications of human impacts on global ecosystems. His books include *Functional Consequences of Biodiversity: Empirical Progress and Theoretical Extensions*.

MICHAEL A. TOMAN was formerly a senior fellow at RFF. He is also an adjunct faculty member of the Nitze School of International Studies, Johns Hopkins University, and the Bren School of the Environment, University of California, Santa Barbara. His current activities include work on economic development, energy, and climate change. Recent publications include *India and Global Climate Change, The Economics of Sustainability* and *Climate Change Economics and Policy: An RFF Anthology*.

Scarcity and Growth Revisited

CHAPTER I

Introduction

The "New Scarcity"

R. David Simpson, Michael A. Toman,
and Robert U. Ayres

O NE OF THE MOST famous and influential books ever published on resources and the human prospect appeared in 1963. In *Scarcity and Growth*, Howard J. Barnett and Chandler Morse interpreted the extensive data assembled by their colleagues Neal Potter and T. Francis Christy Jr. in another seminal work, *Trends in Natural Resource Commodities* (1962). From those data Barnett and Morse made a compelling case that resource scarcity did not yet, probably would not soon, and conceivably might not ever, halt economic growth.

The interplay between scarcity and growth is an issue of perennial concern, however. Only a decade after Barnett and Morse published their work, pundits, politicians, and activists announced the arrival of an "energy crisis." Consumers accustomed to decades of the declining resource prices Barnett and Morse had documented found themselves waiting in long lines and paying skyrocketing prices to purchase gasoline. Academic researchers dusted off the writings of scholars such as Harold Hotelling (1931) and M. King Hubbert (1949), looking for insights into the causes and implications of rediscovered scarcities.

At roughly the same time, other types of scarcity were being recognized. For much of human history, sky, water, and land were employed for waste disposal with little thought about the consequences. When people were few and unspoiled territory plentiful, the consequences of waste were manageable. As these circumstances changed, the consequences of pollution mounted. A human population that stood at less than a billion in 1800 had climbed to 2.5 billion by 1950. By the 1970s some commentators were making apocalyptic projections for continued growth to be followed shortly by catastrophic decline (Ehrlich 1968). Humanity was finding itself increasingly living cheek-by-jowl with its own refuse.

That refuse was a more potent witch's brew than our pre-industrial ancestors were capable of producing. The concentrated wastes of humans and animals have always been a breeding ground for disease, and packing ever-larger populations into cities compounded these risks. With the industrial revolution, however, new poisons came into broader circulation. Smoke from her "dark Satanic mills" stained "England's green and pleasant land." In the early 1950s, thousands of Londoners died in the "killer fog" of coal smoke. Another fossil fuel, petroleum, caused further problems. From Los Angeles to Athens to Tokyo, cities were increasingly smothered by the exhaust of their vehicles. Paper, metalworking, chemical, and other industries fouled air, water, and land with toxic cocktails.

So, in addition to renewed concern regarding the scarcity of energy resources, the early 1970s also saw some of the first broad manifestations of concern with what we will call in this volume the "New Scarcity"—the limitations on the environment's capacity to absorb and neutralize the unprecedented waste streams humanity looses on it. From celebration of the first Earth Day on April 22, 1970, to the first United Nations Conference on the Environment in Stockholm in 1972, to the enactment of broad-reaching Environmental Protection Acts in the United States and other nations, citizens expressed their concerns over environmental degradation, and governments responded.

Scholars also weighed in. The Club of Rome's controversial 1972 volume *Limits to Growth* predicted that such limits were fast approaching, and global society ignored them at our collective peril (Meadows et al. 1972). Others were quick to fault the analysis in *Limits to Growth* (e.g., Nordhaus 1974). New scholarship appeared on the economics and management of natural resources (e.g., Dasgupta and Heal 1974, 1979; Solow 1974; Stiglitz 1974; Clark 1976).

In the fall of 1976, Resources for the Future held a conference to again investigate the topic of *Scarcity and Growth*. In many respects *Scarcity and Growth Reconsidered* (Smith 1979), the volume of collected papers and commentaries from that conference, echoes the optimism of its predecessor. That optimism had been buffeted somewhat by the events of the 1970s, however. Moreover, as one might expect from a volume combining the contributions of a dozen different authors, *Scarcity and Growth Reconsidered* had a less synoptic perspective than its predecessor. Not only did some contributors raise doubts about whether Barnett and Morse's relatively rosy perspective was justified, several also raised concerns regarding the limitations of received economic theory as a tool for analyzing scarcity and growth. And, it must be said, *Scarcity and Growth Reconsidered* evidenced some groping on the part of the practitioners and innovators of "received theory" to determine what their models could, in fact, say on the topic. Several of its chapters focused on determining what it was that available data could tell us about scarcity and growth, rather than on making pronouncements regarding long-term prospects from such data.

More than a quarter of a century has passed since the chapters constituting *Scarcity and Growth Reconsidered* were collected. New modes of analysis have been developed in many disciplines. New empirical understandings have come to the fore. New questions have arisen. Even if we are no closer to a final resolution of questions of scarcity and growth, we are in a position both to bring new tools and facts to bear on them and to consider nuances that must now be addressed in answering them.

To investigate these issues, Resources for the Future, with generous assistance from the Vera I. Heinz Endowment, the Netherlands Ministry of the Environment, and the European Commission, again assembled a panel of distinguished economists, natural scientists, and others to discuss scarcity and growth. On 18 and 19 November 2002, authors presented their draft papers and discussed them with other participants.[1] Following these discussions, the authors revised their contributions extensively.

Before reviewing these contributions, however, let us briefly discuss further the background and motivation for this undertaking.

Scarcity and Growth: The Long View

Before detailing our reasons for again revisiting a classic, we may first take inspiration from it and give the reader a broad overview of the issues, both historical and current, motivating continuing interest in scarcity and growth. Barnett and Morse displayed a genuine erudition that is sadly absent from much of what the economics profession produces today. Many readers remember *Scarcity and Growth* for its substantial quantitative and diagrammatic analyses. However, its first 100-odd pages are dedicated to an extensive, authoritative, fascinating, and in contrast to much that has been written on the subject in the years since, refreshingly jargon- and mathematics-free account of what great thinkers of earlier eras and many disciplines had to say about scarcity and its implications for continuing human well-being. Thus, when Barnett and Morse proceeded to present their quantitative analysis in the last two-thirds of *Scarcity and Growth*, the reader had been afforded a very complete introduction to the topic.

We cannot duplicate Barnett and Morse's introduction. We commend it to the reader for its careful scholarship and as an introduction to the perennial questions of *Scarcity and Growth* that remain as relevant as they were in the year their book was written. We will also, however, in the space of the few pages available now for the purpose, attempt to radically condense and to slightly update their introduction for readers of this volume.[2]

Perhaps the most important point the original *Scarcity and Growth* established is that world views have changed remarkably in the last two centuries, particularly among economists. Thomas Carlyle bestowed the sobriquet "the dismal science" on economics in response to the writings of classical economists.[3] It is ironic that some two hundred years later, economists have come to be seen by some as Panglossian apologists for "business as usual" scenarios in which the invisible hand of the market will solve all problems: in fact, some early economists argued that our problems are insoluble.

Of course, there have always been optimists as well as pessimists. Quintus Tertillianus, writing in 200 A.D., was surely not the first to suppose that the world was in decline. "We are burdensome to the world," he wrote. "The resources are scarcely adequate to us.… Truly, pestilence and hunger and war and flood must be considered as a remedy for nations, like a pruning back of the human race" (quoted in Johnson 2000).[4]

Others, of course, had rosier perspectives, and such perspectives could claim an even more ancient pedigree. According to the Old Testament, God is supposed to have been a "resource optimist":

Let us make man in our image, after our likeness: and let them have dominion over the fish of the sea, and over the fowl of the air, and over the cattle, and over all the earth....

And God blessed them, and God said unto them, Be fruitful, and multiply, and replenish the earth, and subdue it: and have dominion over the fish of the sea, and over the fowl of the air, and over every living thing that moveth upon the earth. (Genesis 1:26, 28)

God promised Abraham, "I will make thee exceeding fruitful, and I will make nations of thee" (Genesis 17:6). The Almighty did not seem worried about the limitations of the earth's resources for Abraham's descendents, as God also said, "I will multiply thy seed as the stars of the heaven, and as the sand which is upon the sea shore" (Genesis 22:17).

Yet a very different view of the human prospect dominated many of the writings of classical economists. In *An Essay on the Principle of Population* (1798), the British economist and cleric Thomas Malthus advanced his well-known theory that population tends to increase geometrically, while food production can be expected to grow at best arithmetically. As a consequence, humanity tends inexorably to approach what came to be dubbed almost two centuries later "the limits to growth," often with catastrophic consequences.

The early economists were a dismal bunch.[5] Even Adam Smith (1776), the founder of the modern discipline, opined, "Every species of animals naturally multiplies to the means of their subsistence, and no species can ever multiply beyond it." Malthus contemporary David Ricardo is remembered for his theory of rent: Resources are in short supply. Those lands favored by location or other attributes command high prices ("rents") and are quickly appropriated and exploited. This leaves latecomers to the economic scene with meager pickings from which to choose. As Robert Heilbroner remarked, "Anyone who was not sufficiently depressed by Malthus had only to turn to David Ricardo" (1967, 86).

Ironically, one of the fundamental insights of economic analysis led the classical economists to their misguided, or to give them the greatest benefit of the doubt and to suppose with today's pessimists that the case is not yet closed, *premature* "dismal" pronouncement. The principle of diminishing returns explains a tremendous amount in economics. The more there is of something, the less productive is still more of it. If natural resources are limited, we can expand the human workforce and manufactured capital employed in combination with them. Without more resources, though, our gains from employing more of other factors of production will be progressively smaller. It is the principle of diminishing returns that underlies the use of marginal analysis in economics, and it is marginal analysis that resolves the greatest puzzle in economics: the paradox of value. Why are some things, like water, both so useful and so cheap relative to things like diamonds, which are neither? Because of diminishing returns. There is so much water relative to diamonds that a little more water is of little incremental, or marginal, value.

The principle of diminishing returns is, then, central in economics. Yet its straightforward implication is that economies relying on fixed stocks of land and other resources are, at best, destined for stagnation. At worst, population growth in combination with profligate resource use and a lack of foresight spell doom for the majority of mankind.

Many once believed this. The great figures of early economics were often inaccurate prognosticators. William Stanley Jevons, who first formalized marginal analysis, is famous for his predictions concerning the calamities that awaited when coal was exhausted. His dire predictions now have a faintly comic tone: Jevons stockpiled reams of writing paper in anticipation of its eventual shortage.

The following might have been written by a modern conservationist:

> The world is really a very small place, and there is not room in it for the opening up of rich new resources during many decades at as rapid a rate as has prevailed during the last three or four. When new countries begin to need most of their own food and other raw produce, improvements in transport will count for little.

This quote is not taken from any consensus report of the recently concluded World Summit on Sustainable Development. It was, rather, the opinion of Alfred Marshall (1907), among the most important economists working at the turn of the twentieth century.

Marshall did, however, also point the way out of the resource-constrained cul-de-sac. He went on to write that "the Law of Diminishing Returns can be opposed only by further improvements in production; and improvements in production must themselves gradually show a diminishing return." Marshall was adopting the wisdom expounded by John Stuart Mill, the greatest economist between his generation and that of Malthus and Ricardo. Mill (1848) amended Smith, Malthus, and Ricardo by noting that the law of diminishing returns might "be suspended, or temporarily controlled, by whatever adds to the general power of mankind over nature; and especially by any extension of their knowledge."

Subsequent generations have yet to determine whether the implications of diminishing returns have, in fact, been "suspended" or merely "temporarily controlled" by the "extension of their knowledge." As economics evolved, however, statistical evidence was marshaled that seemed to show that knowledge was accumulating at an impressive rate (we will discuss a little later whether the evidence *does* in fact show this). Robert Solow, one of the leading economists of the post-World War II generation, is associated with the "residual" to which his name is often attached. This residual is the share of output growth that cannot be explained by an increase in the use of inputs.

The gross domestic product of the United States grew at an average annual rate of some 3.4 percent between 1950 and 2000. We can decompose the growth of output from one year to the next as follows. It is the rate at which output grows in response to increases in labor hours times the change in labor hours, plus the rate at which output changes with changes in equipment used times the change in equipment used, etc., for all inputs. Economic theory shows that the rate at which output changes with changes in the quantity of an input is proportional to the price of that input. We can, then, decompose changes in output into constituent elements: the change in output due to change in labor plus the change due to change in capital equipment and so on, weighting each by its price. Solow noted that these changes fall short of adding up. There is a residual left over—a difference between observed rates of overall growth and measurable changes due to changes in input use. In the decomposition we have just described, there is a missing term for the rate at which output changes with changes in unmeasured inputs times the proportional change in unmeasured inputs. This residual, commonly called multifactor productivity growth, averaged about 1.2 percent per year between 1950 and 2000.

The reason the residual is missing from the calculation we have described above is, of course, that it cannot be measured. In Moses Abramowitz's (1956) memorable phrase, the residual is "a measure of our ignorance." It is, by definition, what is left over when the effects of all measurable explanatory variables have been calculated. Despite this fundamental uncertainty, the productivity residual is often interpreted as the effect of technological progress. To foreshadow an issue we will discuss below, however, we should note other possibilities. The "missing input" the residual represents could also be an increase in the unmeasured consumption of resources that are not traded in markets, a category in which we might also include degradation of the environment.

What does productivity growth mean? If we take the rosy view that it can be extrapolated into the indefinite future, it means that we have little to worry about. Recent work by Martin Weitzman suggests that, if we assume that productivity growth will continue at its historical level *ad infinitum*, conventionally measured national income *understates* true welfare by as much as 40 percent (Weitzman 1997). If we take the view that the residual reflects profligate waste of unmeasured resources, productivity growth would represent an ominously mounting account payable.[6]

The former view has dominated the economics profession in recent decades. The "dismal science" has reversed direction since the dire predictions of Malthus and many who followed him through at least the early years of the last century. Diminishing returns remains a fundamental element of the economics canon, but many believe that its long-run implications have been more than offset by an equally fundamental element. Majority opinion is that even in relatively short periods—years, even months—substitution possibilities obviate resource scarcity. Diminishing returns still provide reasonably robust explanations of many microeconomic phenomena (e.g., why consumers buy less when prices go up). Over any appreciable sweep of time, however, the implications of the scarcity of any particular resource are obviated by the abundance of potential substitutes, an abundance that grows as technology evolves.

In writing the original *Scarcity and Growth,* Barnett and Morse noted that among the classical economists "the doctrine of diminishing returns per capita became embedded in economic theory as a self-evident fact, requiring neither precise formulation nor analytical investigation" (1963, 51). The tables may have turned in the modern era. Some regard the fact that dire predictions have not come true as almost a proof by induction that they never will.

One of the reasons why the classical economists' emphasis on diminishing returns may have been misplaced is found in the passage quoted from Alfred Marshall above. He asserted that "improvements in production must themselves gradually show a diminishing return." On this point he is explicitly contradicted by J. M. Clark, whose remark is more often quoted: "Knowledge is the only instrument of production not subject to diminishing returns" (Clark 1923).

It is difficult to verify Clark's assertion in the absence of an operational measure of a notion as intangible as "knowledge."[7] Knowledge is, however, different than most other economic goods in one important respect. It is *nonrival.* There is no physical reason why my possession of certain knowledge precludes your use of the same knowledge (though there may be *legal* reasons: intellectual property law governs patents, copyrights, and the like). In this respect, there is an interesting parallel between knowledge—often touted as the ultimate solution to problems of scarcity—and the New Scarcity that provides much of our motivation for again revisiting *Scarcity and Growth.* Environmental pollution and the depletion of global ecological assets are also nonrival. The same smoke that makes my eyes water and my lungs burn will have a similar effect on you, and the accumulation of atmospheric carbon dioxide and diminution of global biodiversity may affect people all over the world.

Markets do not allocate nonrival goods efficiently. Too little knowledge is likely to be produced, as innovators often generate "spillovers" that others can appropriate and benefit from.[8] Too much pollution is likely to be produced as polluters generate wastes that "spillover" into the public domain. One might say, then, that the ultimate resolution of scarcity and growth depends on the resolution of two policy issues. First, can we as a global society do enough to restrict the negative spillovers we impose on ourselves through pollution, and second, can we do enough to promote innovation that generates positive spillovers through the augmentation of knowledge?

Scarcity vs. Growth Today

There are now—as there no doubt always will be—a range of opinions concerning the adequacy of resources for a growing population seeking a higher standard of living. A generation ago, *Silent Spring* (Carson 1964), *The Population Bomb* (Ehrlich 1968), and *Limits to Growth* (Meadows et al. 1972) announced the dangers of pesticides, overpopulation, and resource depletion, respectively. Such works have been succeeded by tomes detailing evidence of climate change (IPCC 2001) and biodiversity loss (Wilson 1992). And just as the late Julian Simon (1981) disputed the pessimism of earlier writers, commentators such as Gregg Easterbrook in *A Moment on the Earth* (1995) and Bjorn Lombourg in *The Skeptical Environmentalist* (2001) have argued that things are not so bad and are arguably getting better.

If there is a trend to be noted in the debate between optimists and pessimists, it is not so much that one or the other is prevailing as that the terms of the discourse are deteriorating. Still, for all the vitriol loosed in recent debates among the optimists and pessimists, it is difficult to say from a review of the evidence whether our generation should be more or less concerned than our predecessors over the state of the planet we have inherited from our parents and will pass along to our children.

It is a classic case of a glass that might be seen as either half full or half empty. Let us begin by reporting economic statistics. It has been estimated that the aggregate value of global economic production was something less than $700 billion in 1800. It had increased to about $2.5 trillion a century later, $5.3 trillion by 1950, and was nearing $35 trillion at the turn of the new century (Maddison 2002).

One would, of course, want to know something about how these figures translate into per capita terms before making any judgments concerning what they say about the human prospect. World population has also increased at a dizzying pace. It was not until the early nineteenth century that the population reached one billion, and not until after the First World War, nearly 120 years later, that it passed 2 billion. Since then, during the space of a single unexceptionally long lifetime, world population has tripled.

Economic performance has, however, kept pace with this explosive growth and has itself increased at an increasing rate. On average, each of the billion people on the planet two centuries ago got by on an income that, estimated as best we now can, totaled between $600 and $700 per year. By the turn of the last century this had increased to well more than $1,000 a year. It was more than $2,000 per year in 1950 and now stands at nearly $6,000 per year.

It is, of course, extraordinarily difficult to translate into contemporary dollars the income of someone who lived without electric lights, motorized transportation, even the most rudimentary of medical treatment, or any of the host of other conveniences many of us have come to regard as necessities. A measure of per capita income is intended to convey some notion of the standard of living a person can afford, and it is difficult to make a comparison between such different consumption possibilities as those our ancestors faced and our own.

Several authors have suggested a compelling thought experiment, however. It demonstrates that the pace of economic development has (*on average*, it is important to point out) accelerated markedly in recent decades. While many of the same problems of measurement bedevil comparisons over shorter periods, we might have more confidence in saying that average per capita income has roughly tripled in the last 50 years. Extrapolate that rate of growth to the time of Christ, and one arrives at a per capita income of far less than a penny (15 zeroes before the first significant digit). Even allowing for huge problems of measurement, a human being simply could not survive on so little. The fact that humans did survive the poverty of our race's first thousand millennia establishes that the pace of economic progress has increased dramatically.

Let us return to the population growth figures for a moment. Extrapolation of growth at the rates of the last half-century quickly leads to surreal results. If population were to continue growing at 1.5 percent per year for another two millennia, the human population would weigh more than does the earth itself!

This would, of course, be a physical impossibility. Demographers and others have long recognized that continuing faster-than-exponential growth would sooner or later—but probably "sooner"—lead to a catastrophic collapse. There can be no serious dispute of this conclusion. What reasonable people may disagree upon, however, are the point at which collapse is likely to occur and the extent to which humanity is likely to regulate its own expansion. Perhaps the only answer to the first question is "nobody knows for sure," although many commentators seem to concur that the answer is probably "more than there are now." As Joel Cohen, who wrote the definitive tract *How Many People Can the Earth Support* (1996), notes, the issue may not be how many people can the earth support, so much as what kind of earth is consistent with a human population of any given size?

There also seems to be emerging evidence that the rate of population growth is slowing. While population rose over the last decade at roughly the same average annual rate of 1.5 percent per year, the United Nations *World Population Prospects* (UNPD 2002) predicts a population of some 8.9 billion in 2050. Although this increase of almost three billion people over the current total represents more *additional* people than the *total* number who were alive in 1950, it also projects a significant reduction in growth rates: it implies an average annual rate of growth over the next half-century of less than half of the current growth rate. This average annual growth rate overstates the long-term trend, as most experts agree that population growth will slow to a halt at some point in the next century or two (if, that is, numbers do not first decline).

The prediction of a stable (as opposed to a collapsing) population reflects some optimism concerning the ingenuity or self-discipline of humanity. This optimism arises in large part from the experience of the world's wealthier nations. There is a pronounced, if imperfect, correlation between nations' wealth and their population growth rates. Poor nations tend to have high rates of population growth—despite often appalling rates of infant and overall mortality. In contrast, some rich nations are now facing potential fiscal crises as their aging populations reach retirement age with few young workers to take their place. Immigration will surely be required if the aging populations of today's wealthy nations are to retain their standard of living.

So, taken in aggregate, world economic performance has increased, this perception of improvement survives translation into per capita figures, and there is reason for hope that we will not overwhelm the planet with our sheer numbers. Even these observations do not point to entirely rosy prospects, however. As we have just noted, population growth rates tend to be higher among the poorer nations, and the plight of the poor remains heartbreaking. Average figures mask the fact that the distribution of income remains highly skewed. Well more than a billion people now eke out the best living they can on an income of about a dollar a day. Recall that this is about half of world per capita income in 1800.

The plight of the poor raises two disquieting worries: either the poor may become wealthy, or they may not. While it may seem a wonderful thing for the world's poor to become wealthy, the prospect brings the issues of scarcity and growth into stark relief. A billion or so wealthy people are now exposing the earth to unprecedented environmental threats. Can the planet tolerate *eight* billion people living the lifestyle of the wealthy?

Or, as no less a humanitarian than Jesus is reported to have said, "Ye have the poor with you always" (Mark 3:8). Millennia ago, a few thousand wealthy nobles enjoyed greater material wealth than did a few million struggling peasants. Now a billion-odd denizens of the wealthy industrial countries enjoy far greater health, wealth, and prospects than do the billions who live hand-to-mouth. Philosophers might debate whether current inequities are more troubling than historical ones, but no one would suggest either is ideal. It is troubling that there is little compelling evidence that the world's poor are catching up to the wealthy (Pritchett 1997). Would we be satisfied living in a "sustainable" world if most people were frozen in a status in which they struggle to achieve subsistence?[9]

Such questions beg the larger question of what we mean by "economic growth." If economic growth were *necessarily* linked with increases in the physical insults we impose on the earth, it seems clear that indefinite growth would not be desirable or even possible. If economic growth among the poor meant that tomorrow's poor would exactly replicate the production processes and consumption possibilities of today's wealthy, we might also conclude that such growth is either illusory or a contradiction in terms.

Yet economic growth has, in fact, been as much a qualitative as a quantitative phenomenon. The optimistic view of the improvements in productivity we have discussed above mean essentially that every year we acquire the ability to make greater physical quantities of outputs using smaller physical quantities of inputs. Moreover, every year we acquire the ability to make *different* outputs. Automakers may have made different choices than environmental advocates might have wanted them to as they refined their designs, but cars today are, in general, lighter, stronger, faster, safer, and more efficient than they were 50 or 75 years ago. They are qualitatively different, and certainly if we impose the standard of value per dollar spent or per hour worked to acquire, qualitatively better than their predecessors. The same may be said of airplanes, communications, fabrics, electric lights, office equipment, and virtually any other purchased good or service we might name.

A problem is captured in the limitation "any other *purchased* good or service," however. Any number of *unpurchased* goods and services deteriorated as incomes rose, populations increased, and the activities captured in the measured economy increased. Progress has undoubtedly been made in controlling the production and release of many pollutants. Researchers have identified an "Environmental Kuznets Curve," under which countries' emissions of some pollutants first increase, then decrease, as their per capita wealth increases (Grossman and Krueger 1995). One might hope that airborne particulates, organic wastes in water, and other "common" pollutants would be controlled as incomes increase.

This may be too optimistic a prognosis in several respects, however. First, there may be a "composition effect" in the income–pollution relationship. Wealthier societies often concentrate their own economic activity in high-technology and service industries, relegating—or perhaps even "exporting"—dirtier industries to less developed countries. Moreover, such composition effects may be more powerful in explaining the observed pattern of industries and pollution across countries at different income levels than they will prove to be in explaining aggregate global emissions over time. It seems unlikely that *every* country will devote its economy to low-emissions manufacturing and services: someone will need to specialize in the industries in which pollution is more difficult to control.

Second, not all pollutants can be expected to follow the "inverted-U" relationship of the Environmental Kuznets Curve. It is not clear exactly how dependent modern industrial economies are on fossil fuels, but it is clear that wealthier countries generally consume more of them. Such fuels are the primary contributors to greenhouse gas emissions, which are, in turn, the driver of global climate change. It seems unduly optimistic to suppose that the world as a whole will simply grow its way out of concern with climate change.

Third, greenhouse gases are an example of a process that cannot be quickly reversed. They are cumulative pollutants. It has been estimated that carbon dioxide

(CO_2)—the most important of the greenhouse gases) concentration in the atmosphere was more or less stable around 280 parts per million (ppm) for thousands of years.[10] It began to increase following the industrial revolution, reaching 315 ppm in 1957 and some 362 ppm now (Vitousek et al. 1997). Even if all further CO_2 gas emissions were to stop tomorrow, it would take quite some while for the atmospheric concentration to return to pre-industrial levels.

Changes in atmospheric chemistry may be slow to reverse. Losses in biological diversity are, on any reasonable scale, completely irreversible; after each of the mass extinctions revealed in the geological record, the time to restore diversity measured in the millions of years (Wilson 1992). It will make little difference to the preservation of biodiversity, then, if in 10, 100, or 1,000 years' time human societies have found ways to reduce the chemical pollution, overharvesting, transport of exotic pests, predators, competitors, and disease, climate change, and most importantly, destruction of indigenous habitats that now threaten the species with which we share the planet.

No one is certain how many species now live on earth. Estimates range from a few million to many tens of millions. New species come into being through a combination of geographical separation and genetic mutation. Existing species are extinguished when they cannot adapt to their habitat or compete with existing or introduced rivals. The geological record suggests a "background level" of species extinctions of about one species per million per year. Some biologists suggest that current rates of species extinction exceed this background rate by a factor of 1,000 (Raven 2002).

These digressions concerning climate change and biodiversity underscore important aspects of the New Scarcity. Both climate and biodiversity provide global public goods. Everyone on earth is affected by changes in climate. Everyone on earth may be affected by changes in biological diversity and the ecological services that diverse natural ecosystems provide. As we will detail, the global scale of the problems and their duration create tremendous challenges for policy. In the original *Scarcity and Growth* and in *Scarcity and Growth Reconsidered,* the general sense conveyed was that the mechanisms of markets were adequate to the task of allocating resources in an efficient and sustainable way. Most of the attention was paid to resources traded in markets, however: to the "Old Scarcity." In this volume the focus has largely shifted to the adequacy of our social mechanisms for coping with the New Scarcity.

Let us now turn to the ways in which our authors have addressed these matters.

The Essays

David Menzie, Donald Singer, and John DeYoung Jr. survey a subject that was of tremendous importance in *Scarcity and Growth*: the physical availability of resources. These authors reach some conclusions that echo those of earlier investigators. They conclude that the supply of minerals per se is not a limit to growth. Yet causes for concern persist. While abundant quantities of many resources remain, they are becoming progressively more remote. It is only natural that mineral (and fuel and biological) resource stocks were most intensively exploited first in the areas closest to where they were used. As demands increased, exploration and eventually extraction took place across oceans, in inhospitable climates, and for minerals and fossil fuels especially, deeper and deeper beneath land and water.

This renews the classic tension between the depletion of (easily accessible, at least) resources on one hand and the reduction of exploration and extraction costs by technological progress on the other. Per capita consumption of many minerals has remained relatively steady in the developed nations, despite talk of the "dematerialization" of economies dominated by high-technology and service industries. As less developed countries become wealthier, their consumption may be poised to increase to levels comparable to those of the industrialized nations. An optimistic view is that resource scarcity has yet to decrease use in an ever wealthier and more populous world. In fact, an optimist might be further buoyed by the observation that wealthy countries have not even been compelled to take recourse to substitutes for common mineral commodities (of which optimists would suppose there to be many). A more pessimistic view is that we simply do not have adequate reserves to suppose that ever growing numbers of people can *all* consume at the level to which the wealthy have grown accustomed.

Recent trends in exploration effort could also motivate similarly dichotomous views. Menzie, Singer, and De Young note with some concern declines in mineral exploration budgets, research expenditures, and training. Here a pessimistic view would be that this lack of preparation could reflect, at best, a lack of foresight and, at worst, a recognition that such effort would be futile given the existing state of depletion. Yet one might as easily take the contrary view that the reason for a decline in exploration is the recognition that known stocks will be adequate, and new ones can be identified as needed.

One thing that does come through clearly, however, is that society is increasingly recognizing the nonmarket costs of mineral extraction. More and more land is being placed off-limits to such activities, either in appreciation of the unique services provided by pristine landscapes or in recognition of the fact that mining competes with residential uses with which it is incompatible. Moreover, similar concerns arise with respect to the residuals arising from mining: society is less willing to tolerate the pollution and degradation that unrestricted extraction can impose.

Here again one might take either an optimistic or a pessimistic view. The pessimist may see the increasing environmental restrictions placed on mining as further evidence of the constraints resource scarcity places on us: both minerals and pristine ecosystems are becoming scarce, and scarcity of one imposes further pressure upon the other. A world confronted with such interlocking constraints simply cannot afford to continue in its ways. The optimist could reply that the fact that the world has, in fact, continued in its ways indicates that we have the means to continue our material consumption while preserving our environment.

Optimists of a different stripe might take another tack, however. Whether or not the preservation of natural landscapes proves to be compatible with continuing physical consumption or spiritual enjoyment, the fact is that something has been done to preserve natural landscapes. Voters have voted, legislators have legislated and, as Menzie, Singer, and De Young note, nongovernmental environmental interest groups have taken action to restrict environmental harms. Reasonable people can differ as to whether such developments represent overreactions or an instance of too little too late, but they do suggest that there is a marketplace of ideas and policies in which the New Scarcity has begun to be discussed and addressed.

The late Jeffrey Krautkraemer surveys some of the same territory as Menzie, Singer, and De Young in "Economics of Scarcity: State of the Debate." He also marshals extensive economic, as well as physical, data on resources. By updating many of the figures presented by Barnett and Morse, he generally confirms their findings after a span of some 40 years. The real prices of *most* commodities have declined since 1960.

Krautkraemer's figures highlight some other important developments, however. Researchers could not have claimed that resource prices were generally falling in 1980. Many mineral and food prices peaked at approximately the same time as those of fossil fuels. While different commentators express different opinions, it again seems that the data may support conflicting conclusions. On one hand, one might assert that the 1970s and early 1980s were an historical aberration in the long-run trend of declining real resources prices. On the other, one might say that the last two decades have been the aberration, a temporary—and perhaps final?—respite from inexorable scarcity.

The pessimism of the latter conclusion might be refuted by reference to futures markets and the prices of underlying assets. If it is generally felt that resource prices will increase over time, rational investors will bid up the current prices of, for example, oil leases and mining sites. While confidence in the rationality of investors is an article of faith among many economists, observers from other disciplines draw different conclusions. The case for the latter has surely been aided by the economy's recent experience with the "irrational exuberance" that appears to have characterized other asset markets.

The evidence from existing markets for assets that have entered the market economy is, then, generally positive but admittedly mixed and subject to the skepticism of those who are reluctant to accept the economists' common postulate that markets reflect the foresight of rational participants. Krautkraemer also surveys some physical evidence concerning the state of resources that are not traded in markets. There is greater cause for concern over the status of air, water, climate, and biological diversity. Again, however, the question remains open as to whether emerging institutions and instruments for the allocation and preservation of these nonmarket resources will be adequate to the task of securing a sustainable future in the face of the new scarcity.

This question is taken up in greater detail by David Tilman and Stephen Polasky. "Ecosystem Goods and Services and Their Limits: The Roles of Biological Diversity and Management Practices" details concerns with what may be one of the most profound but least understood changes humanity is inflicting on the planet. Paleobiologists have identified five episodes of mass extinction in the geological record. Each was likely caused by volcanic or astronomical cataclysm. The most recent occurred some 65 million years ago, when an asteroid impact plunged the earth into darkness and exterminated the dinosaurs, among many other less well known forms of life.

Many natural scientists now warn that we are entering a "sixth extinction crisis." This, in contrast to the first five, is believed to be caused by the increasing dominance of a single species: *Homo sapiens*. The numbers we cite above concerning the growth of human population and our economies may have a darker side—our impact on the other forms of life with which we share the planet. An often-cited article by Peter Vitousek and his colleagues (1997) has estimated that humanity appropriates either directly or indirectly 40 percent of the world's net primary productivity.[11] Other indices of our impact are equally alarming: we note earlier the accelerated extinction of species.

Biodiversity loss stems from many causes. In addition to natural threats from competition and predation, a number of human-induced factors threaten other species. Some have been hunted to extinction (passenger pigeons, dodos), some nearly so (blue whales, otters). Others, particularly on islands, have been out-competed or preyed upon by exotic species introduced by human travelers. Chemical pollution threatens some species. The effects of climate change on biodiversity are difficult to predict, but potentially profound. Many biologists believe, however, that the greatest threat to biodiversity today comes from the conversion of natural habitats to alternative human use. The native forests, prairies, and wetlands of the world are being felled, plowed, and drained for factories, homes, and farms. When natural habitats disappear, the organisms that depend upon them do also.

What does this "cost" us? To many people, either such a question will seem crass ("How can we put a price on life!") or its answer obvious ("It is costing us our soul!"). There are more pragmatic answers to the question, however. While economists and ecologists have not yet been able to provide any concise dollars-and-cents answers, Tilman and Polasky have been at the forefront of efforts to provide quantitative estimates of the effects of biodiversity loss.

Less diverse habitats are demonstrably less "productive" by various measures than both ecologists and economists might think important. Tilman and Polasky report the results of a number of experiments in which less diverse habitats have been shown to produce less biomass, leach more nutrients, and otherwise perform worse than do more diverse natural assemblages. The authors also present some simple conceptual models explaining why such effects might result.

The observation that more diverse habitats are in some sense "better" than those converted to other human uses begs some questions, of course. Obviously, if everyone thought natural landscapes superior to simplified and managed landscapes everywhere, no natural landscapes would be converted to alternative uses. The problem, then, arises if *too many* natural landscapes are converted or used too intensively. That is, there is a problem if those who convert natural landscapes to artificial ones for their own benefit impose additional costs and burdens on others.

This, however, is precisely what the problem with biodiversity loss may be. The people who clear the land that shelters biodiversity derive private benefits from doing so, but most of the costs of biodiversity loss may be "spillovers" accruing to other people at other locations. This would be the case with "ecosystem services," such as water purification, habitat for organisms that control pests and pollinate crops, prevention of erosion, and a host of others.

More generally, everyone in the world may be rendered spiritually poorer for the loss of the other species with which we share the planet. Tilman and Polasky remind us that we ignore our relationship with dynamic creation at peril to both our physical and spiritual well-being. Regrettably, there is probably less evidence that the biological aspects of the New Scarcity are being addressed than are the chemical.

The chemical aspect of the New Scarcity that has attracted the greatest attention in recent years is global warming. It has been known since Svante Arrhenius, a Nobel laureate in chemistry, advanced the hypothesis in 1896 that the anthropogenic release of carbon dioxide from the burning of organic compounds can lead to global warming via a "greenhouse effect." While a potpourri of gasses have been identified as having such effects, carbon dioxide is the most important among them, both because the mass of its anthropogenic releases is greatest and because the circumstances of its release are so ubiquitous. Burning fossil fuels—petroleum-based distillates, coal, and natural gas—releases carbon dioxide (CO_2) into the atmosphere. So does felling forests. Both activities take place on a huge scale.

Christian Azar considers policies to address climate change in "Emerging Scarcities: Bioenergy-Food Competition in a Carbon Constrained World." His message illustrates a combination of both possibilities and concerns that exemplifies the challenges embodied in the New Scarcity. On one hand, it is clearly possible to do something about the problem of climate change. The generation and adoption of new technologies could have profound effects on the concentration of greenhouse gases in the atmosphere.

The challenges are daunting, however. First, let us consider the magnitude of the problem. People in the United States produce approximately 17 times as much CO_2 per capita as do people in the developing nations of Africa and the Asian subcontinent. Stabilizing CO_2 concentrations in the atmosphere would require that the per capita emissions of the wealthy countries plummet toward levels currently encountered in the world's poor nations. If business continues as usual, it seems more likely that the emissions of the poor countries will increase to the levels now encountered among the wealthy.

This need not be the case, however. There are technological alternatives that might yield comparable quantities of usable energy without further increasing greenhouse gas concentrations. The key theme of Azar's chapter, however, is that there are no entirely benign alternatives. Consider, for example, greater reliance on nuclear power. On one hand, this might represent a Faustian bargain: the current generation would be producing highly poisonous spent fuel whose safekeeping would be passed on to future generations. While technological fixes may arise, supposing that they *must* seems unwarrantedly optimistic.

Moreover, nuclear technology raises uncomfortable issues concerning the poorly understood aspects of social relationships. Harold Barnett may have been prescient again when, in 1979, he was asked to contribute his perspective to *Scarcity and Growth Reconsidered*. Among other observations, he noted,

> The nuclear nightmare is rooted in military and political affairs and violence, not in economic growth. The dangers … are societal terrorism, violence, mass destruction, and related political problems…. The solutions for environmental pollution are relatively simple and at hand…. [T]his cannot be said of nuclear dangers.

If the "solutions for environmental pollution" are not in fact "relatively simple and at hand" in an era in which climate change and biodiversity loss create problems of global scope, it is hard to disagree with his assessment of the dangers of "societal terrorism, violence, mass destruction, and related political problems." In 1945 the United States, the world's wealthiest nation, produced its first atomic weapon. Little more than half a century later, India and Pakistan exploded nuclear devices—and made ominous threats to use them against each other. Recent events have led the world to doubt that even "peaceful" nuclear programs can stay entirely benign. Presidents and prime ministers around the world spend nights worrying about nuclear weapons falling into the hands of terrorists.

What may seem a more benign solution to the problem of climate change is to meet our energy needs in a renewable and sustainable way. Azar provides some rough estimates of what would be required if the world were to avert climate change by relying on biomass—crops grown for their energy value. Could it be done? Perhaps. What would be the consequences? Most likely a sizable increase in the amount of the globe placed under cultivation, and with it an appreciable increase in the price of food to the developing world, as well as further reduction in the amount of relatively unspoiled habitat maintained to support biodiversity.

There is perhaps no better source of far-fetched anecdotes than the projections "experts" of previous decades or centuries made of the fate of their descendents. Azar's calculations are, then, offered with appropriate modesty. However, regardless of whether his particular projections come true—or the options he identifies are, in fact, chosen in preference to others—he illustrates a general principle. The New Scarcity is actually a complex of interrelated scarcities that cannot be considered in isolation. If the potential of the atmosphere to absorb greenhouse gases without overheating is, in fact, limited, we must look to alternatives that may themselves impinge on the scarcity of arable land, the scarcity of biological diversity, the scarcity of international oversight of nuclear programs, virtually *ad infinitum*.

Having said all this, it is also true that hope springs eternal and need not always be disappointed by harsh realities. We employ technologies our great-great-grandparents never imagined, and our great-great-grandchildren may regard our circumstances as similarly rustic. Still, Azar's chapter reminds of us several things. One is simply that we are confronted by emerging and overlapping scarcities. A second is that resolution of these scarcities does require careful consideration of social policies, either for addressing them directly or developing technological alternatives.

Third, inventing new technologies may involve unleashing genies from bottles. Nuclear power and weapons are perhaps the foremost examples, but the advent of biotechnology is another. The techniques of genetic modification may allow spectacular advances in medicine, agriculture, and industry. They also spark fears of ecological catastrophe and, in the hands of the insane or disgruntled, could lead to the creation of terrifying epidemics—"the poor man's nukes," as some worried strategists call them. It may be worthwhile remembering that we do not always appreciate the ramifications of our new technologies.

Most of what we have said thus far involves, in one aspect or another, notions of "sustainability." This term has been used increasingly frequently since it entered the popular lexicon at the time of the Brundtland Commission's report in 1987

(WCED). As John Pezzey and Michael Toman note in "Sustainability and Its Economic Interpretations," however, part of the appeal of the term is its lack of precision. It is difficult to be opposed to a concept that essentially means "intergenerational fairness," especially if the person proposing it is not required to say what she means by "fair."

We might all agree that "fairness" means that no one should be treated better than anyone else. Beyond that, however, we run into a number of difficult philosophical problems. Pezzey and Toman distinguish "weak" from "strong" sustainability. The former essentially means that earlier generations are not better off than are later ones. The latter seeks to guarantee that outcome by assuring that later generations do not have to make do with a depleted set of assets relative to that enjoyed by their ancestors.

Such definitions beg many questions, however. "Strong sustainability," if taken to an extreme, admits a *reductio ad absurdem*. Even if a society took the strongest possible measures to preserve all aspects of its natural, technological, and social assets for posterity, it would be impossible to accomplish this end literally. Just as "one cannot step into the same river twice," natural processes alone will ensure that there are *some* differences in the constellation of assets available from one generation to the next. The recourse of defining a sustainable world as one in which "natural" processes are free to run their course begs the question of what is "natural" when humans take conscious steps to intervene or not intervene in "natural" processes.

Much of the debate surrounding sustainability concerns the possibilities for substitution among assets. Advocates of a "strong" form of sustainability generally argue that opportunities for substitution between natural capital, such as that represented by diverse natural ecosystems and the climate moderation and waste disposal services of the atmosphere, and anything that humanity can manufacture are, at least under the conditions in which we now find ourselves, very limited. Further reductions in natural capital could prove catastrophic, as there is no alternative to its use.

These are the types of questions that arise in the tricky navigation among concepts and criteria of sustainability. In Pezzey and Toman's chapter, two other major issues arise. The first concerns when a market economy is sustainable. To some this will seem a rather strange question. Under some restrictive conditions that we will discuss further below, a market economy can be shown to satisfy certain optimality conditions. What would it mean if an economy were "optimal" but not "sustainable"?

Pezzey and Toman suggest that the paradox of an unsustainable market economy can be resolved by postulating dual roles for individuals, as both "consumers" and as "citizens." I can, as a consumer, behave as the textbook *homo economicus* who maximizes his satisfactions by undertaking economic transactions. As a citizen, however, I may have a different outlook and pursue different objectives. I may manifest a concern for future generations that belies the "maximize discounted present value" objective that economists presume and private consumers perhaps pursue.

The idea that we pursue different agendas as consumers than as citizens is not really that unusual and leads into the second major issue in discussing the sustainability of market economies. As citizens, at least those of us fortunate enough to live in democratic societies regularly vote to express our preferences for public policies to restrain our excesses as consumers and producers. A market economy can achieve a socially optimal allocation of resources across time without resorting to any central system of command and control if there are no "externalities." An externality is a cost or benefit generated by one party that accrues to others and for which no payment is rendered. Such externalities are the very essence of the New Scarcity. When there are externalities, there is no guarantee that a free market will allocate goods optimally over time or even across consumers in a given period. It is not, then, necessarily a contradiction in terms, nor does it reflect a philosophical inconsistency, to ask whether our economy achieves socially desirable outcomes or sustains our long-term well-being.

The next several contributions investigate the processes by which technological innovations are generated and applied. Robert Ayres's perspective is largely historical. In "Resources, Scarcity, Technology, and Growth" he traces the development of important technologies of the eighteenth through twentieth centuries. In the cases of coal-fired steam engines, the development of practicable technology for aluminum smelting, and the rise and spread of electrification, he identifies common themes. One is that one innovation begets another, leading eventually to feedback loops. One of the first applications of the steam engine was pumping water from coal mines. This made it possible to mine deeper, leading to greater availability of coal, which led in turn to wider application of this relatively compact source of energy to industry and transportation. Railroads brought coal to steel mills, where it was used to fire furnaces in which were forged more rails and parts for larger and stronger steam engines. Similar feedback loops among innovations characterize the development of other industries.

These cycles of development have interesting attributes. One is that they often drive a "rebound effect" in which scarcity of a resource motivates an innovation that, through cycles of synergies and recombinations with other innovations, eventually results in far more consumption of the resource whose perceived scarcity drove the original innovation. Steam engines were first used to facilitate the recovery of coal. Once the steam engine had been developed, applied to transport and industry, and spurred the creation of innumerable other innovations, far more coal was mined and consumed than before its invention.

Ayres believes that innovation displays a fundamentally episodic character. Major innovations, such as the development of the steam engine, appear as discrete events in the historical record. Such breakthroughs constitute major, discontinuous changes in possibilities. Between breakthroughs, long periods of incremental improvement occur. Eventually, however, the momentum of the breakthrough innovation dissipates and the pace of change slows until the record is punctuated with another breakthrough.

Many of these breakthroughs have to do with the use of energy. This is problematic in light of the fact that energy resources are limited both by their own inherent scarcity and by their environmental consequences. Our age is in need of a breakthrough that will both reduce our dependence on nonrenewable energy and reduce the environmental consequences of its use. Public intervention may be called upon to finance and motivate such a breakthrough.

This last notion provides an apt lead-in to Sjak Smulders' chapter, "Endogenous Technical Change, Natural Resources, and Growth." One of the most important developments in economics since *Scarcity and Growth Reconsidered* has been a closer investigation of determinants and consequences of innovative activity. As we note above, innovation and environmental improvement share a distinctive feature: each typically generates benefits that its providers cannot fully "internalize." Hence, there is a *prima facie* case for public intervention to encourage socially beneficial improvements.

Smulders takes up this theme in his chapter. He makes a number of points relating technological change and resource scarcity. First, one should not automatically jump to the conclusion that technological innovation is a panacea for coping with resource scarcity. Technological advances may create opportunities for substituting away from scarce resources. They might also, however, result in the more rapid degradation of such resources. As an example, one need only think of what has happened with many of the world's fisheries: "better" technologies have led to even more rapid depletion of fish stocks.

This example highlights the importance of policy interventions. In the case of fisheries, steps need to be taken to reduce the "common pool" problems that arise when fish stocks are open to exploitation by whomever can first capture them. As we have already remarked, a similar but opposite problem arises in innovation: since the benefits of innovations may be appropriated by people who do not pay for their generation, innovations may be undersupplied. This observation again underscores the main point. Policy interventions will generally be required, first to prevent the overexploitation of the resources whose degradation is the subject of the New Scarcity, and second to provide the impetus for innovations that will further reduce pressure on such resources.

Smulders notes another interesting issue. What are the long-term prospects for humanity if we are, in fact, confronted with an insurmountable limit on our material resources? As we note above, economists have had very different perspectives on these matters over the years. Classical economists applied the principle of diminishing returns in conjunction with the assumption of fixed resources and concluded that our long-term prospect is for stagnation, if not decline. Later economists, buoyed by evidence that productivity continued to increase even as resources generally declined, came to more optimistic conclusions.

Still, it may be naively optimistic to suppose that technological innovations will fall like manna from heaven in perpetuity. Innovation is, as Smulders points out, an economic activity like any other. Innovative activity responds to incentives. In economic analysis, the returns to one factor of production are an increasing function of the quantities of others. Thus, if natural inputs remain a constraint, even incentives for further innovation could someday dwindle away. This unhappy state of affairs might still be averted even if we do not benefit from an unending stream of innovations, however. Knowledge itself may be a "capital stock" that grows over time to offset the constraints imposed by natural resources. In this scenario the growing economy pulls itself up by its own bootstraps.

The subjects Smulders surveys are very much at the cutting edge of modern economic research, and progress on these topics has had a profound effect on our understanding of the economy. Yet these topics are also very much part of a long-standing tradition in economic thought, in which forward-looking agents respond in rational fashion to economic incentives. In "Evolutionary Analysis of the Relationship between Economic Growth, Environmental Quality, and Resource Scarcity," Jeroen van den Bergh explores a different perspective. Economists often assume that observed behaviors can be explained as those that maximize the welfare of the individuals displaying them. Biologists, on the other hand, generally assume that observed behaviors are those passed along by individuals whom nature has selected to survive and reproduce. Van den Bergh explores a synthesis of these economic and biological paradigms.

Such a synthesis is both natural and problematic. It is natural, as the assumptions economists often make to explain the emergence of "optimal" behavior in economic models may strain credibility. Given human foibles, however, it may seem unreasonable to suppose that we can always immediately and accurately recognize our own best interests or the opportunities to realize them.

Might "optimal" behavior then emerge through learning or emulation? If so, evolution might explain patterns that emerge in social as well as biological interactions. The analogy is imperfect, however. The mechanics of biological evolution can be reduced to molecular genetics. No similar mechanism of inheritance has been proposed to explain economic evolution. The most logical choice—that success spawns emulation—is open to dispute. It is difficult to distinguish between a mechanistic process in which the most successful among a group of randomly distributed variants generates the most progeny and a more directed one in which emulation is a rational choice exercised by followers who may also propose their own enhancements.

Economic evolution cannot then be reduced to, or made strictly analogous to, biological evolution. But by the same token, thinking about social issues from an evolutionary perspective can be edifying. Van den Bergh demonstrates this in his contribution, in which he shows that an evolutionary perspective can be particularly enlightening when we are considering how a group of organisms—humanity in this case—shapes, and is shaped by, its environment.

From such a perspective, certain common notions in economics become less clear-cut. Much of what is written about the virtues of markets concerns the ability of individual "optimizers" to produce "efficient" outcomes. From an evolutionary perspective, however, the environment determines what is "optimal." Evolution toward what is optimal under one set of circumstances may result in modifications that do not prove optimal as circumstances also evolve. The adaptation resulting from selection only converges to a long-term steady state if the natural environment shaping evolution is static. When social and natural systems coevolve, it is questionable to ascribe developments in the former as "progress" when we are necessarily unsure as to the effects social development will have on natural systems.

An evolutionary perspective may also motivate a different view of policy formulation. It is common in economics to suppose that social planners are just another set of rational optimizing agents, defined as maximizers of some social objective (or in more realistic or cynical models, as maximizers of more personal satisfactions). Even if we can speak meaningfully of a role for rationally directed policymakers in a model in which selection occurs by the application of exogenous standards of fitness to agents whose attributes have been randomly generated, such policymakers may have a different role in evolutionary than in traditional models. The maintenance of variability is important in providing for the selection of fit types in rapidly changing environments. Perhaps the objective of policymakers should not be so much to identify "winning" ideas as to retain enough of the "also-rans" as to provide other options should conditions change.

Debates about the capacity of technology to alleviate scarcity of natural and environmental resources normally center around technical feasibility in resource substitution. Sometimes they extend to consider whether more harm to natural resources and environments will result from a new technology than from the old. A prototypical example of such debates is the one involving the application of genetically modified organisms in agriculture, whereby individuals of different persuasions debate the ability of these organisms to provide sustained increases in agricultural productivity and their potential for lessening use of chemicals versus their potential harm to other organisms. Another example is nuclear power—a source of electricity without conventional pollutants or greenhouse gas emissions but with a legacy of waste that is toxic to the environment and to humans and as well is hard to manage.

More recently, concerns about the use of technology vis-à-vis broader questions of social progress and stability and risk management have been on the upswing. For example, both nuclear and genetic recombination technologies can be seen as posing risks not solely because of their unintended environmental consequences, but also because of their potential for deliberate and malicious misuse. As technologies grow more complex, moreover, questions arise as to the ability of the public to understand and compare their risks and even whether the various risks and benefits are comparable. Yet, it seems that in much of the modern literature on scarcity and growth, technologies are simply means by which to expand society's production possibilities. Much less concern is generally expressed over how technology and society relate in a broader sense.

In reflections on these subjects, Sylvie Faucheux brings to bear many of the concepts of strong sustainability that have become central in the European community of ecological economics. She argues in particular that the evaluation of technology's risks and benefits cannot be undertaken apart from the social context in which they are experienced, requiring therefore more place-based analysis and public participation in assessment and decisionmaking. She also argues that with growing complexity or with the technological and other risks faced by citizens, the premise that all risks and rewards can be commonly compared is increasingly open to question. There are reasons to believe that individuals faced with modern and complex risks exhibit a kind of risk aversion that puts a premium on avoiding major uncertain changes to the status quo of complex natural systems. This line of argument puts Faucheux at odds with the mainstream of natural resource and environmental economics, especially in the United States. Even if one does not accept her other conclusions, it is difficult to dispute the point that reduction of the conditions that cause social instability—poverty and injustice—may be as valuable in the portfolio of social investments as expanding technological capacities.

David Pearce and Molly Macauley consider policy choices from more conventional, albeit not entirely "standard" perspectives. In "Environmental Policy as a Tool for Sustainability," Pearce investigates how society should deal with the New Scarcity of environmental resources. He begins by noting the similarities and differences between the old and new scarcities. Concerns over the Old Scarcity revolve around the adequacy of stocks of mineral, fuel, and other resources. Many commentators dismiss such concerns by noting that these resources are traded in markets. As they become scarce, their prices will increase to reflect that scarcity, and three responses will arise: consumers will economize on their use, substitutes will be identified and exploited, and new technologies will be developed to further economize on use and introduce more substitutes.

It may seem natural, then, in addressing the New Scarcity of environmental resources to try to emulate the markets in which other natural resources are traded. If environmental assets are, in fact, growing scarce, the prices assigned to their further degradation should rise, and we would expect the same pattern of economizing on use, employment of substitutes, and investment in innovation to arise.

As Pearce points out, however, such a scenario presumes that certain prerequisites have already been met. Suggesting that, to cite the environmental economist's mantra, we "get the prices right" presumes that society has determined that environmental resources are, in fact, growing scarce and, in a closely related vein, that this scarcity has risen to a level of concern in which public sector involvement is required. Having made these determinations, we can turn to the details of how to structure environmental policy. In many nations, however—particularly the poor countries of the developing world—there is as yet no social consensus that environmental problems do, in fact, motivate the sacrifices required to address them.

It is not surprising that there is a sort of geographical hierarchy determining which problems are addressed first in developing countries. Water pollution, for example, is often a relatively local problem that can be addressed with relatively simple measures, such as improving sanitation systems. Air pollution may be more complex. As we have already noted, emissions of greenhouse gases and losses of biodiversity arise from local action but have global consequences. We would expect—and evidence seems to confirm—that generating the social consensus required for action on these problems will be more difficult than addressing environmental problems at more local levels.

There is a sense in which the stages of development of environmental policy that Pearce proposes overlap. A society is more likely to recognize that the scarcity of environmental resources has reached critical proportions if it recognizes relatively low-cost ways to ameliorate such scarcity. Moreover, we might suppose that a society otherwise committed to the principles of a market economy might be more willing to address its environmental issues if it could employ the apparatus of the market economy to them.

These observations emphasize the central paradox that Pearce's essay investigates: why is it that "market-based instruments" (MBIs) are, as yet, so rarely employed? Pearce is careful not to overstate the case. There are enough instances

of the use of MBIs, such as fuel and effluent taxes and tradable permits in air pollutants, to establish them as more than "experimental" approaches to environmental policy. Still, given the enthusiasm expressed for them among economists (and,
increasingly, elements in the environmental advocacy community), MBIs remain
conspicuous by their relative absence.

The reasons for this paucity are myriad. First and perhaps most obviously, many
countries, particularly those of the developing world, have few environmental
regulations of any kind. Once sufficient social pressure mounts to implement *some*
such regulations, newly empowered regulators often find themselves in urgent
circumstances. It is then time to do something clear and definite, rather than
fumbling with the niceties of guessing what economic measures will yield which
physical results. The certainty attached to requiring specific measures may often
be preferred to what may seem the less tangible benefits of implementing more
flexible incentives.

Once an environmental administration is established with an emphasis on technical, as opposed to economic, concerns, it may be difficult to reverse the culture
of the organization. It is only natural that bureaucrats will seek to perform their
duties in such a way as to maximize the demand for their own expertise. One
advantage often claimed for MBIs is that they "decentralize" environmental decisionmaking by leaving the specifics of compliance to private agents. Viewed from
a bureaucrat's perspective, however, eliminating the need for her own management
expertise may seem less of an advantage.

Other factors in determining public policy toward the environment may be
still more malign. While much environmental policy aims to improve upon policies that are already "good," in the sense that they tend to improve environmental performance, great progress might also be made by reforming policies that
are demonstrably "bad." Many commentators bemoan the continuing existence
of perverse subsidies—payments that, by encouraging profligate use of resources,
result in their greater degradation. While lamentable, it is, perhaps, not surprising that perverse subsidies continue. Large economic players wield political clout.
Environmental degradation is a problem in large part because important elements
of society realize private gain—albeit at the expense of losses to the broader public—by engaging in activities that degrade the environment. It should not be
surprising that such major players find ways to underwrite their activities with
public funds.

Conventional wisdom holds that environmental improvement tends to be a
popular cause among the wealthy. But this constituency may be more often motivated by fervor than by careful economic analysis. Efficiency *per se* has a limited
constituency (Stroup 2003), a view that Pearce echoes in noting that, aside from
academicians, there is limited lobbying for MBIs.

The *form* in which environmental policy is implemented may seem a matter of
less importance in responding to the New Scarcity than is the question of whether
to institute some form of environmental policy. However, Pearce suggests that
activists will not begin to achieve their goals until they come to some accommo-
dation with economic interests. Halting economic development is simply not an
option. Overcoming resistance to cost-effective environmental regulation is, then,

in some respects tantamount to overcoming resistance to environmental regulation generally.

Molly Macauley's "Public Policy: Inducing Investments in Innovation" addresses the other aspect of public policy that bears strongly on the resolution of scarcity: technological innovation. Governments have provided incentives for innovation for almost as long as there have been governments. The motivation for such incentives is found in the nature of innovation to which we have alluded above: new ideas often generate spillovers that can spread far beyond the industry or sector for which the original innovation was applied (as noted, for example, in Robert Ayres's chapter detailing the ways in which important technological innovations tend to reinforce each other). As we also note, the incentives for public underwriting of innovation are greater when innovations ease the constraints imposed by the New Scarcity of environmental resources.

The generality of the possible effects of far-reaching spillovers can create wide opportunities for abuse, however. If we cannot predict the ultimate applications to which *any* particular innovation might be put, should society subsidize *every* innovation? For that matter, can we even distinguish "innovation" from other undertakings? While Macauley states the conventional case for public support of innovative activity, she also points out the pitfalls of taking this principle too far. Publicly funded research may ease environmental problems, but it may also induce rent-seeking among would-be recipients of public funds and generate wasteful "pork."

Macauley buttresses these conclusions with a number of examples drawn from case studies. Governments have underwritten any number of research ventures in energy, resources, and environment (as well, of course, as many more in other fields) with decidedly mixed results. There have been some successes, but there have also been spectacular failures, such as the United States synthetic fuels program. In that case, it appears that private industry more accurately perceived long-term price trends in fuels than did the government, which initiated a slapdash program to combat an energy "crisis" that, a decade later, had largely disappeared. It remains to be seen if other "crises" turn out to be more serious or enduring, but the synfuels program illustrates the pitfalls that can occur when large sums of public money are devoted to poorly conceived ventures.

Public policy can motivate innovation in other ways, however. Public research subsidies provide "carrots," but innovation may also be motivated by "sticks," such as regulatory restrictions, taxes on pollutants, and the like. Conventional wisdom has it that MBIs are superior to other forms of regulation because they leave open to the regulated parties the issue of how best to reduce their costs of polluting. If regulated parties can accomplish this most efficiently by devising better modes of production and effluent control, MBIs will allow them to do so.

This observation brings us back to the theme of Pearce's paper, that MBIs, although perhaps becoming more common, are still thin on the ground. Macauley adds another important cautionary note to Pearce's recitation of the reasons: the regulator who seeks purely to "maximize social welfare" in the absence of market failures is as much a fiction as is the perfect market of the introductory economics textbook. Idealized depictions of economies are useful as pedagogical devices and benchmarks against which to compare actual performance. The same might be said of idealized depictions of regulatory programs. Macauley reminds us that policies must be made on the basis of comparisons of *actual* economic conditions and *actual* regulatory capacity and performance. Comparisons between ideals are generally irrelevant; those between the *actual* performance of an unregulated economy and an unobtainable regulatory *ideal* are deceptive and can prove counterproductive.

While the authors of the contributions we have surveyed thus far make some mention of the importance of environmental resources in poor countries, their emphasis is largely on economic, social, and physical conditions in the wealthier countries. Yet there is widespread agreement both that the consequences of environmental degradation and resource scarcity are greatest for the poor and that the greatest environmental challenge of our era is to raise the living standard of the poorest of the poor without further degrading the global environment.

The final two chapters extend the analysis of scarcity and growth to the developing world. In "Intragenerational versus Intergenerational Equity: Views from the South," Ramón López elaborates on the theme of "sustainability." He encapsulates a very telling point in a question: "Will governments that systematically neglect the welfare of the vast majority of the current population … consider the interests of those not yet born?"

López argues that much of the existing debate on sustainability ignores two crucial issues, both of which appear in his question. First, sustainability is essentially a question of equity, and one cannot consistently espouse concern for *inter*generational equity without also showing a corresponding concern for *intra*generational equity. We should be at least as concerned with the plight of today's poor as we are with that of tomorrow's.

López's second observation is that it is "*governments* that systematically neglect the welfare of the vast majority" (our emphasis). Understandably, economists writing on the prospects for sustainable development emphasize economic policies and forces in their analyses. Yet, in much of the world, the institutional underpinnings for well-functioning market economies are scarce or absent. Markets facilitate trade in goods and services, but the efficiency- and welfare-enhancing properties claimed for trade in goods and services do not arise from their appropriation or theft. Elite minorities in many developing countries appropriate their nations' resources for their own benefit, while employing the private wealth they acquire to insulate themselves from the consequences of environmental degradation.

To López, then, effective policy intervention to institute sustainable development cannot be limited solely to working within the economic system to "get the prices right." We must first get the institutions of governance right. Until we do the latter, there will be little incentive to do the former.

Partha Dasgupta begins the final contribution by noting a revealing semantic distinction. Environment is, in the wealthier countries, an "amenity," or a "luxury good," that comes to be in greater demand as income increases. This view leads to a one-sided policy prescription: since wealthier people demand cleaner environments, the solution to the developing world's environmental problems lies in promoting its economic growth.

This perspective and prescription are problematic in a couple of respects. With regard to the semantic point, Dasgupta argues that natural resources and environmental "amenities" are not luxuries, but rather, necessities. Poor people have no substitutes for the filthy water and smoky air of their communities. Second, in prescribing economic growth as an antidote for environmental degradation, we need to be very careful about measuring "growth."

Dasgupta has titled his contribution "*Sustainable Economic Development* in the World of Today's Poor" (our emphasis), and it raises themes introduced in Pezzey and Toman's chapter on sustainability. An appropriate measure of yearly income is one reflecting the consumption possible at present *without compromising future prospects*. In both theory (Weitzman 1976) and practice, national income is calculated by adding together the market value of consumption and that of net investment.[12] Suppose that the latter of these were markedly negative—that the country whose national income was being calculated were liquidating its capital stock or allowing it to depreciate without replacing it. In such circumstances one should subtract the value of the lost capital stock from the value of consumption in calculating income.

Amending accounts in this way is straightforward when the "value of consumption" and "value of investment" can be measured by simple accounting exercises. The value of consumption is typically described as the sum of all consumption goods produced, with each weighted by its market price. The value of investment is typically calculated in a parallel fashion, summing quantities weighted by prices.[13]

Problems arise in attempting to do accounting in the absence of prices. Yet this is the situation, by definition, when nonmarket goods are involved. In the absence of actual prices, one must infer "shadow" prices from other markets or information. Dasgupta acknowledges the many practical impediments to estimating such prices, but making some heroic assumptions in the interest of illustration, presents figures that offer a sobering perspective on the true economic performance of the developing world. Nations that appear to be making substantial progress in rising from poverty may, on closer reflection, be achieving only illusory growth by degrading their natural assets. In some of the world's poorest regions, the outlook is dismal indeed. Official economic statistics indicate stagnation. When such assessments are augmented to reflect environmental degradation, the scenario is worse yet.

Dasgupta also demonstrates that similar concerns dog estimates of productivity. Imagine, for example, a nation that fuels its economy by felling ever larger portions of virgin forest. If there is no official market for trees ("stumpage" in the forestry vernacular), it may appear that the economy is making more with less each year—until the trees run out. Thus, whether measured as high income growth rates or growing productivity, official statistics may not represent economic progress so much as the imminence of a bill coming due.

Dasgupta notes that institutional changes can induce progress as much as can innovations in physical processes. Resources will be used more efficiently when institutions evolve that require users to pay the full costs of their use, that is, for example, the private costs of their current provision, the costs that may fall on future generations if such resources will no longer be available, and the costs that users of the resources may impose on other members of society because of the pollution, biodiversity loss, and so on, they may occasion. While it is clear that a society that makes more efficient use of its available resources will better serve the interests of its citizens in the long run, it may appear to be performing less well in the short run. This is because some resources are underpriced. The measured costs of production must increase when the prices of resources increase.

This leads us again to the critical question of when and how society—or societies—will decide to acknowledge the New Scarcity and take steps to offset it. This is a fitting point from which to begin to sum up our main themes.

Scarcity and Growth in the New Millennium: Conclusions

The contributors are an eclectic group, and they come at a host of issues from a variety of perspectives. Each has, however, addressed the essential question of how scarcity affects prospects for growth.

To organize our conclusions, let us consider how scarcity might constrain growth and which scenarios seem most likely:

- In the worst-case scenario, absolute limits on physical capacities might condemn humanity to stagnation—or worse. If an essential resource were available in fixed supply we would, in the fullness of time, be doomed.
- Slightly more favorable would be a situation in which an essential resource were renewable, but only at a relatively slow rate. While a "sustainable" path might be open to us, even an economy in which this essential resource were privately owned might not choose to follow it.
- If essential but renewable resources are not privately owned, disaster could ensue unless public action is taken to address problems of pollution and degradation.
- If no resources are essential, or if essential resources reproduce at a sufficiently rapid rate, a market economy could follow a sustainable path with no public oversight.

None of our authors espouse the extreme views represented in the first and last possibilities. None is saying that humanity is doomed, or even that a human population of the current size is doomed, nor is any saying that a *laissez-faire* policy in which all resource allocation decisions are left to private actors will assure us of a rosy future. While John Pezzey and Michael Toman consider the prospect raised in the second scenario, that a market economy could follow a path that is both "optimal" and "unsustainable," they and most other authors devote most of their attention to the third scenario.

In short, the most pressing question of scarcity and growth is What public action is required to address the New Scarcity of nonmarket environmental resources?

A natural inclination—among economists, at least—is to mimic markets. If environmental resources are becoming scarce, government action can lead private actors to reflect this scarcity in their decisions by taxing emissions, setting quotas on them, or otherwise "making markets" where they did not exist before. As David Pearce shows, however, market-based incentives remain relatively rare.

Why is this? In addition to the reasons Pearce cites in his chapter, we might consider other factors. Ramón López reminds us that the institutions of the economy are subsidiary to those of governance more generally. We should not be surprised to observe market "failures" when such "failures" may, in fact, represent tremendous "successes" for the governing elites that perpetuate them. David Tilman and Stephen Polasky's work on biodiversity reveals that part of the problem of crafting incentives, even in an otherwise well-structured society, is complexity. Exactly *what* would we place a tax or quota on in order to preserve the world's species? This complexity is amplified by the fact that scarcities interlock in a global society that has grown so large. As Christian Azar points out, we could put a sizable tax on carbon dioxide emissions that could well lead farmers to plant renewable energy crops, and in the process expand land under cultivation, reduce biodiversity, and further compound the challenges facing the world's poor. Even the resolution of the Old Scarcity raises issues with regard to the New Scarcity. David Menzie, Donald Singer, and John DeYoung Jr. note that compliance with environmental restrictions is likely to be as important to the mining industry's performance in coming years as is physical depletion. A similar message emerges from Jeffrey Krautkraemer's chapter: markets seem to have been adequate to the task of allocating exhaustible natural resources, such as oil and minerals, over time. The more difficult it becomes to establish ownership over resources, however, the less satisfactory has been performance. Fisheries—which are inherently more difficult to "own" because the fish themselves can move—are declining, and a host of environmental issues remain intractable, especially those being played out at a global scale.

Is the answer to the New Scarcity to launch into bold programs to achieve major technological breakthroughs? Robert Ayres notes that the historical record shows several major episodes in which radical new technologies have transformed the world through spillovers and feedback loops. He argues that new breakthroughs are required in order to achieve similar breakthroughs to overcome the scarcity of energy and the environmental consequences of its use. Sjak Smulders provides conceptual support for this notion by reviewing an influential literature that emphasizes the similarities between environmental and technological issues. Each may motivate public involvement to counteract the "nonrival" aspects of the goods in question. Yet Molly Macauley cautions that the actual performance of many public programs intended to spur breakthroughs in environmental and resource technology has left much to be desired. As she notes, real bureaucrats may be a good deal less efficient in achieving socially desirable ends than are their representations in idealized models.

Of course, it is only to be expected that the realization of emerging scarcities and the evolution of policies to counter them will involve fits and starts. While many credit private markets with resolving issues pertaining to the Old Scarcity of marketed natural resources, the history of metals, minerals, and fuels markets is replete with countless bankruptcies, bubbles, and miscalculations. We chose the phrase *"evolution of policies"* in the first sentence of this paragraph deliberately, as we mean to evoke the principles covered by Jeroen van den Bergh. If a thriving and, one would hope, increasingly equitable economy is to survive a brush with the limits of the planet's finite carrying capacity, there must be some experimentation by which "fit" strategies for survival are winnowed from those that will not see us through. As Sylvie Faucheux points out, the criteria by which we predict "fitness" ought not be limited to those factors that are easily incorporated in analytical models. Although economists often suppose either that technology evolves exogenously or can be treated as an outcome of economic activity no different than any other, Faucheux suggests that our long-term prospects may be determined by the complex, multidimensional relationship between society and its creations.

In closing, we consider Partha Dasgupta's observations on the experience of communities around the world that lived close to nature. Survival was, for them, a matter of learning the limitations of their environment and living within them. It would be comforting to know that this is, in fact, the natural progression of mankind. Regrettably, the historical record suggests that it is not inevitable. There is a "selection bias" in that the traditional societies that survived to modern times are those that solved the problem of living within their ecological means. Societies now known only by their archaeological artefacts may not have.

It remains an open question whether, as the globe turns into a single, integrated society, it musters the technological, economic, and perhaps most importantly, social wherewithal to address the implications of the New Scarcity. We have not yet, as a race, run into the limits imposed by reserves of the things we regularly buy and sell. The ultimate question, our authors agree, is whether we will be as successful in managing resources that have, to date, largely remained outside the realm of the market economy. While "economic" prescriptions might be suggested for addressing the question, we suspect that the ultimate deciding factors will include the wisdom, compassion, and vision we can bring to our collective choices.

Notes

1. We would especially like to acknowledge the insights offered in the course of the workshop by Cutler Cleveland, Linda Cohen, Frank Dietz, Faye Duchin, Rod Eggert, Charles Kolstad, Toni Marechaux, Richard Newell, Thomas Sterner, Timothy Swanson, Frans Vollenbroek, and Cees Withagen.

2. The scale of our debt to Barnett and Morse in preparing this introduction is reflected in part by the fact that most of the passages quoted from famous economists in this section are taken from *Scarcity and Growth*.

3. Somewhat surprisingly, in light of the credence given the contrary notion, Carlyle was commenting on the racial views of John Stuart Mill, rather than Thomas Malthus's theory of population, when he wrote the remark. See Dixon (n.d.) and Levy and Peart (n.d.).

4. D. Gale Johnson (2000) notes that Quintus "includes nearly all the modern complaints about the effects of excessive population on environment—deforestation, loss of biological diversity, farming unsuitable land, drainage of the natural refuges for wildlife—as well as the massing of people in cities." Joel Cohen (1996) provides an illuminating review of perspectives on the perils of overpopulation from Assyrian times to the present.

5. It is interesting to note, in fact, that Darwin credited Malthus with the fundamental insight motivating his theory of evolution: The scarcity of resources leads to the survival of the fittest.

6. We should note some logical limitations on this pessimistic view, however. To suppose that measured productivity growth is predicated on deterioration of natural resources would require that such deterioration continue year after year. After a certain point it would be implausible to maintain that degradation was fueling productivity, as a sufficiently long stretch of productivity growth would have to mean that the resources in question were fully depleted.

7. This, however, is not impossible. Many modern approaches to economic growth theory have employed the Dixit-Stiglitz (1977) model of differentiated products. If "knowledge" is defined as "the number of products we know how to make," then there can, in fact, be increasing returns to scale in "knowledge."

8. It is, however, possible that a "common pool" effect can result in the opposite result. If one innovator's product can supercede or obviate another's, one rival's investments in innovation can be stranded by another's success. If the successful innovator does not take account of this effect, he or she may devote too much effort to producing new and better products.

9. An interesting historical question arises: although the numbers clearly show that people are *on average*, far better off now than they were a millennium ago, *who* is better off? One might argue that Nefertiti of Egypt or Louis XIV of France enjoyed a more opulent lifestyle than do, say, Jennifer Lopez or Roman Abramovitch. The general penury of their subjects assured the historical figures of less expensive labor-intensive goods! While such comparisons are difficult, one tremendous difference is in life expectancy and medical services. Life expectancy in medieval times was only about 30 years. Of course, infant mortality and death in childbirth were appallingly high. Even a man who had survived his first 30 years could expect to live only about another 22 years, however (Nofi and Dunnigan 1997; figures are for England in 1300). By contrast, worldwide life expectancy is now about 67 years, and averages almost 60 years in low-income countries (although, tragically, there are still areas of Africa in which poverty, strife, and, in recent years, AIDS combine to hold life expectancy below 40). In wealthy countries, life expectancy is more than 78 years (UNDP 2003). Thirty-year-old males in the United States can now expect to live another 46 years (NCHS 2002). Inasmuch as death did not respect wealth or station in earlier centuries, it seems we may reasonably conclude that today is a good time to be alive on average and for the wealthiest.

10. This can be inferred from its concentration in polar ice caps. The snow that fell millennia ago has been insulated from the atmosphere by subsequent falls.

11. Net primary productivity is the accumulation of biomass in plants.

12. The formula can be further elaborated to include government expenditure and net exports.

13. Year-to-year variations in price levels are accommodated by dividing aggregates by price indices, although the construction of such indices can be problematic.

References

Abramowitz, M. 1956. Resource and Output Trends in the United States since 1870. *American Economic Review* 46 (May): 5–23.

Barnett, H. 1979. Scarcity and Growth Revisited. In *Scarcity and Growth Reconsidered,* edited by V.K. Smith. Baltimore: Johns Hopkins University Press for Resources for the Future, 163–217.

Barnett, H., and C. Morse. 1963. *Scarcity and Growth: The Economics of Natural Resource Availability.* Baltimore: Johns Hopkins University Press for Resources for the Future.

Carson, R. 1964. *Silent Spring.* Boston: Houghton Mifflin.

Clark, C. 1976. *Mathematical Bioeconomics: The Optimal Management of Renewable Resources.* New York: John Wiley.

Clark, J.M. 1923. Overhead Costs in Modern Industry. *Journal of Political Economy* 31 (December): 606–636.

Cohen, J. 1996. *How Many People Can the Earth Support?* New York: W. W. Norton.

Dasgupta, P., and G.M. Heal. 1974. The Optimal Depletion of Exhaustible Resources. *Review of Economic Studies* 42 (Symposium): 3–28.

———. 1979. *Economic Theory and Exhaustible Resources.* Cambridge: Cambridge University Press.

Dixit, A.K., and J. Stiglitz. 1977. Monopolistic Competition and Optimum Product Diversity, *American Economic Review* 67(June): 297–308.

Dixon, R. n.d. The Origin of the Term "Dismal Science" to Describe Economics. http://www. economics.unimelb.edu.au/TLdevelopment/econochat/Dixonecon00.html (accessed August 14, 2003).

Easterbrook, G. 1995. *A Moment on the Earth.* New York: Viking.

Ehrlich, P.R. 1968. *The Population Bomb.* New York: Sierra Club–Ballantine Books.

Grossman, G.M., and A.B. Krueger. 1995. Economic Growth and the Environment. *Quarterly Journal of Economics* 110(May): 353–377.

Heilbroner, R. 1967. *The Worldly Philosophers.* 3rd edition. New York: Simon and Schuster.

Hotelling, H. 1931. The Economics of Exhaustible Resources. *Journal of Political Economy* 39(April): 137–175.

Hubbert, M.K. 1949. Energy from Fossil Fuels. *Science* 109(4 February): 103–109.

IPCC (Intergovernmental Panel on Climate Change). 2001. *Third Assessment Report: Climate Change 2001.* New York: Cambridge University Press.

Johnson, D.G. 2000. Population, Food and Knowledge. *American Economic Review* 90(March): 1–14.

Levy, D.M., and S.J. Peart. n.d. The Secret History of the Dismal Science: Economics, Religion and Race in the 19th Century. http://www.econlib.org/library/Columns/LevyPeartdismal. html (accessed August 14, 2003).

Lomborg, B. 2001. *The Skeptical Environmentalist.* Cambridge: Cambridge University Press.

Maddison, A. 2002. *The World Economy: A Millennial Perspective.* Paris: Organisation for Economic Co-operation and Development.

Malthus, T.R. 1798. *An Essay on the Principle of Population.* Library of Economics and Liberty. http://www.econlib.org/library/Malthus/malPop1.html (accessed December 10, 2004).

Marshall, A. 1907. Social Possibilities of Economic Chivalry. *Economic Journal* (17): 65, 7–29.

Meadows, D.H., D.L. Meadows, J. Randers, and W.W. Behrens, III. 1972. *Limits to Growth: A Report for the Club of Rome's Project on the Predicament of Mankind.* New York: Potomac Associates.

Mill, J.S. 1848. *Principles of Political Economy.* Library of Economics and Liberty. http://www. econlib.org/library/Mill/mlP12.html (accessed December 10, 2004).

NCHS (National Center for Health Statistics). 2002. National Vital Statistics Reports. http:// www.cdc.gov/nchs/fastats/pdf/nvsr51_03t11.pdf (accessed June 4, 2004).

Nofi, A.A., and J.F. Dunnigan. 1997. *Medieval Life and the Hundred Years War.* http://www.hyw. com/books/history/1_Help_C.htm (accessed June 4, 2004).

Nordhaus, W.D. 1974. Resources as a Constraint on Growth. *American Economic Review* 64(May): 22–26.

Potter, N., and F.T. Christy Jr. 1962. *Trends in Natural Resource Commodities.* Baltimore: Johns Hopkins University Press for Resources for the Future.

Pritchett, L. 1997. Divergence, Big Time. *Journal of Economic Perspectives* 11(Summer): 3–17.

Raven, P.H. 2002. Science, Sustainability, and the Human Prospect. *Science* 297: 954–958.

Simon, J.L. 1981. *The Ultimate Resource.* Princeton: Princeton University Press.

Smith, A. 1776. *An Inquiry into the Nature and Causes of the Wealth of Nations.* Library of Economics and Liberty. http://www.econlib.org/library/Smith/smWN1.html (accessed December 10, 2004).

Smith, V.K. (ed.) 1979. *Scarcity and Growth Reconsidered.* Baltimore: Johns Hopkins University Press for Resources for the Future.

Solow, R.M. 1974. Intergenerational Equity and Exhaustible Resources. *Review of Economic Studies* 42 (Symposium): 29–45.

Stiglitz, J.E. 1974. Growth with Exhaustible Natural Resources. *Review of Economic Studies* 42 (Symposium): 122–152.

Stroup, R.L. 2003. *Eco-nomics: What Everyone Should Know about Economics and the Environment.* Washington, DC: Cato Institute.

UNDP (United Nations Development Programme). 2003. Human Development Indicators. http://www.undp.org/hdr2003/indicator/indic_1_1_1.html (accessed June 4, 2004).

United Nations Population Division (UNPD). 2002. *Word Population Prospects: The 2002 Revision Database*. http://esa.un.org/unpp/ (accessed June 4, 2004).

Vitousek, P.M., H.A. Mooney, J. Lubchenco, and J.M. Melillo. 1997. Human Domination of Earth's Ecosystems. *Science* 277(July 25): 494–499.

Weitzman, M.L. 1976. On the Welfare Significance of National Product in a Dynamic Economy. *Quarterly Journal of Economics* 90(February): 156–162.

———. 1997. Sustainability and Technical Progress. *Scandinavian Journal of Economics* 99(March): 1–13.

Wilson, E.O. 1992. *The Diversity of Life*. Cambridge, MA: Belknap Press of Harvard University Press.

WCED (World Commission on Environment and Development). 1987. *Our Common Future*. http://www.are.admin.ch/are/en/nachhaltig/international_uno/unterseite02330/ (accessed June 4, 2004).

CHAPTER 2

Mineral Resources and Consumption in the Twenty-First Century

W. David Menzie, Donald A. Singer,
and John H. DeYoung, Jr.

MODERN SOCIETIES ARE HIGHLY dependent upon energy and mineral resources to produce and deliver the material goods and even the services of everyday life. Although societies' dependence upon fossil fuels is evident and understood by much of the population, few people are as well informed about their dependence upon a wide variety of nonfuel minerals. This ignorance may result from two interrelated conditions. First, in contrast to fossil fuels, few people directly use nonfuel minerals in recognizable forms because most use is as part of manufactured products. Second, the value of raw ($38 billion) and even processed ($397 billion) nonfuel minerals in the United States in 2002 was small relative to the value the industries that consume these materials contribute to the economy ($1,700 billion). That is, although nonfuel mineral inputs are indispensable to construction and to the manufacture of durable and even nondurable goods (USGS 2003), their value is modest compared with the value of the final products.

Nevertheless, societies' dependence upon minerals has resulted in periodic concern about the adequacy of mineral supplies to continue to support the economy. This concern has been rooted in experiences as diverse as shortages of minerals in wartime, depletion of individual mineral deposits and even mining districts (Hewett 1929), and rapid increases in mineral consumption when countries industrialize. These experiences have sparked periodic debates about the abundance or scarcity of mineral resources.

Addressing such questions has been particularly difficult because the concept of a mineral resource involves both geologic and economic aspects and because knowledge about the earth and future economic conditions is limited. On the one hand, mineral deposits are physical entities that form as the result of geologic processes and are therefore finite and exhaustible. On the other hand, the amount of a mineral commodity that will be produced in the future is dependent on future costs of discovering and processing new mineral deposits, recycling used materials, and developing substitute materials (Tilton 1996). Many debates about mineral scarcity have taken place because participants did not appreciate both the physical and economic aspects of mineral resources.

The debates about the adequacy of future mineral supplies that took place in the 1970s illustrate this lack of understanding. The twentieth century was an era of unprecedented metal availability due to geologic and economic circumstances. The settlement of Australia, Canada, and the western United States, and colonial-era mineral exploration in Africa in the second half of the nineteenth century, led to the discovery of an unprecedented number of mineral deposits that provided abundant, high-quality minerals throughout the twentieth century. In addition, innovations in mining, such as Daniel Jackling's recognition that low-grade porphyry copper ores could be profitably mined by using economies of scale, as well as innovations in mineral processing and metallurgy, have reduced the cost of producing metals and added to the inventory of very large deposits that can be profitably processed. Further, advances in information processing and search technologies used in mineral exploration, including the widespread application of geophysical tools, geochemical methods, satellite imaging, and mineral deposit models, increased the efficiency of mineral exploration.

The twentieth century, however, was also an era of accelerating mineral consumption. Although mass consumption of resources in North America did not begin until the 1950s, and consumer societies did not develop in Europe and Japan until a decade later, when they had recovered from World War II (Rostow 1960), by the late 1960s rapid increases in consumption of minerals in these developed countries were leading some observers to question the adequacy of mineral resources to support continued growth. The peaking of oil production in the United States and the first Arab oil embargo, together with the publication of *The Limits to Growth* (Meadows et al. 1972), initiated a wide-ranging debate about the scarcity of mineral resources. The authors of that Club of Rome report used a computer program that dynamically modeled population, agriculture, industry, pollution, and natural resources to predict future conditions worldwide. Using well-explored economic resources as a measure of mineral availability, they underestimated existing stocks of minerals and failed to allow for replenishment by exploration. As a result the computer model that was the foundation for *Limits to Growth* mistakenly predicted the demise of industrial civilization because of a shortage of resources.

Mineral economists, such as Gordon (1972), wrote that the computer analysis used by Meadows et al. was no improvement on Malthus and pointed out that resource creation is a continuous process controlled by man, not nature. Emphasizing that new technologies, recycling, and market forces would be the solutions to material and energy shortages, Gordon questioned the wisdom of changing public attitudes on material well-being and cautioned against using scarce resources to "undertake dubious policies to prevent the questionable possibility of resource exhaustion." If the authors of *The Limits to Growth* misunderstood the economic aspects of mineral resources, one of their critics misunderstood the physical aspects of resources when he stated that "future quantities of a natural resource such as copper cannot be calculated even in principle, because of new lodes, new methods of mining copper, and variations in grades of copper lodes; because *copper can be made from other metals* (emphasis added); and because of the vagueness of the boundaries within which copper might be found—including the sea, and other planets" (Simon 1980, *1435*). When the deficiencies of the computer model became apparent, and no immediate shortages of minerals developed, the debate over mineral scarcity subsided and society turned its attention to more pressing problems.

One of those problems, also raised by *The Limits to Growth* model, was the concern that increased pollution from industrial activities would overwhelm natural systems. Warnings that the ozone layer was being destroyed by the use of chlorofluorocarbons and that the earth's atmosphere was changing due to the combustion of fossil fuels are perhaps the best-known examples of such concerns. By the end of the twentieth century, concern about the ability of earth systems to absorb the wastes resulting from the use of mineral products had supplanted concerns about mineral scarcity in public debate. Growing awareness of the effects of production and consumption of material goods on the natural environment has led to calls for limits to material consumption (During 1992). The degree to which developed economies are reducing (Wernick et al. 1996) and can reduce (Ausubel 1996) both the level of consumption of material goods and the environmental effects of that consumption have become central questions that will affect the future quality of the environment.

Entry into a new century is an appropriate time to reexamine both the old and new questions of mineral resource scarcity. As the twenty-first century opens, rapid economic development in Asia is increasing the consumption of mineral commodities. Will mineral production be adequate to meet the needs of rapid economic development of large Asian countries? At the same time there are growing concerns that global environmental systems are already being affected by mineral production and use. Can they absorb the wastes from increased mineral production? Can we reduce the levels of such wastes sufficiently to avoid environmental problems?

This paper examines the challenges posed both by mineral scarcity and by environmental effects of mineral production and consumption by

- investigating the physical nature of mineral resource occurrence in order to determine if resource depletion is likely to impose a constraint on mineral production during the next several decades,
- investigating the current state of systems that produce mineral commodities to determine if society faces social and technical constraints that could affect mineral availability,
- investigating emerging changes in the consumption of mineral resources to determine the extent to which increases in consumption may pose constraints on mineral availability, and
- investigating the extent to which increased demand may result in an increase in environmental effects that may constrain growth.

Copper is used as an example for the present study because rates of consumption of metals are more likely to show effects of resource scarcity than are nonmetallic mineral commodities; growth in copper consumption is a good indicator of economic development. Recent studies (DeYoung and Menzie 1999; Menzie et al. 2001) that examine the growth in consumption of aluminum, cement, copper, and salt, demonstrate that copper is a good indicator of mineral consumption in developing countries. Although consumption of aluminum also grows with economic development, it trails copper consumption.

The Physical Nature of Mineral Resource Occurrence

That the earth contains vast quantities of every metal used by man cannot be denied. Even an order-of-magnitude decrease in estimates of crustal abundances still implies immense quantities of metals. Most uses of metals do not reduce their amounts, they only relocate and reconstitute them. The question of how much metal is available in concentrations and forms of interest to man is more difficult to address. As long as costs of extraction of metals are above zero, man will seek out the concentrations and forms of metals that have cost-reducing physical attributes (DeYoung and Singer 1981). Both form, or mineralogy, and concentration, or grade, are critical to consider because either can make some of the total amount of metal inaccessible due to excessive costs of extraction and processing. Current concepts of cost go beyond traditional views of direct production costs to include, for example, chemical attributes that increase environmental costs, whether they are internalized or not.

Earth's less common metals, such as copper, zinc, lead, and nickel, are usually found evenly distributed in common silicate minerals (Skinner 1976). As Skinner points out, metals in this form cannot be significantly concentrated to reduce volumes to be processed, and the entire mineral must be broken down chemically to separate the desired metal from the other atoms. Because the chemical bonds in most common minerals are strong, this is a complicated and very energy-intensive process. Hence mining of these scarce metals is focused on rare deposits that contain metal in compounds with elements such as sulfur or oxygen. The metals in such deposits are more easily extractable and occur in significant concentrations or grades (Figure 2-1). Total amounts of copper, zinc, lead, silver, and gold contained in all discovered mineral deposits range from 0.015 percent to 0.002 percent of the amount contained in the upper one kilometer of continental crust (Singer 1995). Thus, presently identified mineral deposits represent a very small proportion of the total amount of metal.

Several researchers have constructed plots of the metals in mineral deposits or ore deposits versus metals in rocks, usually the upper kilometer of the earth's crust (McKelvey 1960; Erickson 1973; Brooks 1976; Barton 1983). Barton (1983, 6) points out that the large amounts of metals that are not accounted for in identified conventional metal deposits offer great opportunities for undiscovered deposits of these metals. These and similar analyses renewed speculation about the promise in the observed relation between decreasing grades acceptable for mining and resulting increases in mineral availability (Lasky 1950). The mathematical model used as a basis for such extrapolations has been demonstrated to produce physically impossible results outside the limited range of observed mineral deposit grades (DeYoung 1981). In the case of titanium, which is already mined from deposits where its grade is below average crustal abundance, miners seem to have already reached the promised land (Brooks 1976, *149*).

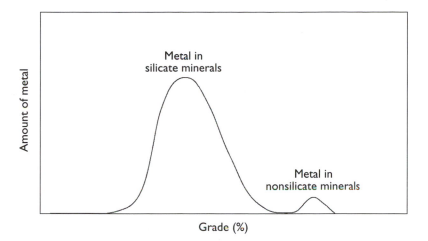

Figure 2-1. *Probable Distribution of Geochemically Scarce Metals in the Earth's Crust (after Skinner 1976)*

It is quite difficult to estimate the total amounts of the earth's metals that are in undiscovered deposits that might become available for mining. Based on an extrapolation of copper in one of the world's most heavily mineralized areas in the southwestern United States, the National Academy of Sciences (COMRATE 1975) estimated that the upper limit on the quantity of copper in ores with grades greater than 0.1 percent that can be produced in the United States is 0.9 billion metric tons (Gt). In a detailed probabilistic assessment of undiscovered mineral resources, the U.S. Geological Survey estimated that about as much copper remains to be found in the United States as has been found to date (USG Survey Minerals Team 1996; USG Survey National Mineral Resource Assessment Team 1998). The total amount of copper discovered in mineral deposits in the United States is 0.4 Gt—about one-fourth of which has already been mined. These independent estimates of the total amount of copper in mineral deposits in the United States are remarkably consistent—an upper limit estimate of 0.9 Gt versus an estimate of 0.8 Gt in known deposits plus estimated undiscovered deposits. Thus, assuming that the deposits are of acceptable grade and size, and discoverable, about seven-eighths of the total resources of copper in the desirable sulfide or oxide form in the United States should yet be available for mining.

But are these resources in deposits with acceptable grades? Worldwide, at least 74 percent of all discovered gold, silver, zinc, or lead is in deposits with grades above the respective median grades, and 44 percent of copper is in deposits with grades above the median grade of all deposits (Singer 1995). Comparison of the proportion of deposits with the proportion of metals grouped in increasing grade classes (average grades of whole deposits) provides a way to examine both frequency of grades and tendency of metals to concentrate at certain grades. All of these data are available because private or public enterprises have expended funds to drill these deposits in the belief that they might be economic to mine (Figure 2-2). Contrary to the common misconception, there is no indication that lower-

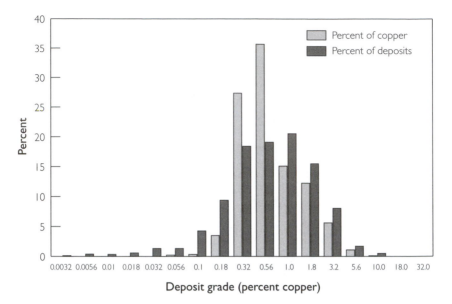

Figure 2-2. *Percent of 2,265 Copper-bearing Deposits by Average Grade of Mineralized Rock Class and Percent of 2.16 Billion Metric Tons of Copper by Copper Grade Class*

grade deposits are more common than higher-grade deposits or that lower-grade deposits contain large amounts of metals; the evidence points to the reverse. Even when summed, lower-grade deposits do not contain significantly more total metal than higher-grade deposits.

Figure 2-3 illustrates frequency of deposit sizes and propensity of metals to concentrate in deposits of certain sizes. Size class values represent lower tonnage limits for the classes except for the lowest class, which also includes any lower values and the highest class, which is unbounded. For each size class, the number of deposits was counted and the associated amount of copper summed; each of these numbers was converted to its percentage of the total number of deposits or metric tons of copper. Most of the copper is in the few largest deposits. A cumulative plot of deposits and copper metal by deposit size (Figure 2-4) shows that the largest 10 percent of copper-bearing deposits contain 76 percent of all copper known. Deposit size is an excellent predictor of contained metal. More than 96 percent of total amounts of copper, zinc, lead, silver, and gold resides in deposits having greater than median deposit size—between 47 percent and 79 percent of all metal is contained in the largest 10 percent of deposits (Singer 1995). Only the largest deposits can have a profound effect on supply. The large size of these deposits makes them amenable to economies of scale. Mineral deposits in the earth's crust are rare, and large ones are especially uncommon. Of fundamental concern in considering future supply is estimation of how many of these large deposits remain to be found and where they might exist.

About 50 percent of the earth's land surface is covered by apparently barren rocks and unconsolidated sediments. Because most mineral deposits exposed at the

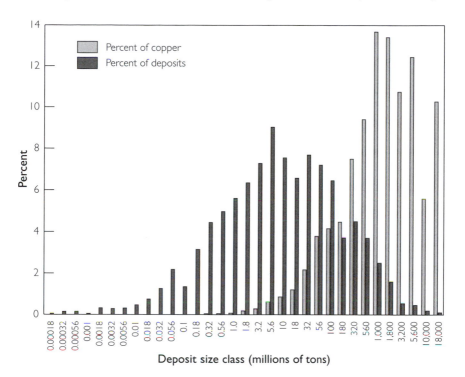

Figure 2-3. *Percent of 2,265 Copper-bearing Deposits by Tonnage of Mineralized Rock Class and Percent of 2.16 Billion Metric Tons of Copper by Tonnage Class*

surface in well-explored regions are believed to have already been found, as in the United States (USG Survey Minerals Team 1996; USG Survey National Mineral Resource Assessment Team 1998), a prime concern is the nature of and depth to possible mineral deposits under this cover. Cover has a profound effect on the likelihood and cost of discovery and the cost of extraction.

Mining engineers have continually lowered mining and processing costs for deposits near the surface. Avoiding mineral scarcity, that is, maintaining current prices, will require comparable low-cost methods for deposits under cover, as well as improved exploration methods.

As demonstrated for copper, probably only the few largest deposits of most mineral resources are capable of supplying the bulk of society's needs. The likelihood of periods of mineral "scarcity" depends on when these few deposits are discovered and whether they will be economically accessible.

In a recent analysis of long-term trends in copper prices, Tilton (2002) states that increases in the supply of copper, especially from Chile, together with reduced costs of production through new technologies have increased copper supplies more rapidly than mineral supplies were reduced by mineral depletion and increased costs of complying with environmental regulations. The resource development pattern for copper is typical of most metals. It is unlikely that the physical availability of mineral resources will be a constraint on mineral produc-

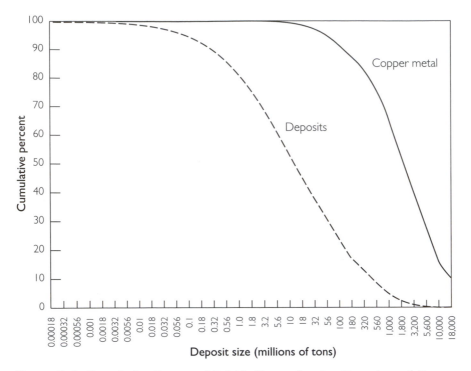

Figure 2-4. *Cumulative Percent of 2,265 Copper-bearing Deposits and Percent of 2.16 Billion Metric Tons of Copper versus Tonnage of Mineralized Rock*

tion in the next 20 years. Even regions, including the United States, that have been intensively explored and have yielded significant mineral production have produced only a fraction of the easily processed metal that they are estimated to contain. Although no assessment of global mineral resources comparable to that done for the United States is available, it seems reasonable to assume that at least a similar proportion of global resource remains elsewhere. Certainly, copper resources are abundant.

The State of Mineral Production

If future episodes of "mineral scarcity" occur, they will likely be the result of temporary imbalances between production and consumption. Because significant lead times are needed to discover and develop new deposits, rapid increases in consumption could outstrip mineral production. In addition, mineral exploration and processing are highly technical fields that require research and development to effectively function. Finally, mineral exploration and production require favorable social and legal frameworks to operate effectively. Thus, the states of mineral exploration and processing, technological developments that support mineral production, and the social and legal environment in which these activities take place are critical to maintaining the flow of minerals used by societies.

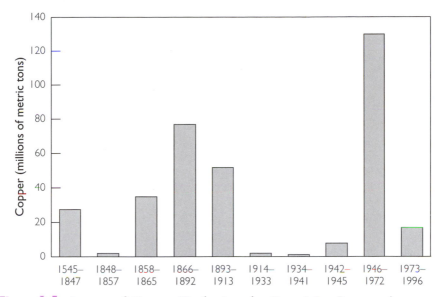

Figure 2-5. *Amount of Copper (Production plus Remaining Resources) Contained in Significant Deposits Discovered in the United States by Historic Period (after Long, DeYoung, and Ludington 2000)*

Exploration

During the 1980s and 1990s, new deposits of copper were developed in Chile and Indonesia. The discovery and development of these deposits during the past several decades is one of the factors that have led to recent decreases in copper production costs (Tilton 2002, 26). However, many of these deposits were the result of grass-roots exploration in the 1960s and 1970s, periods of intense exploration for copper in the United States (Long, DeYoung, and Ludington 2000) and in South America (see Figure 2-5). Exploration since the early 1970s may not have been extensive enough to provide the additional deposits needed to support future consumption.

In recent years, reported exploration budgets have been directed mainly toward precious minerals, such as gold, platinum-group metals, and diamonds; base metals like copper have drawn a smaller proportion. Mineral exploration budgets (Figure 2-6) have declined since 1997 (Wilburn 2002); although the percentage of funds has varied by region, that for exploration has dropped in all regions. Further, doubts have been raised about whether an adequate inventory of deposits has been identified to meet projected needs. Recently, an industry geologist pointed out that although five large porphyry copper deposits must be developed to replace the production from deposits that will be depleted by 2014, no one in industry can identify the replacement deposits (Smith 2001), and the study did not consider increases in consumption that will occur. Finally, industry exploration programs are being restructured. Consolidations of mining companies and changing exploration philosophies have increased the role of junior mining companies in exploration. The objective of this change is to produce smaller, more efficient exploration teams. It is not yet clear,

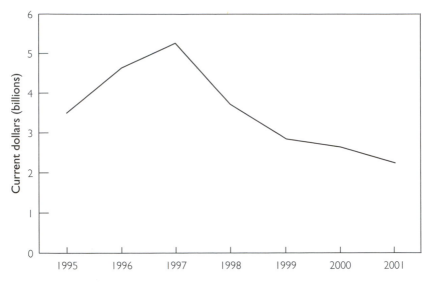

Figure 2-6. *World Exploration Budgets (after Wilborn 2002)*

Source: Metals Economics Group.

however, whether these changes will furnish adequate exploration budgets and a sufficient number of such teams to discover needed new resources. Many university geology and mineral economics programs in the United States have been restructured and reduced in size in response to declining demand for graduates. This could limit the number of newly trained professionals available to staff new exploration teams unless foreign universities expand their programs in these disciplines.

Other developments point to a more positive evaluation of recent exploration efforts. Deposits are still being found at or near the earth's surface using conventional mineral exploration techniques (geologic prospecting, geochemical surveys, remote sensing, and other geophysical methods) in regions that have not drawn mineral exploration in the past. The Batu Hijau deposit in eastern Indonesia and the Bajo de la Alumbrera deposit in Argentina are among the deposits discovered in the early 1990s and recently brought into production. In addition, recent exploration has located promising porphyry copper prospects now undergoing further exploration in Ecuador, Mongolia, and western China.

Although the geographic broadening of exploration to regions that have previously drawn limited exploration has resulted in the discovery of new deposits at or near the surface, in the longer term, an increasing fraction of the remaining resource, especially in extensively explored regions, will be in deposits that are concealed by younger rocks. These resources will certainly be more difficult to discover and more costly to mine.

Technological Developments

The second key to the increase in copper supplies during the 1980s and 1990s was the adoption of new, more efficient technologies to mine and to process copper ores (Tilton 2002). Many mining professionals expect technology to continue to increase the efficiency of exploration, mining, and mineral processing (Ward 1996). Perhaps the most obvious new technologies are mining equipment, especially the larger trucks used in open pit mines. The larger scale of open pit operations has permitted surface mining of ground that had heretofore been considered only for underground operations if grades were high enough. Open pit mines are particularly suited to gains through economies of scale. However, increased scrutiny of the environmental effects of mining and the fact that a greater proportion of discoveries will lie beneath cover imply that more deposits will be processed by underground methods which benefit less from economies of scale.

A second gain in efficiency has been the result of the use of hydrometallurgical methods to process ores. Large-scale leaching of copper ores and tailings has dramatically lowered the costs of producing copper, though at the cost of creating large leach piles. New technologies, such as the in-situ leaching of sulfide ores and the use of robotics in mining, offer the hope of future cost reductions. But reduced public and private support for research in mining and mineral processing are reasons for concern.

The Social and Legal Environments

Changing societal attitudes during the last 40 years in the United States have greatly reduced the amount of land open to mineral exploration. Beginning with the passage of the original Wilderness Act in 1964, large amounts of public lands have been designated as Wilderness Areas, National Parks, and National Monuments. Limiting the amount of land available for exploration significantly reduces the likelihood that the critical few large deposits will be discovered (Singer and DeYoung 1980; Singer 1995; Allais 1957). In addition, growth of the population, especially in urban areas in the mountain west (Arizona, Colorado, Idaho, Montana, Nevada, New Mexico, Utah, and Wyoming) has been accompanied by increased scrutiny of mineral projects, restrictions on certain methods of production, and increased costs of permits for mineral producers. This has led mining companies to allocate more of their budgets to foreign exploration (Ward 1996). Scrutiny of mineral projects outside of the United States, however, is also increasing. Nongovernmental organizations that oppose mineral extraction are increasing their international presence, as evidenced by the dispute over proposed mining of the Tambo Grande deposit in Peru.

Increasing societal concern about large-scale alteration of landscapes could dramatically affect resource production because open-pit mining and hydrometallurgical processing of ores both significantly change landscapes. Because technological developments related to these two methods of producing metals have been responsible in large part for reductions in production costs over the last several decades, limits on landscape alteration could dramatically increase the costs of mineral production.

In addition, societal concern about production residuals, especially metals such as mercury and arsenic, and use of mineral commodities and products is increasing. The technology to reduce the release of these elements to the environment is also likely to increase costs.

Mineral Consumption

The physical existence of minerals and the ability and willingness to produce minerals only partially determine whether minerals will be abundant or scarce. The level at which minerals are consumed is also important. Rapid economic development in Asia is increasing consumption and raises the question of whether mineral production will be adequate to meet the needs of society over the next 20 years. The answer to this question will in part depend on how fast developing countries increase their mineral consumption and whether developed countries are able to reduce theirs.

In theory, consumption of minerals can be measured at different stages of use: as raw minerals by mineral processing industries, as processed minerals by industries, and as minerals in final goods by consumers. Consumption is most commonly measured at the industrial stage because data on production and trade in primary metal products are easily obtained. Measuring mineral consumption in final goods is difficult because it requires disaggregating the finished goods into component materials, a process that can require a large investment of time and strong assumptions about the compositions of classes of goods.

Patterns of Industrial Consumption

The effects of development on mineral consumption have been examined by recent studies of changes in per capita apparent consumption of aluminum, cement, copper, and salt. One covered Korea, Japan, and the United States from 1965 to 1995, another included the 20 most populous countries from 1970 to 1995 (DeYoung and Menzie 1999; Menzie et al. 2001).

The former study found that in the United States per capita consumption of cement, copper, and salt did not significantly change over the period of study. In Korea, per capita consumption of aluminum, cement, and copper increased dramatically, while the consumption of salt grew slowly. In Japan, per capita consumption of all four commodities grew at slow rates. See Figure 2-7. By the end of the period, Japan and the United States had achieved a similar level of development as measured by per capita GDP, and they consumed similar levels of these metals per capita, copper (10 kg) and aluminum (30 kg), whereas their per capita consumption of cement and salt, a construction material and industrial mineral, respectively, differed by a factor of two. The order in which growth began in Korea may be significant. Cement consumption grew earliest, followed by copper and aluminum. This could reflect development of basic infrastructure, followed by development of workplaces and light manufacturing, followed by manufacture of heavier structural goods.

The second study included additional developed or high-income countries (Germany, the United Kingdom, and France), which, like the United States and Japan, did not show significant changes in per capita consumption of cement and copper over relatively long periods, but did show increasing per capita consumption of aluminum. This study also included countries that, like Korea, were undergoing significant development (increases in income) and significant increases in per capita consumption of aluminum, cement, and copper. China, Thailand, and

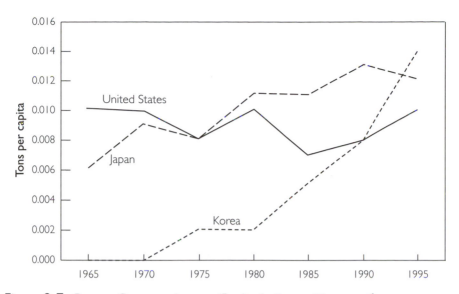

Figure 2-7. *Copper Consumption per Capita in Japan, Korea, and the United States*

Turkey show significant increases in per capita consumption of all three commodities, although Thailand's consumption of copper declined slightly in 2000 and Thai consumption of cement declined significantly in response to the Asian economic crisis of the late 1990s. Several other developing countries—Mexico (aluminum and cement), Egypt (cement), and Iran (copper)—show significant increases in one or two commodities. Finally, two additional groups of countries show other patterns of consumption. India and Indonesia also showed increases in per capita consumption of several commodities, but because their initial level of consumption was low, growth is only beginning to appreciably increase their total consumption of minerals. Another group of countries, including Bangladesh and Nigeria, show no significant consumption of mineral commodities.

Together, the two studies show a consistent pattern of per capita consumption of minerals with increasing income. At very low income levels, consumption is very low. During the initial states of income growth, it increases slowly. After income has reached a threshold level, per capita consumption increases very rapidly. Finally, when countries reach high levels of income, the per capita consumption is essentially constant. The pattern of use of these commodities is consistent with the phenomenon of dematerialization, or declining consumption per unit of economic activity (Wernick et al. 1996). The pattern, however, does not imply decreasing per capita consumption. The results of the two studies also suggest that levels of per capita consumption of metals in developed countries may be similar, while the per capita consumption of construction materials and industrial minerals may differ from country to country. Taken together the pattern of increasing per capita consumption of minerals with increasing incomes, and the similar absolute levels of use of metals form a basis for predicting future consumption of metals.

A Model of Copper Consumption

This pattern defines a growth curve that may be modeled by a logistic function:

$$C = \left(\frac{K}{1 + e^{-r \log(i)}} \right) P$$

where

C is consumption of a commodity
K is a constant representing the saturation level of the per capita consumption of the commodity in an economy
r is a constant
i is per capita income, or per capita GDP
P is population

Consumption in the future may be calculated by adjusting per capita GDP for real growth of GDP.

Because the consumption of copper in all high-income countries became stable at about the same level, copper was selected to develop a model of consumption as a function of income and population. The equation was used to estimate future levels of copper consumption in the 20 most populous countries in 2020. The model forecasts that copper consumption of the 20 countries will be 19 Mt in 2020. If the 20 countries consume the same proportion of world copper in 2020 as they did in 2000, world copper consumption in 2020 would be 24 Mt. This is 1.8 times 2000 consumption and represents an average rate of growth of world copper consumption of 3.1 percent, which is about 10 percent faster than the annual growth rate (2.8 percent) of per capita consumption of copper between 1980 and 2000.

Developing countries have been increasing their proportion of world copper consumption. In 1980, France, Germany, Japan, the United Kingdom, and the United States accounted for 68 percent of world copper consumption. By 2000, these five countries accounted for only 51 percent. The model predicts that they will account for only 30 percent of the world's consumption of copper by 2020. One way to comprehend the magnitude of change implied by the model is to consider which countries consume more than 1 Mt of copper. In 1980, the Soviet Union consumed 1 Mt, Japan 1.2 Mt, and the United States 2.2 Mt. In 2000, Germany and Japan each consumed 1.3 Mt, China 2 Mt, and the United States 3 Mt. The model estimates that in 2020, Brazil will consume 1.2 Mt, Japan 1.4 Mt, India 1.6 Mt, the United States 3.5 Mt, and China 5.6 Mt. Such estimates may appear remarkable until one recalls the growth that Korea achieved in 20 years.

This projected growth in consumption has important implications about the need to develop new reserves of copper and other minerals. The U.S. Geological Survey reports world copper reserves in 2001 of 340 Mt (Edelstein 2002). In 1995, world copper consumption was 10.5 Mt. Copper reserves are equal to about 30 years of supply at the 1995 level of consumption. The logistic model predicts world consumption of more than 24 Mt per year by 2020. If consumption reaches this level and reserves are kept proportional to consumption, at about 30 years of consumption, and allowance is made for intervening draw down of reserves, more

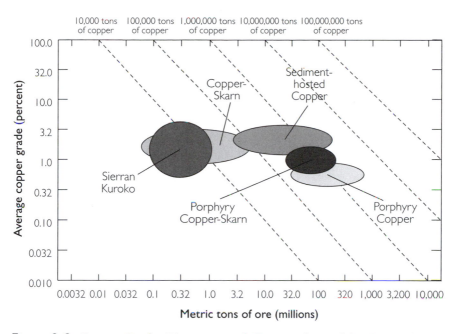

Figure 2-8. *Copper Grades, Tonnages, and Contained Metal by Deposit Type*

Note: Each circle represents 45 percent of the deposits.

than 700 Mt of copper must be added to world reserves, assuming that 10 percent of consumption will come from secondary sources. Porphyry copper and sedimentary copper deposits currently contain the bulk of the world's copper reserves. The ability to develop these additional reserves depends on new discoveries of large porphyry copper or sedimentary copper deposits or significant deposits of a new type. On average, porphyry copper deposits contain more metal than sedimentary copper deposits (Figure 2-8). The five largest porphyry copper deposits contain on average about 68 Mt of copper. To maintain the predicted growth in consumption and retain a 30-year reserve will require the discovery of 1.1 Gt of copper in new deposits by 2020. To put this immense amount into perspective, it means the discovery and development of 16 additional deposits, each of which is as large as the average of the 5 largest porphyry copper deposits known from a century of exploration.

Embodied Consumption

Although the stabilization of industrial consumption of many minerals in high-income countries seems to lend support to the argument that high-income countries can sever the link between economic growth and material consumption, a number of recent studies, including Roberts (1987) and Al-Rawahi and Rieber (1990), have noted that industrial consumption of metals may not be a good indication of final consumption of some minerals because many developed countries import significant quantities of metals embodied in manufactured goods.

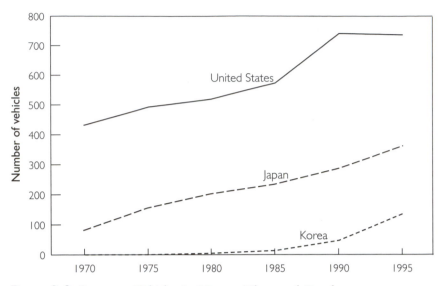

Figure 2-9. *Passenger Vehicles in Use per Thousand People*

Al-Rawahi and Rieber estimated the consumption of copper embodied in manufactured goods from 1965 to 1985 and compared estimates of industrial consumption with total domestic consumption of copper. Although their data did not show an increase in consumption across the period of study, it did indicate that total domestic consumption of copper was about 16 percent higher than industrial consumption of copper.

Embodied consumption of nonmetallic mineral commodities is largely unstudied. Although it is easy to imagine significant embodied consumption, especially of those minerals used as fillers and extenders, the embodied consumption of mineral materials such as cement is probably negligible.

Growing levels of embodied flows could call into question the conclusion that rates of mineral consumption are stabilizing in developed countries and increasing in developing countries, which could instead be exporting minerals to developed countries as manufactured goods. One way to examine the growth of materials consumption in developed and developing countries is to examine a time series of the number of passenger vehicles in use per thousand people. Such data substantiate recent reports that consumption of personal vehicles is rising rapidly in China (Murphy and Lague 2002; *Washington Post* 2002). Cross-sectional data indicate that vehicles in use per capita increase with income; Figure 2-9 shows the number of passenger vehicles in use per thousand people in the United States, Japan, and Korea from 1970 to 1995. Over this period, the number of passenger vehicles per thousand people increased an average of 2.1 percent per year in the United States, 6 percent per year in Japan, and 18 percent per year in Korea. The rates of growth correspond to economic development in the three countries during the period. Korea's increase in passenger vehicles appears to lag the increase in aluminum consumption by about 10 years, the increase in copper consumption by about 15 years, and the increase in cement consumption by about 20 years.

Although these data are crude and ignore important information, such as weight and material composition of the vehicles, the fact that the United States was still increasing its use of passenger vehicles per capita provides some reason to question whether rates of material consumption have stabilized in developed countries. The number of passenger vehicles per capita increased an average rate of 2.1 percent in France, 2.3 percent in Germany, and 2.3 percent in the United Kingdom. These data reinforce the impression that, at least for some uses, material consumption may still be rising in developed countries.

Environmental Residuals

Although rapid increases in mineral consumption could result in temporary scarcity, environmental problems posed by increased mineral consumption are likely to be more difficult to solve. Such problems could include increased use of land for mineral production, increased wastes related to mining, mineral processing, and manufacture of goods, and increased disposal of obsolete goods. Concerns about the environmental effects of material production and use in modern economies has led to studies of the flows of materials that take place between sectors of industrial economies and to efforts to develop a system of physical accounts that measure these flows in order to develop measures of the efficiency with which economies use materials (Matthews et al. 2000). One of the benefits of such a system of flows would be to associate wastes generated by early stages of the mineral production cycle in one country with their end use, which because of the global nature of mineral production and consumption may take place in other countries.

Matthews et al. (2000) have estimated material flows that are *outputs to* the environment *from* industrial economies. In addition to measuring those materials that become additions to the nation's stock of infrastructure and durable goods, they quantified two types of materials: those used and then discarded in the production process and the "hidden flows" that are not measured in economic accounting (overburden in mining, soil erosion related to farming and other human activities). Together, these amounts represent the total quantities of material outputs and material displacement within national borders. In the United States more than 25 tons per person of domestic process output and more than 60 tons per person of hidden flows were returned to the environment in 1996. Coupled with prospects of significantly increased consumption of mineral commodities in developing economies, these additional quantities of moved materials add complexity to the task of achieving sustainable development. More than 30 tons per person were generated as overburden and waste from mining. More than 20 tons of carbon dioxide emissions were generated as a result of burning of fossil fuels.

The magnitude of wastes that will be generated by increasing copper production will be enormous. On average, about 350 tons of waste rock and 147 tons of tailings are generated for each ton of copper produced (Kesler 1994, *211*). Assuming that grades of ore produced do not change, these data and the results of the model of copper consumption imply about 130 Gt of waste rock and 56 Gt of tailings from the mining and milling of copper ores will be generated between 2000 and 2020. Wastes from converting copper concentrate to copper metal are somewhat more difficult to calculate because copper metal may be made either by smelting copper sulfide ores or by chemically leaching copper oxide and carbonate ores. Copper smelting can release sulfur dioxide and arsenic to the atmosphere and hydrosphere. The amount released depends on the proportions of copper and arsenic that are smelted and the technology employed in the smelter. Estimation of wastes from the stages of the mineral production cycle allows the identification and prioritization of efforts to reduce waste. Some wastes will by their nature be difficult to reduce. For example, the amounts of waste rock and tailings are unlikely to be reduced unless copper ores are leached in place, and in-place leaching would create other environmental challenges. The amounts of arsenic and sulfur dioxide released into the environment are more amenable to reduction by the adoption of new technologies. New copper smelters currently are able to recover 99.99 percent of the sulfur in the ore.

New strategies of reducing waste, such as materials flow and industrial ecology (Graedel 1996), have shown some success in reducing environmental releases of some toxic elements such as lead (Socolow and Thomas 1997) by reducing the use of products that contain toxic materials and instituting mandatory recycling of products, such as car batteries, that contain toxic materials. If wastes from material production and use are to be reduced, additional studies of materials flow within individual industries, such as a recent study of environmental challenges and opportunities presented by the manufacture of cement (van Oss and Padovani 2003) will be needed. Such studies can identify opportunities for industrial symbiosis in which waste from one industry can become an input to another industry.

Besides the residuals that are the direct result of producing a commodity, there are residuals that enter the environment due to use and final disposal of the goods made from the commodity. Although recycling of material such as steel, aluminum, and even aggregate from road construction has increased significantly in recent years, studies question whether recycling rates have nearly peaked (Barrett 2000; Fialka 2002; Hamilton 2002, Logomasini 2002). Remanufacture and reuse of materials offer additional opportunities for reducing material consumption, but legal and regulatory changes may be required to make them common practice.

One of the barriers to recycling, remanufacturing, and reuse is the cost of disassembling finished goods and sorting the component parts. These barriers may be lowered in the future through the use of radio chips, which are being developed to control product inventory. If this technology were applied to component parts and if products were designed for easy disassembly, the costs of sorting used parts for reuse, remanufacture, or recycling could be significantly reduced. Rising costs of primary materials and competition from developing countries could provide an incentive to manufacturers in developed countries to undertake the re-engineering of products that will be necessary if these new strategies are to be adopted.

Conclusions

Rapid economic development in Asia and the resulting increase in consumption of mineral commodities have revived concerns about resource scarcity. But the effects of mineral production and use on global environmental systems are equally troubling.

Estimates of undiscovered mineral resources suggest that mineral supply per se is not a limit on growth. In addition, mineral deposits that occur at or near the surface are still being found in some regions. Over the next 20 years, however, an increasing proportion of unexploited resources are likely to be in deposits concealed deep beneath covering rock or sediments. Such deposits will be more difficult to find and probably more costly to mine.

Although undiscovered mineral resources are abundant, rapid increases in mineral consumption could create temporary supply problems. The decline in mineral exploration budgets since 1997, together with structural changes within corporations engaging in mineral exploration, and in universities that teach economic geology and mineral economics, are reasons to doubt that a ready inventory of deposits will be available to meet increases in mineral consumption implicit in the growth of developing economies. In addition, reduced funding from governments and private sources for research in mining engineering and mineral processing, combined with staffing cuts in associated departments of universities, makes it unclear whether technical innovations will continue to reduce the costs of mineral production as they have over the past 40 years. Finally, social commitment to resource extraction is declining. In the United States, environmental legislation and population growth have impeded exploration. Similar trends can be seen around the world.

Increased mineral production and use will exacerbate known problems—appropriation of land, waste from mining and manufacturing, and disposal of obsolete goods—that demand new strategies. Recycling rates seem to be reaching their limits and, despite talk about dematerialization, per capita consumption of a number of mineral commodities in developed countries has remained roughly constant for the last 30 years. In addition, a number of developing countries with large populations are reaching levels of economic development where per capita mineral consumption is likely to rise rapidly.

References

Allais, M. 1957. Method of Appraising Economic Prospects of Mining Exploration over Large Territories. *Management Science* 3(4): 285–347.

Ausubel, J.H. 1996. Can Technology Spare the Earth? *American Scientist* 84(2): 166–178.

Barrett, R. 2000. Can Recycling Run Out of Steam? *Metal Bulletin Monthly* 351(March): 12–17.

Barton, P.B., Jr. 1983. Unconventional Mineral Deposits—A Challenge to Geochemistry. In *Cameron Volume on Unconventional Mineral Deposits*, edited by W.C. Shanks III. New York: Society of Mining Engineers of the American Institute of Mining, Metallurgical, and Petroleum Engineers, 3–14.

Brooks, D.B. 1976. Mineral Supply as a Stock. In *Economics of the Mineral Industries*, 3rd edition, edited by W.A. Vogely. New York: American Institute of Mining, Metallurgical, and Petroleum Engineers, Inc., 127–207.

COMRATE. 1975. Resources of Copper. In *Mineral Resources and the Environment*. Washington, DC: National Academy of Sciences, 127–183.

DeYoung, J.H., Jr. 1981. The Lasky Cumulative Tonnage–Grade Relationship: A Re-Examination. *Economic Geology* 76(5): 1067–1080.

DeYoung, J.H., Jr., and W.D. Menzie. 1999. The Changing Uses of Minerals Information—A Government Perspective. In *Proceedings of the Workshop on the Sustainable Development of Non-*

Renewable Resources towards the 21st Century, New York, October 15–16, 1998, edited by James Otto and Hyo-Sun Kim. New York: United Nations Development Programme, 111–127.

DeYoung, J.H., Jr., and D.A. Singer. 1981. Physical Factors That Could Restrict Mineral Supply. In *Economic Geology 75th Anniversary Volume, 1905–1980*, edited by B.J. Skinner. Lancaster, PA: Economic Geology Publishing Co., 939–954.

During, A. 1992. *How Much Is Enough?* New York: W.W. Norton.

Edelstein, D.L. 2002. Copper. In *Mineral Commodity Summaries 2002*. Washington, DC: Government Printing Office, 54–55.

Erickson, R.L. 1973. Crustal Abundance of Elements, and Mineral Reserves and Resources. In *United States Mineral Resources*. U.S. Geological Survey Professional Paper 820, edited by D.A. Brobst and W.P. Pratt. Washington, DC: Government Printing Office, 21–25.

Fialka, J.J. 2002. Recycling Faces a Heap of Trouble. *Wall Street Journal*, July 9, A2.

Gordon, R.L. 1972. The Revival of Exhaustion of [sic] Malthus Rents a Computer. *Mining Engineering* 24(7): 322.

Graedel, T.E. 1996. On the Concept of Industrial Ecology. *Annual Review of Energy and the Environment* 21: 69–98.

Hamilton, M.M. 2002. Is Recycling Being Canned? Budget Constraints, Other Factors Slowing Progress. *Washington Post*, Sept. 22, H1, H5.

Hewett, D.F. 1929. Cycles in Metal Production. *American Institute of Mining and Metallurgical Engineers Transactions* 85: 65–93.

Kesler, S.E. 1994. *Mineral Resources, Economics and the Environment*. New York: Macmillan College Publishing Co.

Lasky, S.G. 1950. How Tonnage and Grade Relations Help Predict Ore Reserves. *Engineering and Mining Journal* 151(4): 81–85.

Logomasini, A. 2002. Forced Recycling Is a Waste. *Wall Street Journal*, Mar. 19, A22.

Long, K.R., J.H. DeYoung Jr., and S. Ludington. 2000. Significant Deposits of Gold, Silver, Copper, Lead, and Zinc in the United States. *Economic Geology* 95(3): 629–644.

Matthews, E., C. Amann, S. Bringezu, M. Fischer-Kowalski, W. Huttler, R. Kleijn, Y. Moriguchi, C. Ottke, E. Rodenburg, D. Rogich, H. Schandl, H. Schütz, E. Van Der Voet, and H. Weisz. 2000. *The Weight of Nations—Material Outflows from Industrial Economies*. Washington, DC: World Resources Institute.

McKelvey, V.E. 1960. Relations of Reserves of the Elements to Their Crustal Abundance. *American Journal of Science* 258-A (Bradley volume): 234–241.

Meadows, D.L., D.H. Meadows, J. Randers, and W.W. Behrens III. 1972. *The Limits to Growth: A Report for the Club of Rome's Project on the Predicament of Mankind*. New York: Signet Books.

Menzie, W.D., J.H. DeYoung Jr., and W.G. Steblez. 2001. Some Implications of Changing Patterns of Mineral Consumption. Paper presented in the World Mining Policies session of World Mining Congress XVIII at MINExpo INTERNATIONAL 2000®. October 9–12, 2000, Las Vegas, NV. Available online as U.S. Geological Survey Open-File Report 03-382 at http://pubs.usgs.gov/of/2003/of03-382.

Murphy, D., and D. Lague. 2002. As China's Car Market Takes Off, the Party Grows a Bit Crowded. *Wall Street Journal*, July 3, A9.

Al-Rawahi, K., and M. Rieber. 1990. Embodied Copper—Trade, Intensity of Use and Consumption Forecasts. *Resources Policy* 17(1): 2–12.

Roberts, M.C. 1987. The Consumption of Metals and International Trade. *Materials and Society* 11(3): 391–406.

Rostow, W.W. 1960. *The Stages of Economic Growth—A Non-Communist Manifesto*. Cambridge: Cambridge University Press.

Simon, J.L. 1980. Resources, Population, Environment—An Oversupply of False Bad News. *Science* 208(4451): 1431–1437.

Singer, D.A. 1995. World-Class Base and Precious Metal Deposits—A Quantitative Analysis. *Economic Geology* 90(1): 88–104.

Singer, D.A., and J.H. DeYoung Jr. 1980. What Can Grade–Tonnage Relations Really Tell Us? In *Ressources Minerals—Mineral Resources*. Bureau de Recherches Géologiques et Minières Memoire 106, edited by Claude Guillemin and Philippe Lagny, 91–101.

Skinner, B.J. 1976. A Second Iron Age Ahead? *American Scientist* 64(3): 258–269.

Smith, L.D. 2001. Informal Presentation by Manager, Project Evaluations, Billiton Base Metals, Toronto, from a 2000 Brook Hunt study at a Resources for the Future Workshop on the Long-Run Availability of Minerals. April 23, Washington, DC.

Socolow, R.H., and V.M. Thomas. 1997. The Industrial Ecology of Lead and Electric Vehicles. *Journal of Industrial Ecology* 1(1): 13–36.

Tilton, J.E. 1996. Exhaustible Resources and Sustainable Development—Two Different Paradigms. *Resources Policy* 22(1–2): 91–97.

———. 2002. Long-Term Trends in Copper Prices. *Mining Engineering* 54(7): 25–32.

USGS (U.S. Geological Survey). 2003. *Mineral Commodity Summaries 2003.* Washington, DC: Government Printing Office.

USGS (U.S. Geological Survey) Minerals Team. 1996. *Data Base for a National Mineral-Resource Assessment of Undiscovered Deposits of Gold, Silver, Copper, Lead, and Zinc in the Conterminous United States.* Open-File Report 96-96. Reston, VA: U.S. Geological Survey.

U.S. Geological Survey National Mineral Resource Assessment Team. 1998. *1998 Assessment of Undiscovered Deposits of Gold, Silver, Copper, Lead, and Zinc in the United States.* U.S. Geological Survey Circular 1178. Washington, DC: Government Printing Office.

van Oss, H.G., and A.C. Padovani. 2003. Cement Manufacture and the Environment, Part II—Environmental Challenges and Opportunities. *Journal of Industrial Ecology* 7(1): 93–126.

Ward, M.H. 1996. Mining in the Twenty-First Century—Who, How and Where? 5th Annual G. Albert Shoemaker Lecture in Mineral Engineering. *Earth and Mineral Sciences.* University Park, PA: Pennsylvania State University. 65(2): 14–18.

Washington Post. 2002. In China, a Rush to Get Behind the Wheel. June 6, A1, A24.

Wernick, I.K., R. Herman, S. Govind, and J.H. Ausubel. 1996. Materialization and Dematerialization—Measures and Trends. *Daedalus (Journal of the American Academy of Arts and Sciences)* 125(3): 171–198.

Wilburn, D.R. 2002. Exploration—Annual Review 2001. *Mining Engineering* 54(5): 26–36.

CHAPTER 3

Economics of Scarcity
The State of the Debate

Jeffrey A. Krautkraemer

W HETHER ECONOMIC GROWTH CAN be sustained in a finite natural world is one of the earliest and most enduring questions in economic literature. In essence, the issue is whether technological progress and capital accumulation can overcome diminishing marginal returns to finite natural resources. The debate began with the birth of economics as a separate discipline and continues to this day. Its intellectual roots still play a prominent and significant role. It is the topic of the two previous volumes on scarcity and growth published by Resources for the Future. While the general nature of the debate is unchanged, the focus and topics of discussion have evolved.

The past two centuries have seen unprecedented growth in human population and economic well-being for a good portion of the world. This growth has been fed by equally unprecedented natural resource consumption and environmental impacts, including conversion of large portions of the natural world to human use, which have prompted recurring concern about whether the world's natural resource base is capable of sustaining such growth. To some degree, this concern is supported by simple mathematics: exponential physical growth in a finite world eventually generates absurd results. For example, any positive population growth rate eventually has the population completely covering the face of the earth and expanding rapidly into space; any positive growth rate for petroleum consumption eventually results in annual production that is greater than the mass of the earth.

While exponential growth can be expected to lead to increasing resource scarcity, human creativity can ameliorate increased scarcity. Humans have been quite adept at finding solutions to the problem of scarce natural resources: finding more abundant substitutes for various natural resources, exploration for and discovery of new reserves, recovery and recycling of materials, and, perhaps most importantly, the development of new technologies that economize on scarce natural resources or that allow the use of resources that were previously uneconomical.

These responses are not automatic but are the result of purposeful activity in response to signals of increased scarcity. Successful outcomes are not guaranteed. Consequently, there has been a persistent tension between impending scarcity and technological progress. Recent decades have seen increasing concern about the environmental impact of population and economic growth. Because environmental resources—ecosystem services and "resource amenities"—are not generally traded on markets, scarcity signals for these resources may be inadequate, and appropriate policy responses are difficult to implement and manage.

This chapter reviews the extensive scope of the debate over the economic scarcity of natural resources and assesses its current state. One significant change in recent years is a greater focus on the ecosystem services and the resource amenities yielded by natural environments—a shift from food, timber, coal, iron, copper, and oil to air and water quality, global climate, and ecosystem preservation.

The distinction between resource commodities and resource amenities is an important one. The answers to central questions regarding natural resource scarcity differ across the two types of natural resource goods and services. This chapter discusses the nature of resource amenities and the significant management challenges they present. Then it reviews current empirical and theoretical findings. The general conclusion is that technological progress has ameliorated the scarcity of natural resource commodities, but resource amenities have become more scarce, and it is unlikely that technology alone can remedy that.

Brief Historical Overview

The scarcity and growth debate began in earnest with Thomas Malthus's observations on the fecundity of human nature and the relative stinginess of Mother Nature (Malthus 1798). Diminishing marginal returns was a cornerstone of classical economics and played an important role in his pessimistic view of the prospects for economic improvement. In this view, as more capital and labor inputs were applied to a fixed amount of land, the marginal product of capital and labor combined would eventually decrease, as would output per capita. Expansion of agricultural activity to previously uncultivated land was not a solution, because the best agricultural land would have been put into production first. Productivity could increase with technological improvements, but the pace of technological progress up to that time had been slow so this was not given great weight by Malthus and other classical economists of his era.

The other half of the Malthusian dilemma was mankind's propensity to reproduce. If wages were above a subsistence level, Malthus argued, then family size would increase. Population growth combined with diminishing marginal returns would reduce wages back to a subsistence level, or even below, and stem population growth through malnutrition, famine, and delayed marriage. Because population tended to increase geometrically while agricultural output increased arithmetically, the demand for food would necessarily bump up against the ability to produce food, the end result being a subsistence standard of living for most of the population.

Malthus wrote at a time of great social upheaval. The English population was growing rapidly and the rising prices of basic foodstuffs were kept high by restrictions on grain imports. The enclosure movement had moved thousands from their traditional agricultural lands to cities, where many were unable to find work and lived on relief. Malthus could not have foreseen the rapid technological progress and the decline in fertility rates that would allow large portions of the world to avoid the Malthusian population trap. Some would argue that this is because human society has been living off its natural capital endowment, while others would argue that humankind's ingenuity in finding solutions to resource constraints has allowed it to prosper.

The nineteenth and twentieth centuries were punctuated with misgivings that adequate natural resources were not available to sustain economic growth. The mid-nineteenth-century British economy was heavily dependent on coal for energy, and Stanley Jevons (1865) argued that as the cost of coal increased it would undermine economic activity. But shortly thereafter, petroleum displaced coal for many purposes.

In the United States, the conservation movement of the late nineteenth and early twentieth centuries opposed depletion of a broad range of natural resources, including minerals, forests, soil, and fisheries. Its focus was efficiency in a technological sense, based on the belief that private interests could not make the best use of current natural resources and conserve them for future generations. Instead, scientific management in the public interest was necessary to achieve "the greatest good for the greatest number."

The rapid growth of the U.S. economy following World War II spurred further concern about natural resources, particularly for defense purposes as the Cold War began. Resources for the Future was born, in part, as an outgrowth of the 1950–1952 President's Materials Policy Commission.

RFF's *Scarcity and Growth* (Barnett and Morse 1963), the first systematic empirical examination of historical trends, analyzed scarcity hypotheses for a variety of natural resources for the period 1870–1958. Except in the case of forests, the empirical evidence—discussed in more detail below—supported decreasing rather than increasing scarcity.

Scarcity and Growth Reconsidered (Smith 1979), RFF's second volume on scarcity and growth, appeared at a time of heightened concern about natural resources and the environment. Natural resource prices, especially energy prices, were increasing, and deteriorating air and water quality and other environmental problems had led to the enactment of a profusion of environmental legislation. The price increases began before 1970 and were exacerbated in the 1970s by the oil embargo and OPEC price increases. The Club of Rome published *The Limits to Growth* (Meadows et al. 1972), which predicted dire consequences by the early part of the twenty-first century unless population and economic growth were significantly curtailed. The tone of *Scarcity and Growth Reconsidered* was less hopeful than that of *Scarcity and Growth*, although overall it was cautiously optimistic, at least with respect to the availability of natural resource commodities. The values of environmental amenities and basic life support systems were mentioned as important, but the focus remained on productive resource inputs.

The most recent renewal of concern over natural resource scarcity began in the mid-1980s under the rubric of "sustainability" or "sustainable development." The term "sustainability" has powerful connotations even if its exact meaning cannot be pinned down. It has become a catch phrase in the current debate about the ability of the natural world to support both current and future population and economic activity. One key element in this renewal of the debate is the much greater focus on the resource amenities provided by the natural environment. The effects of current economic activity on basic environmental life support systems now seem more critical than the availability of particular natural resource commodities.

Ecosystem Services and Resource Amenities

Natural resource commodities used to produce material goods and services are not the only economic services provided by the natural world. Other services include the basic life support systems of the earth: the air, fresh water, carbon, nitrogen, and nutrient cycles; the climate in which we live and to which the flora and fauna have adapted; the sinks where we deposit the waste products of production and consumption; and the ecosystems that support our agricultural and other economic activities. The natural world serves as a storehouse of genetic information and the original source of many of the world's pharmaceutical products. It provides the "playgrounds" where many of us recreate and which we often observe with wonder. These goods and services are known by the term "resource amenities." While "amenities" may not be the best appellation—they do include fundamental services—it does distinguish them from the use of natural resources in the production of the commodities more commonly treated as economic goods and services.

These amenity resources have played a role in the economic growth debate at least since the time of John Stuart Mill, who observed,

> Nor is there much satisfaction in contemplating the world with nothing left to the spontaneous activity of nature; with every rood of land brought into cultivation, which is capable of growing food for human beings; every flowery waste or nature pasture ploughed up, all quadrupeds or birds which are not domesticated for man's use exterminated as his rivals for food, every hedgerow or superfluous tree rooted out and scarcely a place left where a wild shrub or flower could grow without being eradicated as a weed in the name of improved agriculture. (1848)

Environmental concerns are raised in both *Scarcity and Growth* and *Scarcity and Growth Reconsidered,* but they are secondary in both volumes.

The very title of John Krutilla's seminal paper, "Conservation Reconsidered" (1967), highlighted a new focus of conservation themes. Although the initial research concerning resource amenities had concentrated on recreation and wilderness preservation rather than ecosystem services, the latter could easily fit within the same analytical framework. Krutilla made a compelling argument that technology was much better able to provide substitutes for resource commodities than for resource amenities; as a result, the relative value of resource amenities would increase over time. This, in turn, would have important implications for development decisions, particularly when future values were uncertain and the loss of preserved environments was irreversible.

Many economic activities, from the extraction of resource inputs to the emission of wastes, damage resource amenities. Dams and water diversion projects provide water for irrigating crops. This greatly enhances agricultural production—40 percent of crop production occurs on the 17 percent of cropland that is irrigated (WRI 2000). However, upstream water withdrawals for irrigation reduce water availability downstream with potentially disastrous effects. The most extreme example may be the Aral Sea, whose volume dropped precipitously before the 1990s as a result of irrigation diversions; 20 of the 24 fish species in the lake disappeared as a result (WRI 2000). It is difficult to imagine any extractive use of natural resources that does not in some way affect natural resource amenities—from the potential environmental impacts of oil drilling on the pristine wilderness of the Arctic National Wildlife Refuge to the more general impact of carbon dioxide emissions on global climate.

A key aspect of ecosystems and the provision of resource amenities is the inextricable connections between the elements of an ecosystem. Commercial exploitation for natural resource commodities generally considers at most a few of the elements in the ecosystem. But the extraction of one element or the addition of excessive amounts of another can disrupt the entire balance of the ecosystem with unforeseen consequences. Our understanding of ecosystems is incomplete, and there is much uncertainty about how they are affected by different uses and their ability to provide resource amenities over the long run. This complexity raises important questions about how property rights to the various elements of the ecosystem can be assigned and how all of the externalities of commercial exploitation can be internalized when some may not be identified in advance.

Resource amenities present significant management challenges for social institutions. The natural resources that provide these amenities are often open-access resources, and many of the goods and services are public goods. Consequently, one can expect far different outcomes for these than for natural resource commodities. The interdependence between natural resource commodities and natural resource amenities implies that natural resource and environmental policy cannot be concerned with single resources but must look at complete ecosystems and, indeed, the environment as a whole.

Empirical Considerations

In the "race" between technological progress and diminishing marginal returns in a finite natural world, the prospects for future generations depend upon which trend is proceeding at a faster pace. Many issues, then, boil down to a seemingly simple empirical question of whether technological progress can overcome diminishing marginal returns. Over the years, there have been technological pessimists and optimists, and that pattern continues to this day.

The empirical evidence to date for natural resource commodities is largely in favor of technological progress. The many predictions of impending doom have not come true—at least not yet. The discovery and development of new reserves, the substitution of capital, and technological progress in resource extraction and commodity production have led to generally downward sloping price trends for many natural resource commodities. If there is any systematic bias to past predictions of the future, it is an underestimation of the ability of technological progress to overcome natural resource scarcity. For example, petroleum supply forecasts have persistently overestimated the future price of oil and underestimated oil production (Lynch 2002). The picture is less clear for the amenity goods and services derived from the natural environment.

Resource Commodities

Three economic measures have been used as indicators of resource scarcity: price, extraction cost, and user cost. These three indicators are related through a basic first order condition for optimal resource extraction:

$$P = C_q + \lambda$$

where

P denotes the extracted resource price
C_q denotes marginal extraction cost
λ denotes the user cost

The user cost captures the nonextractive economic cost of current depletion, including the forgone regeneration for a renewable resource and the forgone future use of a nonrenewable resource. It also includes any contribution of the resource stock itself to the net benefit of extraction—for example, a more abundant resource stock may decrease extraction or harvest cost.

Barnett and Morse (1963) focus primarily on extraction cost. Extraction cost is computed as the amount of labor and capital needed to produce a unit of output. This measure is founded on the classical economics view that with diminishing marginal returns and finite natural resources, the cost of natural resource use should increase as demand increases and depletion occurs. The tendency toward increasing extraction cost can be offset by technological progress.

Data from the United States for the period 1870–1958 for agriculture, minerals, forestry, and commercial fishing are examined. During this period, population increased by a factor of 4, annual output increased 20 times, and the output of the extractive industries increased about 6 times. This period of rapid population and economic growth should furnish a good test of the relative impacts of diminishing marginal returns and technological progress.

Agricultural output increased four times over this period, and the unit cost declined by one-half when both capital and labor are included and by one-third when only labor is included. The cost measure for agricultural production actually declines more rapidly after 1920. The economy became more mineral intensive over this period, with mineral resource use increasing 40 times. Even so, the unit extraction cost measure for minerals production declined significantly, with an increase in the rate of decline after 1920. Commercial fishing also saw a decrease in extractive cost. Only forestry unit extraction cost increased, although both output and unit cost tended to level out after 1920. The conclusion is that the data do not support the strong scarcity hypothesis of increasing resource scarcity (Barnett and Morse 1963).

Table 3-1. *Unit Extraction Costs, 1870–1958*

	1870–1899	1900	1910	1919	1929	1937	1948	1958
Labor-capital cost per unit output								
Non-extractive GNP	136	126	115	118	100	102	80	68
Agriculture	132	118	121	114	100	93	73	66
Minerals	211	195	185	164	100	80	61	47
Agriculture relative to GNP	97	94	105	97	100	91	91	97
Minerals relative to GNP	155	155	161	139	100	78	76	69
Labor cost per unit output								
Nonextractive GNP	162	137	121	126	100	103	83	69
Agriculture	151	130	130	115	100	92	66	53
Minerals	285	234	195	168	100	96	65	45
Sawnlogs	59	65	67	108	100	104	88	90
Agriculture relative to GNP	93	95	107	91	100	89	80	77
Minerals relative to GNP	176	171	161	133	100	93	78	65
Sawnlogs relative to GNP	36	47	55	86	100	101	106	130

Source: Barnett and Morse (1963), Tables 6, 7, and 8.

A weaker scarcity hypothesis is that economy-wide technological progress would make it difficult to discern increasing scarcity in the natural resource industries. This weaker hypothesis is tested by examining the movement of unit costs in the resource sectors relative to unit costs in nonresource sectors. The minerals sector shows a decline in unit cost more than one-half the decline in nonresource sectors. Agricultural cost declines from 1929 if only labor is used and is roughly constant if both capital and labor are used as input measures. Forestry, of course, still shows increasing resource scarcity. Table 3-1 provides a summary of the unit cost estimates of Barnett and Morse (1963).

The cost estimates for natural resource industries were updated to 1970 for *Scarcity and Growth Reconsidered,* and results for other regions were also included (Barnett 1979). The results were essentially the same: the agricultural, mineral, and extractive sectors continued to show a strong decline in labor per unit extracted. Labor plus capital declined, but at a slower rate. Others have found a statistically significant increase in extraction cost of U.S. coal and petroleum in the 1970s, although this could be due to the exercise of market power rather than changes in scarcity (Hall and Hall 1984). The decline in extraction cost for metals continued in the 1970s (Hall and Hall 1984). There is weak evidence that some natural resource prices increased relative to nonresource commodity prices during the period 1960–1970 (Barnett 1979).

Barnett also observes that *Scarcity and Growth* viewed environmental impacts as a more significant concern than increasing scarcity of resource commodities. Pollution abatement costs were about 2 percent of output at the time, and that was viewed as relatively small. A more aggressive abatement policy, projected to increase costs to 3 percent of output by the year 2000, would reduce projected annual economic growth by less than 0.1 percent (Barnett 1979).

A shortcoming of the use of labor or labor–capital as the input measure is that it does not include other inputs that may be significant, including energy and environmental services. Some of the decline in labor and capital costs per unit of output occurred because energy was substituted for capital and labor. The output per unit of energy input in mining increased from 1919 to the mid-1950s and then declined to one-half of its peak (Cleveland 1991). In agriculture, output per unit of energy input decreased between 1910 and 1973 and then increased after energy prices increased from the mid-1970s to early 1980s. A similar pattern occurred in forestry.

The period between *Scarcity and Growth* and *Scarcity and Growth Reconsidered* saw considerable study of the dynamics of natural resource use. These theoretical developments pointed out the shortcomings of extraction cost as an indicator of resource scarcity. Extraction cost is an inherently static measure; it does not capture future effects that are important for indicating natural resource scarcity. In addition, extraction cost captures information about only the supply side of the market. If demand is growing more rapidly than extraction cost is declining, then extraction cost will give a false indication of decreasing scarcity. The opposite also is possible—extraction cost could increase even as technological progress develops substitutes for most of the uses of a particular resource.

The other two economic measures of resource scarcity—price and user cost—do incorporate information about the demand for the resource and, at least to the extent possible, expectations about future demand and availability. For this reason they are generally preferred as indicators of resource scarcity (Brown and Field 1979; Fisher 1979). The resource price would "summarize the sacrifices, direct and indirect, made to obtain a unit of the resource" (Fisher 1979), because the price would capture both user cost and the current extraction cost. User cost would be the best measure of the scarcity of the unextracted resource. For most of the twentieth century, natural resource commodity price trends have been generally flat or decreasing. This is particularly true for mineral prices. Since these are nonrenewable resources, one might expect they would be more subject to increasing scarcity and therefore increasing prices. However, mineral prices generally declined during the twentieth century (Sullivan et al. 2000). Figures 3-1 through 3-4 show the long-term price curves for copper, lead, petroleum, and natural gas.

An exception to this downward trend for nonrenewable resource prices is the period from 1945 until the early 1980s. During much of this period, prices for many nonrenewable resources—copper, iron, nickel, silver, tin, coal, natural gas, and a mineral aggregate—show an upward trend (Slade 1982). Almost all mineral prices rose in the 1970s, particularly after the 1973 oil embargo. This seems to match the U-shaped price curve that would occur as depletion exerted enough upward pressure on price to overcome the downward force of technological progress (Slade 1982). By all appearances at the time, it seemed likely that nonrenewable resource prices would continue to increase.

But minerals prices did not continue to increase. The economy responds to price increases in a variety of ways: substitutions are made; research and development produce resource-saving technologies; new reserves are discovered and developed; new methods for recovering resources or reducing the cost of using lower-quality reserves are found, and so on. As a result, most mineral prices have declined since the early 1980s, and some of these declines are substantial (see Figure 3-5). Increases in total factor productivity in mining exceeded the increases in manufacturing as a whole (Humphreys 2001; Parry 1999). The price declines

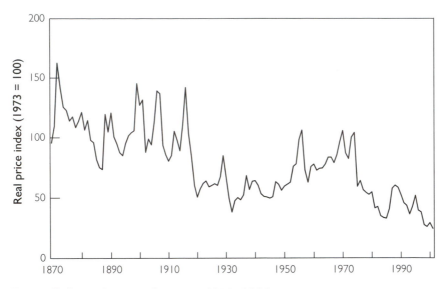

Figure 3-1. *Real Price of Copper, 1870–2001*

Sources: Manthy (1978) for 1870–1973; U.S. Geological Survey, *Mineral Commodities Summaries* for 1967–2001. The two series differ slightly for the period 1967–1973, so an average of the two is used; the general trend is unaffected.

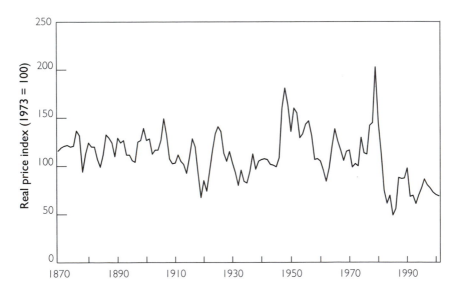

Figure 3-2. *Real Price of Lead, 1870–2001*

Sources: Manthy (1978) for 1870–1973; U.S. Geological Survey, *Mineral Commodities Summaries* for 1967–2001. The two series differ slightly for the period 1967–1973, so an average of the two is used; the general trend is unaffected.

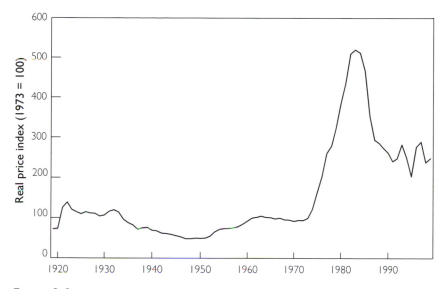

Figure 3-3. *Real Price of Natural Gas, 1919–1999*

Sources: Manthy (1978) for 1919–1973; Energy Information Administration, Department of Energy, *Annual Energy Review* for 1968–1999. The two series differ slightly for the period 1968–1973, so an average of the two is used; the general trend is unaffected.

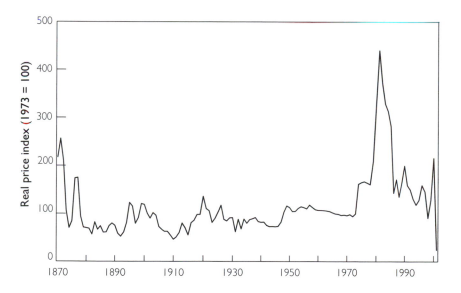

Figure 3-4. *Real Price of Petroleum, 1870-2000*

Sources: Manthy (1978) for 1870–1973; Energy Information Administration, Department of Energy, *Annual Energy Review* for 1949–2000. The two series differ slightly for the period 1949–1973, so an average of the two is used; the general trend is unaffected.

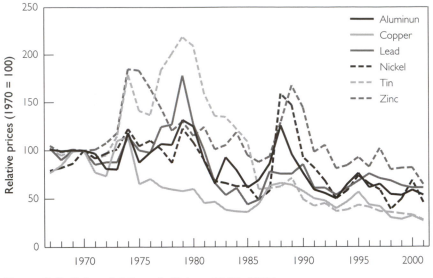

Figure 3-5. *Selected Minerals Prices, 1967–2001*

Source: U.S. Geological Survey, *Minerals Commodity Summaries.*

occurred even as some of the environmental externalities associated with resource extraction were internalized.

The significant decline in the mineral resource intensity of total output after the mid-1970s provides evidence of the ability to substitute away from inputs that become more costly (Tilton 1989). The development of the solvent extraction-electrowinning (SX-EW) method for refining copper ore is one example of the effect of technological progress. This process reduces costs greatly by eliminating smelting and refining, and this allows the economical use of much lower grade copper ore, including material that was left behind by previous mining operations (Tilton and Landsberg 1999; Bunel 2001). The price of copper increased in the late 1980s but has since declined substantially to well below its 1979 level, overcoming previous concerns about the future demand for and cost of extracting copper (Brobst 1979; Goeller, 1979).

The real prices of fossil fuels also declined from peaks in the early 1980s (see Figure 3-6). Petroleum is particularly instructive. Consumption and consumption per capita declined in North America and Europe in the early 1980s following the oil price spike of 1979. Consumption increased slowly after the mid-1980s as real oil prices fell, but consumption per capita remains below its 1980 level in both the United States and Europe. Total consumption in North America is now slightly above its previous peak in 1978, while European total consumption is still below its previous peak in 1979. Consumption declines in the developed world were offset by increasing consumption in the Asian Pacific region, particularly by a doubling of consumption in China in the 1990s.

Developments in computer technology and directional drilling, neither of which was predicted in *Scarcity and Growth Reconsidered*, have substantially lowered exploration and development costs and enhanced recovery from existing reserves.

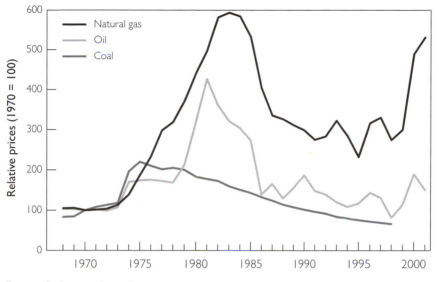

Figure 3-6. *Fossil Fuel Prices, 1968–2001*

Source: Energy Information Administration, Department of Energy, *Annual Energy Review.*

Discovery and development costs in the United States are one-third of what they were 20 years ago (*Economist* 2001). World proved petroleum reserves increased from 660 billion barrels at the end of 1980 to 1,009 billion barrels at the end of 1990. Even though consumption from 1991 to 2000 was approximately 250 billion barrels, proved reserves at the end of 2000 stood at 1,046 billion barrels. The United States produced 28 billion barrels of oil in the 1990s, but its proved reserves dropped by only 4.1 billion barrels (British Petroleum 2001).

Because there is a finite amount of oil in the ground, production cannot increase indefinitely; it must reach a peak, then eventually decline to zero. Hubbert (1969) argued that because of the production technology, the relationship between production and time would trace out a bell-shaped curve with a positive exponential growth rate for production in the early years and an exponential decline to zero. The area under this curve would be determined by the amount of recoverable petroleum. This notion and an estimate of the total availability of reserves led Hubbert (1969) to predict a peak in oil production in the lower 48 states around 1970.

This prediction was right on target, although the peak was higher than forecast. However, the actual decline in production in subsequent years has been slower than forecast. Production in 2000 was about one-half of peak production rather than the predicted one-third. Technological developments have increased the recovery of oil from existing reserves and allowed exploitation of deposits that previously were uneconomical. World production was predicted to peak between 1990 and 2000, depending upon whether the low or high end of estimated reserves was correct. Actual production has not followed a bell-shaped curve. It reached a temporary peak in 1979, fell in the early 1980s, and began a slow rise through the late 1980s and 1990s. A similar pattern holds for U.S. natural gas production, which

Hubbert predicted to peak in 1980. This demonstrates that production is not fixed by the production technology but can be altered by market conditions and technological innovations (Cleveland and Kaufman 1991).

Other major natural resource sectors show increasing productivity and declining prices. Malthus's prediction about population and food supply was inaccurate: food production has exceeded population growth. For much of the last two centuries, the increase in food production was the result of bringing more land under cultivation and farming existing land more intensively. The substitution of tractor power for draft animals made more agricultural production available for human consumption; in the 1920s, the production from about one-quarter of U.S. cropland was used to feed draft animals (Johnson 2002).

Corn yield per hectare was relatively constant from 1800 to 1930, when hybrid corn was introduced. Corn yield had increased about 50 percent by 1950; it tripled between 1950 and 1984. Wheat yields per acre in the United States were relatively constant from 1800 to 1950 and then more than doubled between 1950 and 1984 (Johnson 2002). Average cereal grain yields in the United States from 1996 to 1998 were 22 percent higher than from 1986 to 1988; the increase in cereal grain yields for the world was a little lower, at 17 percent (WRI 2000). The increases in agricultural productivity have increased food availability and lowered prices. The prices of maize, rice, soybeans, wheat, and beef are about one-half of their 1960 levels (WRI 1998). See Table 3-2 and Figure 3-7.

Worldwide fiber production from forests has increased 50 percent since 1960. Timber production in North America and Europe is primarily from secondary-growth forests, and the forested area in developed countries has actually increased in the last two decades (WRI 2000). Forest products have not shown the same general downward price trend, but neither is there a significant upward trend. See Table 3-2 and Figure 3-8.

Marine fishery production has increased six-fold since 1950, primarily through extending fishing to relatively unexploited areas, although aquaculture has increased to more than 20 percent of the total fish harvest (WRI 2000). But many older fisheries are producing much less as a result of overfishing; it is estimated that 75 percent of fisheries have been overharvested. One sign of this is the increase in the catch of low-value species while the catch of some high-value species has declined. The prospects for increasing harvest from existing fisheries are not good (WRI 2000). The harvest from capture fisheries has reached a peak, and growing production from aquaculture threatens capture fisheries, as feedstocks are diverted from natural to commercial production.

The user cost or rental value of a natural resource is the best measure of the marginal value of the resource stock in place. Unfortunately, information about user costs or rental values is not generally available. Many natural resource stocks are not traded or are traded infrequently, and even those that are traded are seldom traded as just resource stocks. Nevertheless, there have been several efforts to construct time series of user cost, and most empirical tests of the behavior of nonrenewable resource prices have found that user cost has fallen rather than increased over time (Krautkraemer 1998). An important exception is the stumpage value of Douglas fir timber from 1940 to 1970 (Brown and Field 1979).

Table 3-2. *Real Commodity Prices in Dollars, 1960–1995*

	1960	1965	1970	1975	1980	1985	1990	1995
Agricultural								
Maize	100	122	111	126	83	78	52	50
Rice	100	106	97	146	110	55	52	52
Sorghum	100	120	113	136	98	82	57	55
Soybeans	100	122	105	109	93	73	56	49
Wheat (U.S.)	100	98	78	118	86	71	48	53
Beef	100	115	146	83	108	88	72	45
Fish meal	100	157	140	97	125	73	74	74
Mineral								
Aluminum	100	90	91	73	83	62	67	62
Copper	100	183	172	84	93	63	81	75
Gold	100	94	83	207	491	269	223	187
Iron ore	100	86	71	69	71	70	56	41
Lead	100	153	126	96	131	59	84	55
Manganese ore	100	83	52	72	52	49	80	40
Nickel	100	102	144	128	115	91	112	88
Silver	100	136	160	221	648	202	109	99
Tin	100	170	138	143	220	159	57	49
Zinc	100	121	99	138	89	96	127	73
Fossil Fuels								
Coal	100	88	100	222	199	157	116	88
Natural gas	NA	NA	100	187	440	532	298	232
Petroleum	NA	NA	100	174	320	273	187	117
Forest Products								
Malaysian logs	90	94	100	87	158	103	103	125
Plywood	NA	NA	100	65	92	75	86	119
Sawnwood	103	104	100	71	79	64	76	89
Woodpulp	NA	NA	100	138	105	86	115	101

Note: Prices are expressed as a percentage of the 1960 or 1970 price, depending upon data availability. NA = not available.

Sources: Food, mineral, and forest product data from WRI (1998); fossil fuel data from Energy Information Administration, Department of Energy, *Annual Energy Review,* www.eia.doe.gov.

Resource Amenities

It is much more difficult to evaluate the scarcity of natural resource amenities. These goods and services are not generally traded on markets, so price and cost data are not available. An alternative is to look at physical measures of scarcity, but even here the data are much sparser than the data for natural resource commodities. In addition, while physical measures of natural resource commodities can be made across relatively homogeneous resources—million metric tonnes of lead, copper, and zinc; barrels of petroleum; trillion cubic feet of natural gas—the same cannot be done with natural resource amenities. A hectare of forestland in the northeastern United States is not the same as a hectare of forestland in the Amazon River basin or the temperate rainforests of Alaska. The use of aggregate data can mask significant local problems.

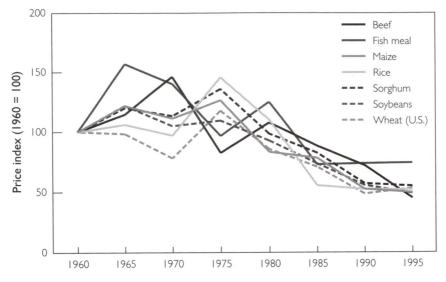

Figure 3-7. *Selected Food Commodity Prices, 1960–1995*

Source: WRI 2000.

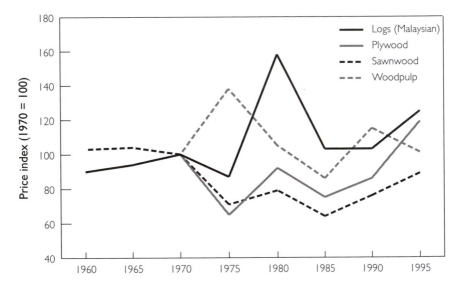

Figure 3-8. *Selected Wood Commodity Prices*

Source: WRI 2000.

Environmental policies have achieved some success in conserving certain amenity values in recent years, particularly in developed countries. The United States has reduced emissions of criteria air pollutants, quite substantially in some cases (WRI 1998), and water quality in the United States and Europe is generally improved (WRI 2000). These successes involve environmental factors that most directly affect human well-being and are more visible than the loss of services from degraded ecosystems. Some natural environments have been better protected and preserved, but there also are reasons for alarm. A commitment to institutional innovation, not just technological innovation, will be crucial for the efficient management of environmental resources.

Conversion of land from its natural state to human use, or degradation of land from human use, is a primary reason for the loss of ecosystem services. The nineteenth and twentieth centuries saw a significant increase in human use of land; 40 percent to 50 percent of land has now been transformed or degraded. In addition, humans appropriate 8 percent of the primary production of the oceans and as much as 35 percent of primary production from the continental shelf in temperate zones (Vitousek et al. 1997). Impacts from human use of the land can extend far from where the use occurs. Nitrogen from fertilizer in runoff water has created a large "dead zone" in the Gulf of Mexico at the mouth of the Mississippi River.

The 40 to 50 percent conversion figure is consistent with data on forestlands. Current forestlands are slightly more than half of the world's original forestland (53.4 percent in 1996); frontier forests (relatively undisturbed intact forest ecosystems) comprise only 21.7 percent of original forestland, and about 40 percent of the frontier forests are threatened, meaning human activities are likely to result in significant loss of ecosystem integrity (WRI 2000). Worldwide, almost 6,000 tree species are threatened. Forestlands provide a variety of ecosystem services. They filter pollutants from air and water; regulate the flow of runoff water, thus controlling floods, preserving soil, and reducing silt in rivers and seacoasts; sequester carbon and buffer temperatures; and provide habitat to a wide variety of species. These services are lost when forestlands are converted to other uses.

Land conversion has reduced biodiversity, although the degree is difficult to measure because the number of species is not known and there is no single measure of biodiversity. Indeed, the rate of loss of species is usually estimated from the rate of loss of habitat. These estimates generally put the loss of species at 100 to 1,000 times the rate that would have naturally occurred (Vitousek et al. 1997). It is relatively easy to identify species whose existence in particular areas has been endangered by habitat conversion. These include in the western United States alone grizzly bears, wolves, wild Pacific salmon, and sage grouse.

Global climate change induced by atmospheric accumulation of carbon dioxide resulting from fossil fuel consumption and deforestation is another avenue through which humans can have a significant impact on ecosystem services. Since industrialization, the atmospheric concentration of carbon dioxide has increased steadily from 286–288 parts per million (ppm) in 1860 to 367 ppm in 1998 (WRI 2000). The current concentration is the greatest in the past 420,000 years. The atmospheric concentration of methane has increased 151 percent since 1750 (IPCC 2001). The global average surface temperature has increased about 0.6 degrees centigrade (1.1 degrees Fahrenheit), and the 1990s were the warmest decade on record (IPCC 2001). While global temperatures and climate vary naturally, a consensus has developed that most of the warming over the last half century has been the result of increased greenhouse gases (IPCC 2001). Changing climate will damage many ecosystems if it occurs more rapidly than they can adapt. A meta-analysis of 143 studies found "a significant impact of global warming is already discernible in animal and plant populations" (Root et al. 2003).

Alaska already seems to be experiencing significant changes. Because the average temperature in much of Alaska is close to the freezing point of water, increased temperatures can have a significant impact. Permafrost is melting and causing buildings and roads to sag, sea ice has thinned significantly, and warmer temperatures have allowed beetles to destroy spruce forests (Egan 2002). Coral reefs are also vulnerable to warmer temperatures because they thrive at temperatures just below the maximum temperature they can survive.

A recent study by the United Nations Development Programme, United Nations Environment Programme, World Bank, and World Resources Institute attempted a comprehensive, qualitative analysis of the state of the world's major ecosystems. The study evaluated the capacity of several types of ecosystems—agricultural, coastal, forest, freshwater, and grassland—to provide a variety of services: food and fiber, water quantity, water quality, biodiversity, carbon storage, shoreline protection, woodfuel, and recreation. Each ecosystem was evaluated for the condition and direction of its changing capacity for providing the various ecosystem services (WRI 2000).

The results, while not bleak, are ominous. With 5 types of ecosystems and 8 types of services, there are 40 possible outcomes each for condition and changing capacity, and it was possible to assess 24 of these. Six of these categories were found to be in good condition, 12 in fair condition, 5 in poor condition, and 1 (biodiversity in freshwater ecosystems) in bad condition. More disturbing, the capacity was declining in 18 of these 24 categories, mixed in 3, and increasing in only 1 (2 were unassessed). Overall, there are considerable signs that the capacity of ecosystems to continue to produce many of the goods and services we depend on is declining (WRI 2000).

Theoretical Considerations

The empirical data for natural resource commodities do not suggest increasing scarcity. However, past successes are no guarantee of future success. Population and economic growth will continue to increase the demand for natural resource commodities and, more importantly, place additional stress on natural environments. Increasing scarcity of a natural resource commodity generally triggers a variety of responses that, at least to some extent, ameliorate that scarcity. By their very nature, the same is not as true for resource amenities; these goods and services are not generally traded in markets so there is no price signal to trigger a response. Detection of the problem is much more difficult, and the response depends upon collective action. Even if one is optimistic about the future availability of resource commodities, it is possible to be pessimistic about the future availability of resource amenities.

Even if natural resource commodities are becoming more scarce, it may be possible to sustain economic production using lower levels of resource inputs to produce equivalent levels of goods and services. This may be achievable through technological progress or the substitution of other more plentiful inputs. The question of what mechanisms can sustain an economy dependent upon an essential nonrenewable resource was examined with highly stylized optimal growth models in the 1970s, and the results were an important theme in *Scarcity and Growth Reconsidered* (Stiglitz 1979; Daly 1979).

In a simple depletion model, if technological progress increases the output obtained from a given resource input, it is akin to having a growing resource stock. If the economy is patient enough to give technological progress the time to increase the effective resource stock, then positive economic growth is sustained (Stiglitz 1974).

The effect of capital accumulation and capital–resource substitution is similar. In a simple capital growth model, capital accumulates as long as its marginal productivity is greater than the rate of time preference; the economy moves to a steady state, where the marginal productivity of capital is balanced against impatience. In a growth model with both capital and a nonrenewable resource, the marginal product of capital depends upon the flow of the nonrenewable resource input. Capital productivity decreases as capital accumulates and the resource input declines. Exactly what happens to capital productivity depends upon how readily capital services can be substituted for the natural resource input. A measure of the substitutability of one input for another is the elasticity of substitution. The elasticity of substitution measures the percentage change in the ratio of the marginal product of the two inputs relative to a percentage change in the input ratio and is captured graphically by the curvature of the production isoquants (Stiglitz 1979).

If the elasticity of substitution between capital and the resource is less than one, the ability to substitute capital for the natural resource input is relatively limited. In this case, the average product of the resource is bounded above, so there is a finite limit on the production that can be obtained with resource depletion: sustained output is not possible. If the elasticity of substitution is greater than one, the substitution possibilities are greater and economic growth can be sustained even as the resource input declines to zero. However, if capital productivity falls below the rate of time preference, the economy will decline. If the elasticity of substitution equals one, then the economy can be sustained only if capital's output share is greater than the resource's output share. In this case, the limiting value of capital productivity is zero, so an economy with a positive rate of time preference is too impatient to sustain growth through capital accumulation (Dasgupta and Heal 1974).

The ability to substitute capital for a natural resource, then, is a critical question in the current scarcity and growth debate. It is relatively easy to find examples where capital can substitute for the use of a natural resource. For example, insulation and thermal pane windows reduce the energy needed to maintain indoor temperatures. The redesign of products like milk and beverage containers that allows the same services to be obtained with less material input can be seen to substitute human capital services for plastic and aluminum. New technologies can replace one resource with another more abundant resource, as fiber optics have replaced copper for telecommunications. The mix of goods produced in the economy can shift from more to less resource intensive commodities. The energy used to produce one dollar of gross domestic product (GDP) was reduced by almost one-half in the United States between 1949 and 2000, with most of that reduction coming after 1970, although total energy use tripled as population doubled and per capita GDP increased (Energy Information Administration 2002). World primary energy use per dollar of GDP has declined by more than 25 percent since 1970 (Smith 2002) and at an annual rate of 1.7 percent during the 1990s (Darmstadter 2002). The use of materials per unit of GDP has declined about one-third since 1970 (Wernick et al. 1996).

The ability to overcome natural resource commodity constraints through substitution and technological progress implies that it is not necessary to extract all productive resource commodities from natural environments in order to sustain a standard of living. The opportunity cost of protecting natural environments is lower the greater the availability of close substitutes for resource commodities. Indeed, when the loss of resource amenity values is taken into account as a cost of extraction, this added cost may warrant using other substitutes or developing new technologies earlier than would otherwise occur. Preservation of the Arctic National Wildlife Refuge is more sensible if petroleum substitutes are readily available than if petroleum is essential to continued economic well-being.

The ability to substitute capital for natural resources is limited by physical laws of nature. It simply is not possible to produce an ever-expanding level of material output from an ever-decreasing quantity of material input. No amount of capital–resource substitution or technological progress can overcome that constraint. The same is true of energy—the amount of work obtained cannot be greater than the amount of the energy expended as an input. Recycling durable nonrenewable resources can increase the life of a given resource stock, but 100 percent recovery and reuse is not practical, so the process cannot continue indefinitely. Consequently, sustaining the economy ultimately must rely on renewable natural resource inputs. It is even more difficult for capital substitution and technological progress to overcome the loss of amenity goods and services from the environment.

The importance of the rate of time preference points to the equitable treatment of future generations as a major impetus for natural resource and environmental conservation. A positive social rate of time preference implies that future generations are not given the same weight in the social welfare function as the current generation. This can be construed as mistreatment of future generations simply because they live in the future, and it can be especially harsh when the economy's only asset is a finite quantity of nonrenewable resources.

However, good things also can happen with the passage of time, and a positive rate of time preference does not necessarily imply that the current generation is better off than future generations. This is clear from a simple capital growth model when the initial capital stock is relatively small. Production and consumption will increase over time as the economy accumulates capital. A zero rate of time preference, or equal weighting of generations, would require the first generation—which has the lowest utility—to make even greater sacrifices to increase the well-being of future generations who already would be better off. Consequently, the same present value criterion can give markedly different relative treatments of earlier and later generations depending upon the technological context. As a result, technological pessimists and technological optimists can differ markedly over the ethical treatment of future generations.

The role of the social rate of time preference becomes even more complex in a world with both physical and natural capital. While the direct effect of a lower discount rate is to increase the accumulation of natural resources, there are indirect effects that can lead to even more rapid resource and environmental depletion. A lower rate of time preference can spur economic activity, increasing the demand for natural resource inputs (Scott 1955), or increase the demand for land development over land preservation (Rowthorn and Brown 1995). Because many extractive industries are capital intensive, a lower discount rate can increase investment in extractive capacity, which allows more rapid resource extraction (Farzin 1984). This can even result in less permanent preservation of natural environments, particularly for open-access natural resources.

Perhaps a more effective way of preserving natural resource and environmental assets for the future is to ensure that all of the contributions those assets make to economic productivity are taken into account. Efficient asset accumulation requires equal marginal rates of return across assets. The marginal return to an asset includes the marginal value of any contribution to well-being. An environmental asset can contribute to economic well-being through the utility function, the production function, and through biological or ecological growth functions. For example, the marginal return to preserved forestland includes the value of water filtration, erosion and sediment control, carbon sequestration, habitat, and recreational and aesthetic services.

The problem for environmental assets is that most, if not all, of their productive contributions are social returns that are not appropriated through the marketplace. If the returns to the environmental asset are not captured, they will be ignored, regardless of the social rate of time preference. Unless market failures are corrected, a lower rate of time preference could lead to more rapid depletion of environmental assets through greater accumulation of other assets. An individual may want to invest in a portfolio of assets, including a healthy environment for their children or grandchildren, but they cannot individually purchase environmental assets for their bequest. The state of the environment depends upon everyone's investment, not just the investment of a particular individual.

An appropriate remedy for this problem is policies that ensure that all of the returns to the natural resource assets are taken into account, not necessarily that the discount rate is reduced. Of course, the proper accounting for environmental services is not a trivial task, and placing quantity constraints on the use of the environment may be a practical alternative. Forest preserves, roadless areas, agricultural conservation reserve programs, areas off limits to mining, habitat conservation areas, shoreline preservation, and development setbacks can all make sense on the basis of efficiency as well as intergenerational equity. Such set-asides also are consistent with optimism about the ability of technological progress to continue to provide an adequate supply of natural resource commodities.

Summary and Conclusions

Empirical evidence does not indicate a significant increase in the scarcity of natural resource commodities. Indeed, the historical evidence is that expansion into previously undeveloped lands and technological progress have enabled the human economy to avoid the Malthusian trap and to maintain adequate supplies of food, forest, and mineral products even as population and economic output have increased substantially.

Population and economic growth into the next century will greatly increase the demand for natural resource commodities. Even though population growth has slowed, a population of 6 billion growing at 1 percent adds the same number of people as 3 billion people growing at 2 percent. The desire for a higher living standard in the developing world places additional demands on technological progress to prevent increasing scarcity of natural resource commodities. The historical success of adaptation to increased demand for these commodities is by no means a guarantee of future success. Little arable land remains to expand agricultural production. Furthermore, the transition from animal to mechanical power has already been made, and a transition from fossil fuels to biofuels would reduce the land available for crops for human consumption. Consequently, the bulk of increased food production will need to come from further increases in yield per hectare. Our understanding of crop production has increased dramatically over the last several decades, and new techniques from biotechnology afford some cautious optimism that the human population can continue to avoid a catastrophic food shortage.

The increased demand for other natural resource commodities will also challenge human ingenuity to continue to overcome impending resource scarcities. Fossil fuel reserves are about as abundant relative to the rate of consumption as they have been over the last century, and the technologies for discovering and recovering these resources have developed substantially over the past decades. However, the world's petroleum supply is finite and cannot last forever—some forecasts place the peak of world oil production within the next decade or two. Coal is more abundant than petroleum, but the environmental evils of energy from coal are generally greater. There are several possible renewable substitutes, including solar thermal and voltaic, wind, geothermal, and biomass energy. Whether these energy sources will allow the same standard of living as fossil fuels depends upon technological advances yet to be made. One can be optimistic or pessimistic about future possibilities, but there does not appear to be a significant shortage on the near horizon.

The same can be said of most mineral resources. The greatest dilemma is the state of ocean fisheries, particularly those species at the top of the marine food chain. Through various institutional failures, many of the oldest fisheries have been overharvested, and their productivity is now well past its peak. Fishing has expanded into new fisheries but, as in agriculture, there is little room left for further expansion. Only better fisheries management will improve the productivity of ocean fisheries as a whole. Freshwater scarcity also threatens many regions, particularly if freshwater supplies are not managed in a more economically rational manner.

The world's economies have not been as adept at preserving natural resource amenities. By their very nature, these goods and services are subject to a variety of market and government failures. The benefits of many of these goods and services are not appropriable, so they are not fully considered when decisions for commercial exploitation of natural resource commodities are made. As a result, many of the world's ecosystems have been degraded. The inability to appropriate the benefits of natural resource amenities reduces the incentive for technological developments that could preserve or restore natural ecosystems. Further population and economic growth will increase the ecological pressures. Without significant improvements in environmental protection, the future availability of natural resource amenities is in jeopardy.

The first step is to correct the institutional failures that result in undervaluation of these goods and services. This step is necessary whether one views sustainability as an efficiency or equity issue, and it is a tremendous task in itself. We are far from understanding how ecosystems function, and the interdependence of their many elements makes it difficult to design simple remedies.

References

Barnett, H. 1979. Scarcity and Growth Revisited. In *Scarcity and Growth Reconsidered,* edited by V. K. Smith. Baltimore: Johns Hopkins University Press for Resources for the Future, 163–217.

Barnett, H., and C. Morse. 1963. *Scarcity and Growth: The Economics of Natural Resource Availability.* Baltimore: Johns Hopkins University Press for Resources for the Future.

British Petroleum. 2001. *BP Statistical Review of World Energy.* London: British Petroleum.

Brobst, D.A. 1979. Fundamental Concepts for the Analysis of Resource Availability. In *Scarcity and Growth Reconsidered,* edited by V.K Smith. Baltimore: Johns Hopkins University Press for Resources for the Future, 106–142.

Brown, G.M., and B. Field. 1979. The Adequacy of Measures for Signaling the Scarcity of Natural Resources. In *Scarcity and Growth Reconsidered,* edited by V.K. Smith. Baltimore: Johns Hopkins University Press for Resources for the Future, 218–248.

Bunel, M. 2001. Estimating the Effects of Technology and Depletion on the Real Price of Copper in the U.S. Using a Cointegration Approach. Working Papers on Environment and Economics 2/2001. Madrid: Fundacion Biodiversidad.

Cleveland, C.J. 1991. Natural Resource Scarcity and Economic Growth Revisited: Economic and Biophysical Perspectives. In *Ecological Economics: The Science and Management of Sustainability,* edited by R. Costanza. New York: Columbia University Press, 289–317.

Cleveland, C.J., and R.K. Kaufmann. 1991. Forecasting Ultimate Oil Recovery and Its Rate of Production: Incorporating Economic Forces into the Models of M. King Huppert. *The Energy Journal* 12(2): 17–46.

Daly, H.E. 1979. Entropy, Growth, and the Political Economy of Scarcity. In *Scarcity and Growth Reconsidered,* edited by V.K. Smith. Baltimore: Johns Hopkins University Press for Resources for the Future, 67–94.

Dasgupta, P., and G.M. Heal. 1974. The Optimal Depletion of Exhaustible Resources. *Review of Economic Studies* 41 (Symposium on the Economics of Exhaustible Resources): 3–28.

Darmstadter, J. 2002. A Global Energy Perspective: Sustainable Development Issue Backgrounder. Washington, DC: Resources for the Future.

Economist. 2001. Into Deeper Water. Dec. 8, U.S. edition, section TQ.

Egan, T. 2002. Alaska, No Longer So Frigid, Starts to Crack, Burn and Sag. *New York Times,* June 16, section 1, page 1.

Energy Information Administration. 2002. *Energy in the United States 1635–2000.* http://www.eia.doe.gov/emeu/aer/eh/frame.html (accessed May 3, 2004).

Farzin, Y.H. 1984. The Effect of the Discount Rate on Depletion of Exhaustible Resources. *Journal of Political Economy* 92(5): 841–851.

Fisher, A. 1979. Measures of Natural Resource Scarcity. In *Scarcity and Growth Reconsidered*, edited by V.K. Smith. Baltimore: Johns Hopkins University Press for Resources for the Future, 249–275.

Goeller, H.E. 1979. The Age of Substitutability: A Scientific Appraisal of Natural Resource Adequacy. In *Scarcity and Growth Reconsidered*, edited by V.K. Smith. Baltimore: Johns Hopkins University Press for Resources for the Future, 143–159.

Hall, D.C., and J.V. Hall. 1984. Concepts and Measures of Natural Resource Scarcity with a Summary of Recent Trends. *Journal of Environmental Economics and Management* 11(4): 363–379.

Hubbert, M.K. 1969. Energy Resources. In *Resources and Man: A Study and Recommendations*, edited by the Committee on Resources and Man, National Academy of Sciences, National Research Council. San Francisco: W.H. Freeman and Company.

Humphreys, D. 2001. Sustainable Development: Can the Mining Industry Afford It? *Resources Policy* 27(1): 1–7.

IPCC (Intergovernmental Panel on Climate Change). 2001. *Summary for Policymakers: A Report of Working Group I of the Intergovernmental Panel on Climate Change*. In *Climate Change 2001: The Scientific Basis*. http://www.ipcc.ch/pub/spm22-01.pdf (accessed May 3, 2004).

Jevons, W.S. 1865. *The Coal Question: An Inquiry concerning the Progress of the Nation, and the Probable Exhaustion of Our Coal-Mines*, edited by A.W. Flux. Reprinted 1965. New York: Augustus M. Kelley.

Johnson, D.G. 2002. The Declining Importance of Natural Resources: Lessons from Agricultural Land. *Resource and Energy Economics* 24(1–2): 157–171.

Krautkraemer, J.A. 1998. Nonrenewable Resource Scarcity. *Journal of Economic Literature* 36(4): 2065–2107.

Krutilla, J.V. 1967. Conservation Reconsidered. *American Economic Review* 57(4): 777–786.

Lynch, M.C. 2002. Forecasting Oil Supply: Theory and Practice. *Quarterly Review of Economics and Finance* 42(2): 373–389.

Malthus, T. 1798. *An Essay on the Principle of Population*. Reprinted 1983. London: Penguin Books Ltd.

Manthy, R.S. 1978. *Natural Resource Commodities—A Century of Statistics: Prices, Output, Consumption, Foreign Trade, and Employment in the United States, 1870–1973*. Baltimore: Johns Hopkins University Press for Resources for the Future.

Meadows, D.H., D.L. Meadows, J. Randers, and W.W. Behrens III. 1972. *The Limits to Growth: A Report for the Club of Rome's Project on the Predicament of Mankind*. New York: Universe Books.

Mill, J.S. 1848. *Principles of Political Economy with Some of Their Applications to Social Philosophy*, edited by Sir W.J. Ashley. Reprinted 1965. New York: Augustus M. Kelley.

Parry, I.W.H. 1999. Productivity Trends in the Natural Resource Industries. In *Productivity in Natural Resource Industries: Improvement through Innovation*, edited by R.D. Simpson. Washington, DC: Resources for the Future, 175–204.

Root, T.L., J.T. Price, K.R. Hall, S.H. Schneider, C. Rosenzweig, and J.A. Pounds. 2003. Fingerprints of Global Warming on Wild Animals and Plants. *Nature* 421(Jan. 2): 57–60.

Rowthorn, B., and G.M. Brown Jr. 1995. Biodiversity, Economic Growth and the Discount Rate. In *The Economics and Ecology of Biodiversity Decline: The Forces Driving Global Change*, edited by T.M. Swanson. Cambridge: Cambridge University Press.

Scott, A.D. 1955. *Natural Resources: The Economics of Conservation*. Toronto: University of Toronto Press.

Slade, M.E. 1982. Trends in Natural-Resource Commodity Prices: An Analysis of the Time Domain. *Journal of Environmental Economics and Management* 9(2): 122–137.

Smith, J.L. 2002. Oil and the Economy: Introduction. *Quarterly Review of Economics and Finance* 42(2): 163–168.

Smith, V.K. (ed.) 1979. *Scarcity and Growth Reconsidered*. Baltimore: Johns Hopkins University Press for Resources for the Future.

Stiglitz, J.E. 1974. Growth with Exhaustible Resources: Efficient and Optimal Growth Paths. *Review of Economic Studies* 41 (Symposium on the Economics of Exhaustible Resources): 123–138.

————. 1979. A Neoclassical Analysis of the Economics of Natural Resources. In *Scarcity and Growth Reconsidered,* edited by V.K. Smith. Baltimore: Johns Hopkins University Press for Resources for the Future, 36–66.

Sullivan, D.E., J.L. Sznopek, and L.A. Wagner. 2001. Twentieth Century U.S. Mineral Prices Decline in Constant Dollars. Open File Report 00-389. Washington, DC: U.S. Geological Survey.

Tilton, J. 1989. The New View of Minerals and Economic Growth. *Economic Record* 65(190): 265–278.

Tilton, J., and H. Landsberg. 1999. Innovation, Productivity Growth, and the Survival of the US Copper Industry. In *Productivity in Natural Resource Industries: Improvement through Innovation,* edited by R.D. Simpson. Washington, DC: Resources for the Future, 109–139.

Vitousek, P.M., H.A. Mooney, J. Lubchenco, and J.M. Melillo. 1997. Human Domination of Earth's Ecosystems. *Science* 277(5325): 494–499.

Wernick, I.K., R. Herman, S. Govind, and J.H. Ausubel. 1996. Materialization and Dematerialization: Measures and Trends. *Daedalus* 125(3): 171–198.

WRI (World Resources Institute). 1998. *World Resources 1998–1999: A Guide to the Global Environment: Environmental Change and Human Health.* Oxford: Oxford University Press.

————. 2000. *World Resources 2000–2001: People and Ecosystems. The Fraying Web of Life.* Washington, DC: World Resources Institute.

CHAPTER 4

Ecosystem Goods and Services and Their Limits

The Roles of Biological Diversity and Management Practices

David Tilman and Stephen Polasky

H_UMANS RECEIVE VITAL GOODS and services from both managed and natural ecosystems (e.g., Krutilla 1967; Daily 1997) and have become, inadvertently or deliberately, managers of essentially all of the terrestrial ecosystems of the world (Vitousek et al. 1994, 1997b; Carpenter et al. 1998; Tilman et al. 2001). In this chapter we consider the forces that influence the flows of ecosystem goods and services and the greater social welfare to be derived from policies that maximize the net value of the flows of these goods and services.

The existence and importance of ecosystem services has been recognized, at least implicitly, for quite some time. Eighteen hundred years ago, a Roman, Quintus Septimus Florens Tertillianus, lamented the dwindling supply of natural capital and services needed by a large and growing human population:

> charming farms obliterate empty places, ploughed fields vanquish forests, sandy places are planted with crops, stones are fixed, swamps drained.... [T]he resources are scarcely adequate to us; and our needs straiten us and complaints are everywhere while already nature does not sustain us. (quoted in Johnson 2000)

Malthus, Ricardo, Faustmann, Thoreau, and their successors, writing about agricultural systems, forestry, fisheries, and wilderness preservation, also recognized the value of ecosystem services. Until fairly recently, much of the work in such fields as agricultural economics and resource economics emphasized the role of nature as a storehouse that provides raw materials for the production of economic goods. Less thought was given to environmental quality, wilderness value, or other ecosystem services. In *The Economics of Natural Environments: Studies in the Valuation of Commodity and Amenity Resources*, Krutilla and Fisher (1975) note that conventional practice in economics at the time they were writing was to treat natural environments as "unimproved lands" that generated no economic value. Economists have since expanded their view of the value of natural systems, thanks in large part to pioneering work by researchers affiliated with Resources for the Future such as Clawson (1959), Krutilla (1967), Hammack and Brown (1974), and Krutilla and Fisher (1975). Subsequent research has articulated the myriad values created by natural systems.

Though the existence of ecosystem sources is widely recognized, only the most preliminary attempts have been made at valuing most of them (e.g., Costanza et al. 1997; Daily 1997; Nordhaus and Kokkelenberg 1999; Daily et al. 2000). We are still far short of accurately quantifying and valuing many ecosystem services (Daily et al. 2000). Controversy over the attempt to do so by Costanza et al. (1997) highlights the lack of reliable and scientifically grounded estimates (e.g., Ayres 1998; Bockstael et al. 2000; Toman 1998).

There are three fundamental challenges in valuing ecosystem goods and services and assuring their supply. The first is estimating nonmarket values. Some ecosystem goods, including foods, fibers, medicines, and combustible sources of energy, are produced and sold in markets, and thus explicitly priced. Most ecosystem services, however, are not; instead they are treated as if they were provided free of charge. These include purification of water, removal and storage of atmospheric carbon dioxide, detoxification and recycling of waste products, generation and renewal of soil fertility, maintenance of both species and genetic diversity, pollination of crops, control of many agricultural pests, moderation of climate and amelioration of flooding, and human recreational opportunities (Wilson 1988; Daily 1997). For all these services, other means must be used to establish relative value. The economics profession has recognized this challenge, and extensive research on nonmarket valuation is underway (Freeman 1993; Smith 1997).

The second challenge is to understand the factors that control the supply of ecosystem services, and the tradeoffs between providing these services or other goods and services. What are "production functions" for ecosystem services, and what do they depend on? Recent work on ecosystem functioning suggests that the ability of ecosystems to provide such services depends on the number and mix of species living in an ecosystem, as well as anthropogenic inputs (Hector et al. 1999; Tilman et al. 1996, 2001; Loreau et al. 2001). This paper explores the nature of "production functions" for ecosystem services, and makes some initial attempts at integrating aspects of ecology and economics.

The third challenge is to devise incentives that will lead those who make decisions affecting the supply of ecosystem services to maintain their supply. The costs of maintaining ecosystem services are typically borne by landowners or other local decisionmakers, whereas benefits typically accrue more broadly. If a farmer reduces nitrogen runoff by changing farm operations, the costs in lower farm productivity are borne by the farmer, but the benefits are shared by others in the region who use groundwater or surface water. Benefits may extend even beyond the river basin drainage; reducing nitrogen runoff in the Mississippi River basin would reduce hypoxia in the Gulf of Mexico. Mechanisms need to be found to reward decisionmakers for providing ecosystem services that generate external benefits or to punish those who degrade such services.

The focus on the provision of ecosystem services is especially appropriate at a time when humanity is causing widespread, fundamental changes in natural systems. Population growth and increased per capita consumption tend to increase the scarcity of ecosystem services, which could inhibit future economic growth and welfare. Several broad forces influence the per capita availability of ecosystem goods and of consumable ecosystem services. Global population increased from less than 2 billion to 6 billion during the twentieth century and is generally expected to reach 9 billion people by the middle of the twenty-first century. This increased population is strongly associated with the massive conversion of natural, multi-species ecosystems to highly simplified, intensively managed ecosystems, such as row-crop agriculture and monoculture forestry (Wilson 1988, 1992). This conversion, while increasing per capita food production, is causing a sharp decline in the per capita availability of natural and seminatural high-diversity ecosystems, and thus in the per capita flow of goods and services from these ecosystems. In this chapter, we discuss how the conversion of natural multispecies ecosystems to simplified managed ecosystems, and other changes, will affect the flow of ecosystem services and why this implies the need for management policy that accounts for the values of both ecosystem services and goods produced in natural and managed ecosystems.

Most of the ecosystem goods and services essential to human life are produced on terrestrial "useable" land, which we broadly define as the 8.6 billion hectares of the terrestrial earth surface that is not ice or rock (1.5 billion hectares), desert (1.8 billion hectares), tundra (1.1 billion hectares), or boreal (1.5 billion hectares) (Schlesinger 1997). About 15 percent of global useable land currently is under intensive, often monoculture, cultivation; about 34 percent is pastoral land containing seminatural multispecies grasslands used for livestock production; and about 30 percent is forested (FAO 2001; Schlesinger 1997). Some forestland is managed intensively, often with monoculture production practices that mimic those of row-crop intensive agriculture. However, the majority of forest ecosystems are seminatural multispecies ecosystems. Thus, although global land use data are spotty, it seems highly likely that well more than half of the useable global lands contain natural or seminatural multispecies ecosystems.

Terrestrial ecosystems that produce marketable goods (food, fiber, and biomass energy) and ecosystem services can be loosely classified as being either highly simplified, intensively managed, often monoculture ecosystems, or as less simplified, multispecies ecosystems that receive fewer inputs of fertilizer, pesticide, seed, and so on. In this chapter, we discuss the underlying ecology of monocultures and of multispecies ecosystems and the implications of their different characteristics for the rate and stability of production of ecosystem goods and services. The majority of lands producing goods and services of value to society are or could be made multispecies ecosystems, and recent research suggests that ecosystem functioning is highly dependent on biological diversity and species composition, both of which are strongly influenced by management practices.

Maximizing the Net Value of Ecosystem Goods and Services

In one sense, ecosystem services are of immeasurable value—the complete loss of all ecosystem services would be catastrophic. In commenting on such an estimate of the global value of all ecosystem services by Costanza et al. (1997), Michael Toman (1998) offered the wry observation that it was "a serious underestimate of infinity." But what about more gradual and less apocalyptic changes to ecosystems? What are the likely consequences for the production of ecosystem services from recent human activities that have resulted in large-scale altering and simplification of ecosystems? How might changes in the provision of ecosystem services affect human welfare? What factors determine the rates of supply of various ecosystem services? Are there tradeoffs between using a managed ecosystem to maximize the economic value of an ecosystem good, such as marketable timber, and maximizing the value of such ecosystem services as flood control, potable water, and the quality of downstream aquatic ecosystems? If so, what management practices would maximize the value of the net benefits produced by such a managed ecosystem, and what policies could lead to adoption of such practices?

Similar questions can be raised about natural and seminatural ecosystems, most of which are affected, often inadvertently, by human behavior. Humans now control essentially all of the earth's land surface, and humans influence the biogeochemistry of such critical limiting resources for ecosystems as nitrogen and phosphate. Human actions change ecosystem composition, diversity, inputs, and disturbance regimes that fall outside the range of natural variation experienced by these ecosystems for thousands of years. How might these actions impact the goods and services that these ecosystems provide? The following hypothetical sketch illustrates the ideas that we will elaborate in this chapter.

Consider a forest ecosystem that is to be intensively managed to maximize the net present value of all goods and services. To simplify this analysis, assume that the ecosystem is managed so that it provides a constant, sustainable annual flow of all goods and services and that there is a single ecosystem good—marketable timber—and a single ecosystem service—perhaps potable water or sequestration of atmospheric carbon dioxide. Further assume that many different tree species could be planted. The net value of the annual sustainable yield of harvestable timber and of the ecosystem service will depend upon which species is planted. Note that the value of the ecosystem service could be negative if the planting of a certain tree species or management practice leads to degradation of environmental quality. By evaluating the net value of ecosystem goods and services for each species that could be grown in monoculture, it is possible to determine which single species would maximize total value. Clearly, it is possible that the total return to society would be maximized not by the timber species that leads to the greatest timber value net of management costs, but by some other species. A tree species that is of lower value as timber but that provides more of a valued ecosystem service would yield greater total value.

The net value of ecosystem goods and services also depends on management practices. Fertilization might increase the annual sustainable rate of timber production, affect the quality of this timber, or alter the quantity and quality of ecosystem services. Fertilization is likely to decrease the value of potable water but might increase the amount of carbon sequestration. Fertilization also incurs costs of purchase and application, soil testing, and personnel to decide on appropriate rates and timing of fertilization. In general, both alternative crop species and alternative management practices have the potential to influence value.

Although the balance sheet for the total costs and benefits of alternative crops and management practices includes the net value for all ecosystem services, a private land manager typically faces a different balance sheet. With no markets for carbon storage, potable water, and other ecosystem services of value to society, no government subsidies for producing services or taxes on activities that diminish such services, and no laws regulating acceptable or unacceptable actions that bear on them, a private land manager will realize only the net value of timber production. Rational private land managers will adopt species and practices that maximize the total net value of timber harvest, choices likely to harm ecosystem services and decrease total return to society. For example, a land manager will continue to use fertilizer until the marginal benefit of fertilizer for timber production just equals the marginal cost. If fertilizer use results in decreased value of ecosystem services, then under standard assumptions about declining marginal benefit, a private land manager will apply fertilizer at a rate greater than is optimal for society.

This simple case is most relevant to highly managed ecosystems, such as agricultural or monoculture forest ecosystems. In the following section we will examine similar problems in multispecies ecosystems.

Methods to align the balance sheets of private land managers more closely with that of society, by internalizing the positive externalities for providing valuable ecosystem services, are well known at least in principle. Subsidies can be paid to land managers for practices that enhance production of ecosystem services. The agricultural policies of most developed countries have some form of "green payments" that farmers are offered if they adopt practices that are considered environmentally benign (OECD 2001). Alternatively, taxes on environmentally detrimental practices can be imposed, but politically it is much harder to institute taxes on "bads" than it is to pay subsidies on "goods." In principle, however, either approach can give the land manager a financial incentive to promote production of ecosystem services.

Another approach is to establish an explicit market for the ecosystem service. Cap-and-trade systems for pollution are an example of creating a market for an environmental good where none existed before. Under cap-and-trade systems, governments set the overall allowable limit on emissions by issuing a certain number of permits, but allow firms to freely trade these permits. The 1990 Clean Air Act Amendments established a cap-and-trade system for sulfur dioxide (SO_2) emissions from electric utilities; it is widely credited with reducing SO_2 emissions at far lower cost than originally expected (Ellerman et al. 2000). International negotiations over climate change have led to proposals to establish a cap-and-trade program for carbon dioxide emissions, as well as other proposals to pay land owners for carbon sequestration (Sandor and Skees 1999).

Two major problems bedevil implementation of schemes that would reward land managers for providing ecosystem services. First, there must be some reasonably accurate and inexpensive method for measuring the amount of ecosystem services provided. This is a serious problem for most ecosystem services. For instance, despite recent advances, there remain large holes in the scientific understanding of the amounts of carbon sequestered in forests and in agriculture, particularly in the soil, under alternative management and species mixes. Second, there is the question of what value to ascribe to ecosystem services. The value of carbon sequestration comes from the reduced impact of future climate change on society, and no one can do more than speculate on this value. In reality these two problems interact—the value of ecosystem services depends upon the level of ecosystem services provided. Large-scale changes in ecosystem productivity would likely have fundamental structural repercussions on human economies. Fully assessing the value of large changes in ecosystem structure and function should be done with a general equilibrium ecological-economic model. Such a model does not exist, though some initial steps in this direction have been taken (e.g., Brock and Xepapadaes 2002; Tschirhart 2000).

Despite these impediments to an ideal or optimal policy, we already know enough to start moving in the right direction. It is almost surely preferable to give some incentive to practices thought to be environmentally beneficial, even though we cannot know exactly the right level of incentive to provide or easily measure the level of ecosystem services provided. Not knowing the precise value of cleaner air and water has not kept the United States from administering the Clean Air Act and Clean Water Act, with significant reductions in emissions whose benefits likely exceed costs by a good margin (Freeman 1990; U.S. EPA 1997).

Production of Ecosystem Services in Biologically Diverse Ecosystems

The significance of diversity is often overlooked in economics literature, but mixtures of species may return greater value than possible in even the most highly valued monoculture. This section will focus on the potential effects of diversity on the supply and value of ecosystem services.

The previous discussion dealt with lands managed for monoculture production of ecosystem goods and services. What of the larger proportion of global lands that contain many more species? A hectare of tropical forest may contain hundreds of plant species, as may a hectare of grazed grassland (Hubbell 2001). How might this biological diversity, or biodiversity, influence ecosystem functioning, and how might it bear on the total value that society receives from all the goods and services produced by an ecosystem? The simple answer is that the many different plant species living in a forest or grassland compete with each other, and thus each species influences the production of the other species and of various ecosystem services. This means that the production of ecosystem goods and services will depend on the suite of species present in the ecosystem. Exploitative competition for a common pool of limiting resources exhibits an interesting property. Where ecosystem services are proportional to ecosystem productivity, as perhaps with carbon sequestration, more diverse ecosystems produce higher levels of ecosystem services. In other cases, the translation from ecosystem productivity to ecosystem service is not as straightforward.

Experimental Studies

Three aspects of ecosystem functioning, all else being equal, should be closely related to the quality or quantity of various services that ecosystems provide to society. The first is net primary productivity, which is the annual net rate of production of living plant biomass by an ecosystem. With 34 percent of useable lands dedicated to grazing of livestock, and with most of these grasslands containing multispecies plant communities, their net primary production is a direct measure of production of livestock forage and thus of their agricultural value. The same is approximately true for forest ecosystems used in the production of timber and pulp.

The second ecosystem service is the removal and sequestration of atmospheric carbon dioxide. Because of combustion of fossil fuels, deforestation, and agriculture, atmospheric carbon dioxide concentrations are increasing, causing global climate change (IPCC 2001). The sequestration of some of this carbon in living biomass is thus a service of potentially great value to society.

The third ecosystem service is the removal of nutrients from groundwater. Fertilization, combustion of fossil fuels, and natural decomposition all release large quantities of biologically available nitrogen and phosphorus into soils. The removal of these nutrients by terrestrial plants is a valuable service because this prevents leaching of these nutrients into groundwater, lakes, streams, rivers, and marine ecosystems. High nutrient concentrations in groundwater decrease its value as a source of potable water. Elevated nutrient loading into aquatic ecosystems causes eutrophication, spoiling fisheries, recreation, and many other aspects of water quality, including use of fresh waters as sources of municipal drinking water.

Productivity. Several field experiments (Tilman et al. 1996, 1997a, 2001; Hector et al. 1999) and related theory (Loreau 1998a, 1998b, 2000; Tilman et al. 1997a; Lehman and Tilman 2000) have shown that net primary production is a strongly increasing function of the number of plant species living in an ecosystem. The first field experiment, performed in Minnesota, was a long-term study of 168 prairie communities that had been planted to contain from 1 to 16 grassland species. The species composition of each plot was determined by a separate random draw of the appropriate number of species from a pool of 18 grassland species. Throughout the experiment, primary production was a significantly increasing function of plant diversity (Figure 4-1A). The dependence of primary production on diversity became stronger throughout the experiment (Tilman et al. 2001). By the seventh year of the study, the 16-species plots had, on average, 2.7 times greater net primary production than the monocultures.

A similar biodiversity experiment performed in Europe had the added power of being replicated at eight different field sites (Hector et al. 1999) in Sweden, Portugal, Ireland, and Greece. Hector and colleagues found a strong dependence of plant community productivity on plant species diversity, much like that observed in the Minnesota experiment. Both studies showed that the composition of the plant community and the number of species it contained were about equally important determinants of ecosystem functioning; this suggests that ecosystem managers need to be equally concerned with both attributes. At present, composition, not diversity, is often the major variable that managers manipulate.

Carbon Storage. In prairie grasslands, the aboveground biomass dies each year, but the belowground biomass is perennial, long-lived, and often 3 to 4 times as

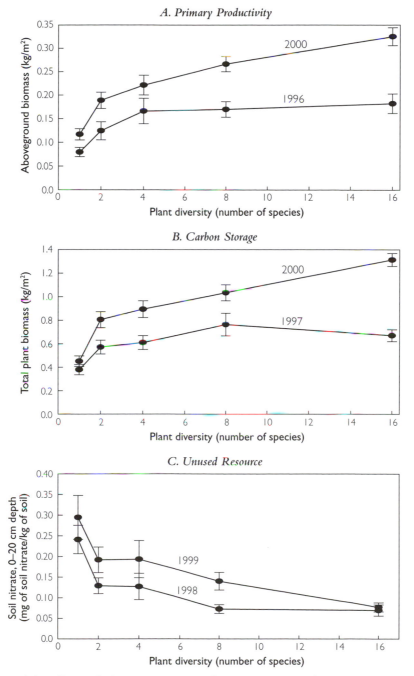

Figure 4-1. *Effects of Plant Diversity in the Minnesota Biodiversity Experiment*

Note: The effects on (A) primary productivity, (B) carbon storage in living biomass, and (C) soil nitrate (inversely related to water quality) are shown for the Minnesota Biodiversity Experiment, for which 2000 was its seventh field season.

Source: Tilman et al. 2001.

large as that above ground. Total biomass (aboveground plus belowground) is thus a direct measure of all carbon dioxide that has been removed from the atmosphere and stored in living plant biomass. For forest ecosystems, much of the stored carbon is in aboveground wood, and thus a potential source of timber and pulp. In the Minnesota experiment, total carbon storage also was a sharply increasing and highly significant function of plant diversity (Figure 4-1B). By the seventh year of the experiment, the 16-species plots had, on average, 2.9 times more stored carbon than the monocultures (Tilman et al. 2001).

Additional carbon is stored in soils. Indeed, in many terrestrial ecosystems, more carbon is stored in soils than in living plants (Schlesinger 1997). However, soil carbon stores change too slowly for current experiments to determine how soil carbon may depend on plant diversity or species composition.

Water Quality. The major limiting nutrient for most terrestrial plant communities is soil nitrate. The concentration of nitrate in soil water is proportional to the rate at which nitrate is leached from the soil into groundwater and thence into wells, streams, rivers, and lakes. The measured levels of soil nitrate were significantly decreasing functions of plant diversity in the Minnesota biodiversity experiment (Figure 4-1C). The threefold lower concentrations of soil nitrate in the 16-species plots compared to the monocultures mean that plant or microbe uptake of nitrate was greater, and nutrient use more efficient, at higher diversity. The lower concentrations also mean that there would be lower leaching of nitrate from the higher diversity communities (Tilman et al. 1996). This improved quality of groundwater at higher diversity could be a valuable service provided by higher-diversity ecosystems.

The Minnesota study suggests that each added species yields the same marginal increase in productivity, at least for increases in this range. Similarly, the quality of the water produced by these grassland ecosystems increased approximately linearly as diversity increased. The surprisingly large effect of diversity on productivity and carbon storage suggests that managing for diversity in grasslands could have significant benefits. The magnitude of the effect in other ecosystem types, such as forests, is uncertain.

Stability and Reliability. Recent experiments have indicated that greater diversity typically leads to greater stability or reliability of the flow of ecosystem services (Tilman and Downing 1994; Naeem and Li 1997; McGrady-Steed et al. 1997). In a grassland experiment in which plant diversity and plant community composition were influenced by the rate of nitrogen addition, primary productivity of plots with lower diversity decreased far more in response to a severe drought. This pattern was unchanged after controlling for such potentially confounding variables as the rate of nitrogen addition and differences in species composition (Tilman and Downing 1994). The year-to-year variability of primary production was also significantly lower in the more diverse plots in this long-term experiment (Tilman 1996). Similar results were observed in two laboratory studies of communities of microorganisms whose diversity was directly controlled. In both studies, greater diversity led to greater repeatability or reliability, that is, the responses of the higher-diversity treatments were more similar to each other than were the responses of the lower-diversity treatments (Naeem and Li 1997; McGrady-Steed et al. 1997). These studies suggest that greater diversity may lead both to greater and more consistent flows of ecosystem services.

Underlying Concepts and Theory

There is a long and controversy-laden history of interest in the potential effects of diversity in ecology. Darwin (1859) was perhaps the first ecologist to suggest that greater diversity led to greater productivity. Odum (1953), MacArthur (1955), and Elton (1958) expanded on the potential links between diversity, productivity, and stability. May (1972) built on the work of Gardner and Ashby (1970) to show that, for a system of equations modeling multispecies competition, individual species become less stable, on average, as diversity increases. McNaughton (1977) disputed the relevance of this theory to nature, but Goodman (1975) concluded that the preponderance of evidence (none of it experimental) failed to support the diversity-stability hypothesis of Elton (1958). Interest in the issues remained dormant until rekindled by Ehrlich and Ehrlich (1981), Wilson (1988), and Schulze and Mooney (1993), and by modern experimental results (Tilman and Downing 1994; Naeem et al. 1994; Tilman et al. 1996) showing that diversity did influence ecosystem productivity and stability.

Recent papers have addressed the mechanisms whereby diversity could affect ecosystem functioning and stability. The theories can be classed as sampling effect models and species complementarity (or niche) models (Huston 1997; Tilman et al. 1997b; Aarssen 1997; Loreau 1998a, 1998b, 2000; Loreau et al. 2001; Loreau and Hector 2001; Lehman and Tilman 2000). Both types of models assume that the species compositions of ecosystems are determined by random draws of species from a pool of potential species; the probability that any given species or combination of species is present in a community increases with the diversity of the community.

Sampling effect models assume that habitats are simple, spatially and temporally homogeneous, and constrained or limited by a single resource. In such habitats, a single species should eventually out-compete and suppress or displace all other species. In such systems, there is a simple hierarchy or ranking of species by their competitive ability. High diversity plots are more likely to contain a superior competitor and thus to have their functioning determined by the traits associated with higher competitive ability. In contrast, many different species could occur in low diversity plots, such as monocultures, so the functioning of low diversity plots typically mirror the average traits of the entire species pool. Thus, as diversity increases, ecosystem functioning should change from that of the average species in the species pool to that of the best competitor. If better competitors are more productive, then greater initial diversity should lead to greater productivity (Huston 1997; Tilman et al. 1997b; Aarssen 1997). Conversely, if better competitors are less productive, then productivity should decrease with diversity. Mathematically explicit models of this variant of the sampling effect reveal its characteristic signature: mean productivity increases with diversity, but the single best monoculture plot has productivity just as great as the best plot initiated with a large number of species.

Figure 4–2A shows that the model predicts great variability in productivity for plots planted to the same initial number of species. This variability is caused by random differences in initial species composition. Note that the greatest variability

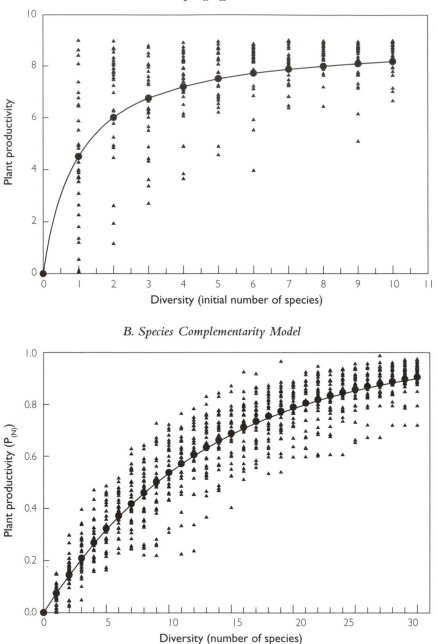

Figure 4-2. *The Dependence of Productivity on Diversity as Predicted by Two Different Classes of Models*

Note: The two models are (A) a sampling effect model and (B) an interspecific complementarity (niche difference) model (with $K = 1$). Both graphs modified from Tilman et al. (1997b). Results for B come from the model of Figure 4–3 and Equation 4–1.

within a given diversity level occurs for the lowest diversity plots and that plot-to-plot variability (comparable to the reliability of McGrady-Steed et al. 1997) declines as diversity increases.

The sampling effect model seems most likely to approximate the conditions of highly simplified, high-input ecosystems, such as agricultural ecosystems. Such managed systems receive sufficient inputs of nitrogen, phosphorus, other limiting nutrients, and water; the only limiting resource is likely to be light. Inputs of pesticides often eliminate other potentially limiting factors. A single species or even a single genotype may lead to the greatest productivity. A manager who plants this species or strain will obtain as great a yield as from any mixture of species. If, as often is the case, crops or strains differ in their values, the return to the manager will be maximized, as demonstrated earlier, by the crop strain and practices that maximize private net returns, and that to society by the crop strain and practices that maximize net returns of all ecosystem goods and services. The value of diversity in such systems derives from the possibility that different strains maximize return in different years, on different soil types, in different climates, in response to different pest pressures, and so on.

Species complementarity models, also called niche models, apply to managed, seminatural, or natural ecosystems with spatial or temporal variability in two or more limiting resources or limiting factors. In these conditions, many species can stably coexist. The interspecific tradeoffs that characterize the diversity of life on earth mean that a species that does well in response to certain combinations of limiting factors responds poorly to others (Tilman 1988, 1990). These tradeoffs mean that species have complementary traits, in that each species has unique resource requirements. This complementarity allows many combinations of species to stably coexist with each other in the same habitat.

Analytical models of communities with various initial numbers of randomly chosen complementary species reveal a consistent result. Systems with greater initial diversity exhibit greater average productivity, greater carbon storage, greater nutrient use efficiency, and greater stability of total production (Tilman et al. 1997b; Loreau 2000; Lehman and Tilman 2000). Moreover, the upper bound of the composition-dependent variation in these ecosystem services is, itself, an increasing function of species diversity (Figure 4-2B). Complementarity allows multispecies coexistence and increased flows of ecosystem services at higher diversity because diverse mixtures of species can, on average, more fully exploit the habitat. The increases in the upper bound of variation for productivity and carbon storage show that some combination of N species outperforms any combination of $N{-}1$ species, for all values of N for these models. The long-term Minnesota biodiversity experiments have an increasing upper bound, which supports the hypothesis that complementarity, not sampling effects, predominate in nature (Tilman et al. 2001), as do analyses of the European biodiversity experiment (Loreau and Hector 2001).

Although numerous detailed, mechanistic models of the effects of diversity have been proposed, a simple model captures their essence. This model assumes that a habitat has spatial heterogeneity in two limiting factors, such as temperature (which varies seasonally) and soil pH. Each species performs optimally (say, in competition for soil nitrogen) at one combination of these two factors (i.e., at a point in the

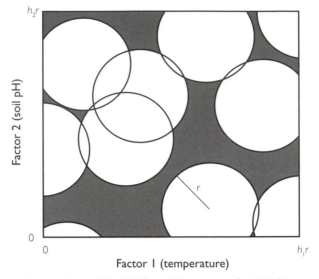

Figure 4-3. *A Simple Niche Model (from Tilman et al. 1997b)*

Note: The habitat conditions in which each species can survive are enclosed within its circular niche, and there is spatial heterogeneity in two niche dimensions (Factors 1 and 2). Simulations of this model led to the results of Figure 2A. For the analytical solution, see Equation 4-1.

plane of Figure 4-3), and perform less well in conditions farther from this point. Thus the region in which each species can maintain a significant population can be represented by a closed curve such as in Figure 4-3 (Tilman et al. 1997b). This circle can be thought of as the "niche" of this species. As a first approximation, the total productivity of a community at a given level of diversity will be determined by the proportion of habitat conditions "covered" by the species present. The black areas of Figure 4-3 show conditions that cannot be exploited by any of the species present. Such areas thus represent resources that cannot be effectively exploited by the suite of species present in that ecosystem.

For a simple analytical version of this model (Tilman et al. 1997b), assume that species have a niche of radius r, and that the habitat has levels of heterogeneity h_1 for niche axis 1 (temperature ranging from 0 to rh_1) and h_2 for niche axis 2 (pH ranging from 0 to rh_2). If N is the number of randomly chosen species, and $P(N)$ is the mean productivity of communities containing N species, then

$$P(N) = K\left[1 - \left(\frac{1-\pi}{h_1 h_2 + 2 h_1 h_2 + \pi}\right)^N\right] \tag{4-1}$$

where K is the maximal productivity of the habitat if completely covered. This equation gives the solid curve shown in Figure 4-2B. The greater the heterogeneity of the habitat, the greater is the effect of diversity on productivity, just as one would expect. Naturally, if there is no habitat heterogeneity ($h_1 = h_2 = 0$), there is no effect of diversity on productivity. This model, though simple, illustrates the mechanism by which species coexist because of niche differentiation (species complementarity) and habitat heterogeneity. It illustrates both visually and analytically the joint effects of habitat heterogeneity and diversity on ecosystem productivity and efficiency of resource use.

This theoretical result suggests that the flow of ecosystem goods and services could be maximized in natural and seminatural ecosystems, and in ecosystems managed without large inputs, by maintaining their diversity. It also illustrates, especially at lower diversity and in spatially more homogeneous habitats, that species composition is an important determinant of performance.

Scaling to Managed Ecosystems

How many species, then, are needed to maximize ecosystem services in 100 or 1,000 hectares of grassland or forest or some other ecosystem? At first glance, the results presented here might seem to suggest that 8 to 16 species would be sufficient. However, the experimental and theoretical results summarized above apply to the immediate neighborhoods within which individual plants interact with one another. Such neighborhoods are on the order to 1 to 10 m^2 for grassland habitats and 10 to 100 m^2 for forest ecosystems. In contrast, most land managers deal with sites that are 10 to 10,000 or more hectares, a thousand to a million times as large. Larger areas are, on average, more heterogeneous. This effect of habitat size on habitat heterogeneity is thought to scale according to the classic island biogeographic species–area relationship (Rosenzweig 1995). If empirically observed species–area relations hold, each 10-fold increase in area would be about 1.5 times as heterogeneous, meaning that it would require about 1.5 times as many species to "cover" it and ensure the same level of ecosystem functioning (Tilman 1999). Thus, if 8 species were required to ensure near-maximal flow of ecosystem services in a 1 m^2 plot of grassland, approximately 11 times this number—about 90 species—would be needed for a 100 hectare field. This greater species diversity is needed to provide species with traits that can efficiently exploit the much wider range of conditions that occur, on average, in this larger habitat. However, such analysis is clearly preliminary because it involves extrapolation both from grassland ecosystems to other types and from small scales to much larger scales.

Policy Implications

We have reviewed both theoretical and empirical evidence that systems with greater diversity of species are more productive ecologically and provide more of a variety of valuable ecosystem services. These services, such as carbon sequestration and removal of nutrients from groundwater, are public goods: benefits to society well beyond those that accrue to the private land manager (Balvanera et al. 2001). Carbon sequestration is a case of a global public good, while water quality improvement benefits accrue within watersheds (though they may accrue far from the private land, such as in the case of nitrogen removal in the Mississippi River watershed).

On the other hand, private land managers typically receive such full benefits of harvesting ecosystem goods as agricultural crops or timber. To maximize net benefits of harvest, private land managers may simplify ecosystems to provide controlled, uniform conditions conducive to the productive growth of a single species. To correct this imbalance toward ecosystem simplification in the interest of an optimum harvest, policy needs to adequately reward private land managers for stewardship. As discussed in the previous section, "green payments" for provision of ecosystem services, or "brown taxes" on practices that harm them, are one promising tactic.

In virtually all respects, the problem faced here is analogous to problems of industrial pollution. Steel, automobiles, and petrochemicals are goods valued by society, like agricultural crops and timber. Industrial production also results in unwanted byproducts like air and water pollution. Societies have instituted a variety of pollution abatement policies to curb the production of unwanted byproducts, while not necessarily curbing production of the goods themselves. A similar approach could work here. Policies that encourage conservation of diversity and production of ecosystem services, or discourage their destruction, could correct the current imbalance that rewards private land managers only for production of harvestable ecosystem goods.

Land Management Practices and Ecosystem Services

Managed lands receive deliberate inputs, such as nitrogen and phosphorus fertilizer, irrigation, tilling, species plantings, and pesticides, that influence both the production of desired ecosystem goods and the rate and value of ecosystem services. High-diversity natural and managed ecosystems also receive such inputs, most of which are inadvertent, but nonetheless influence the production rates and values of ecosystem goods and services.

More than half of the useable lands of the earth are dedicated to natural or seminatural grassland and forest ecosystems that provide goods and services of value to society. Increasing demand for agricultural and forest products is likely to increase the proportion of terrestrial useable lands under direct human management. Many management activities can impair the flow of ecosystem services, thus incurring the costs of either producing alternatives or foregoing these services. Furthermore, the flow of services from the declining area of remaining intact ecosystems is divided among an ever-increasing human population.

Management practices in the most intensively used lands—cultivated lands and intensively grazed lands—release into other ecosystems globally significant quantities of biologically available nitrogen and phosphorus and various pesticides (Vitousek et al. 1997a; Carpenter et al. 1998; Tilman et al. 2001). These substances can significantly impair the services provided by these and other ecosystems. Phosphorus and nitrogen can drastically reduce the quality of water, including potable water, because they stimulate growth of nuisance algae (phosphorus mostly in lakes, nitrogen more in the oceans) and, at high loading rates, can cause anoxic conditions that kill many fish species. High nitrite and nitrate concentrations in groundwater used for human consumption are also a serious human health threat. Human actions now cause an annual production of fixed nitrogen that exceeds that fixed by all natural terrestrial processes. Most of this nitrogen comes from agricultural fertilization. Likewise, mainly because of agriculture, humans now add as much biologically available phosphorus to terrestrial habitats as all natural processes. About half of fertilizer nitrogen is leached from cultivated fields into groundwater, lakes, streams, and rivers, and thence into the world's oceans (Howarth et al. 1996). Much of the rest of the agricultural nitrogen is incorporated into crops, then consumed either by livestock or humans, then released into the waste streams of these consumers. Much of the nitrogen in livestock waste is recycled as fertilizer, but much of that in human waste is not removed in sewage treatment but added to aquatic habitats. Although phosphorus is not nearly as readily leached as nitrogen, leaching, erosion, and transfer via human and animal wastes transfers much fertilizer phosphorus to aquatic ecosystems, thus degrading water quality.

Human disruption of the nitrogen cycle also affects nonagricultural terrestrial ecosystems. A portion of global fertilizer nitrogen is atmospherically deposited in other terrestrial ecosystems, as is additional biologically active nitrogen generated by combustion of fossil fuel and other human actions. Because nitrogen is the most important limiting resource of most terrestrial ecosystems, these increased nitrogen supplies can significantly decrease the flow of some services from these ecosystems over time (Vitousek et al. 1997a). Atmospheric nitrogen deposition has been linked to forest decline and excess loss of soil calcium, which can also cause long-term harm to forest productivity (Likens et al. 1998).

Because of increasing population and per capita wealth, the global rates of application of fertilizer nitrogen and phosphorus may increase 2.5- to 3-fold within 50 years (Tilman et al. 2001). Inputs from combustion of fossil fuels are also likely to increase. Clearly, agricultural lands should be managed to enhance the total return to society, not just to private land managers.

Conclusions

This paper has dealt with the need to incorporate into the determination of public policy, first, the value of ecosystem services, especially as they are affected by management practices for agriculture and forestry, and second, the effects of ecosystem diversity on the quality and flow of ecosystem services of value to society. Policy and the actions of land managers must reflect the range of potentially valuable services that ecosystems are capable of producing.

We have discussed only a few ecosystem services, chosen because their dependence on diversity has recently been explored. There are many additional ecosystem services (Daily 1997), perhaps of much greater value to society, which have been studied much less thoroughly. One such service is pollination, critical to many agricultural crops, and most often provided by wild pollinators living in the vicinity of a farm. Another service, control of agricultural pests, is provided by parasites, parasitoids, and other agents that live in natural and seminatural lands adjacent to agricultural fields or, with appropriate management, in agricultural fields themselves (e.g., Settle et al. 1996; Daily 1997).

Human society has been and will remain highly dependent on an unpredictable but large and changing subset of the earth's biodiversity, which serves as a source for new medicines, new crops, and new genetic variation in existing crops. The global food supply depends mainly on three crops, rice, corn, and wheat, which provide 60 percent of food. Because a single pathogen could effectively eliminate a crop (or a major lumber species, such as Douglas fir), it is essential to maintain as much intra-crop genetic diversity as possible and to maintain a suite of alternative crop species. This insurance value of biodiversity, though hard to quantify, is likely immense.

The value to sociey of ecosystem services, including biodiversity, should give added legitimacy to a process in which the total value of all goods and service produced is the criterion that determines how land is to be managed. This will require a major shift in land management policy, a shift toward ecosystem management, with an emphasis on an array of ecosystem services and away from a focus on a single management objective. In current practice, incomplete knowledge of ecosystem function and the difficulty of providing proper incentives to landowners impede meaningful ecosystem management. We are still far more adept at managing ecosystems for a single market objective, such as crop or timber production, than for simultaneous production of an array of goods and services, only some of which are currently priced in markets.

While diversity is of great importance, other changes and pressures on ecosystems are also of great potential consequence. The twentieth century marked a vast increase in the scale of human activity and its impact on ecosystems around the globe. Human activity has caused major changes in geochemical cycling, not only in the global carbon cycle, which has attracted much research attention because of climate change, but also for the global nitrogen and phosphorus cycles (Vitousek et al. 1997a; Carpenter et al. 1998). Increased demand for food production in the twenty-first century will drive further additions of nitrogen and phosphorus to terrestrial ecosystems, perhaps as much as threefold if past practices continue (Tilman et al. 2001). The consequent changes in the composition of species and ecosystem function will endanger valuable ecosystem services. Increased nitrogen application has been linked to eutrophication of surface waters, nitrate leaching into ground water, and the spread of weed species at the expense of other plant species (Vitousek et al. 1997a). Emissions of nitrogen oxides from agricultural soils increase ground level ozone, which undermines human health and agricultural crops (Delmas et al. 1997).

The exact form, impacts, and costs of future global changes in biogeochemical cycles, climate, and other variables are uncertain, as are the best societal adjustments to them. In the face of uncertainty over future conditions, there is value in preserving options (Arrow and Fisher 1974; Henry 1974). Maintaining diversity is both an important form of preserving options and a way to better maintain the flow of ecosystem services. It is hard to know exactly which species or genetic material will be crucial in the future, but losing diversity lowers the probability that the right options will be available. In agriculture, preserving the right germplasm can make the difference between success and failure in developing crop varieties capable of withstanding a novel pest or pathogen. Given the dominance of wheat, rice, and corn in the global diet, the ability to withstand newly evolved pests and pathogens for these crops is of vital importance. So, too, is there great potential value in the ability of diversity to increase the flow of ecosystem services from both managed and natural ecosystems.

References

Aarssen, L. W. 1997. High Productivity in Grassland Ecosystems: Effected by Species Diversity or Productive Species? *Oikos* 80: 183–184.

Arrow, K., and A.C. Fisher. 1974. Environmental Preservation, Uncertainty, and Irreversibility. *Quarterly Journal of Economics* 88: 312–319.

Ayres, R. 1998. The Price–Value Paradox. *Ecological Economics* 25(1): 17–19.

Balvanera, P., G.C. Daily, P.R. Ehrlich, T.H. Ricketts, S.-A. Bailey, S. Kark, C. Kremen, and H. Pereira. 2001. Conserving Biodiversity and Ecosystem Services. *Science* 291: 2047.

Bockstael, N., A.M. Freeman, R.J. Kopp, P.R. Portney and V.K. Smith. 2000. On Measuring Economic Values for Nature. *Environmental Science and Technology* 34: 1384–1389.

Brock, W., and A. Xepapdaes. 2002. Optimal Ecosystem Management When Species Compete for Limited Resources. *Journal of Environmental Economics and Management* 44: 189–220.

Carpenter, S.R., N.F. Caraco, D.L. Correll, R.W. Howarth, A.N. Sharpley, and V.H. Smith. 1998. Nonpoint Pollution of Surface Waters with Phosphorus and Nitrogen. *Ecological Applications* 8: 559–568.

Clawson, M. 1959. Methods of Measuring the Demand for and Value of Outdoor Recreation. Reprint no. 10. Washington, DC: Resources for the Future.

Costanza, R., R. d'Arge, R. de Groot, S. Farber, M. Grasso, B. Hannon, K. Limburg, S. Naeem, R.V. O'Neill, J. Pareulo, R.G. Raskin, P. Sutton and M. van den Belt. 1997. The Value of the World's Ecosystem Services and Natural Capital. *Nature* 387: 253–260.

Daily, G. (ed.). 1997. *Nature's Services: Societal Dependence on Natural Ecosystems.* Washington, DC: Island Press.

Daily, G.C., T. Söderqvist, S. Aniyar, K. Arrow, P. Dasgupta, P.R. Ehrlich, C. Folke, A. Jansson, B.-O. Jansson, N. Kautsky, S. Levin, J. Lubchenco, K.-G. Mäler, D. Simpson, D. Starrett, D. Tilman and B. Walker. 2000. The Value of Nature and the Nature of Value. *Science* 289: 395–396.

Darwin, C. 1859. *The Origin of Species by Means of Natural Selection.* Reprinted by The Modern Library, 1998. New York: Random House.

Delmas, R., D. Serca, and C. Jambert. 1997. Global Inventory of NO$_x$ Sources. *Nutrient Cycling in Agroecosystems* 48: 51–60.

Ehrlich, P.R., and A.H. Ehrlich. 1981. *Extinction: The Causes and Consequences of the Disappearance of Species.* New York: Random House.

Ellerman, A.D., P.L. Joskow, R. Schmalensee, J.-P. Montero, and E.M. Bailey. 2000. *Markets for Clean Air: The U.S. Acid Rain Program.* Cambridge: Cambridge University Press.

Elton, C.S. 1958. *The Ecology of Invasions by Animals and Plants.* London: Methuen & Co. Ltd.

FAO (Food and Agriculture Organization of the United Nations). 2001. http://apps.fao.org (accessed January 12, 2005).

Freeman, A.M., III. 1990. Water Pollution Policy. In *Public Policies for Environmental Protection,* edited by Paul R. Portney. Washington, DC: Resources for the Future.

———. 1993. *The Measurement of Environmental and Resource Values: Theory and Evidence.* Washington, DC: Resources for the Future.

Gardner, M.R., and W.R. Ashby. 1970. Connectance of Large Dynamic (Cybernetic) Systems: Critical Values for Stability. *Nature* 228: 84.

Goodman, D. 1975. The Theory of Diversity–Stability Relationships in Ecology. *Quarterly Review of Biology.* 50: 237–266.

Hammack, J., and G.M. Brown Jr. 1974. *Waterfowl and Wetlands: Toward Bioeconomic Analysis.* Baltimore: Johns Hopkins University Press for Resources for the Future.

Hector, A., B. Schmid, C. Beierkuhnlein, M.C. Caldeira, M. Diemer, P.G. Dimitrakopoulos, J. Finn, H. Freitas, P.S. Giller, J. Good, R. Harris, P. Högberg, K. Huss-Danell, J. Joshi, A. Jumpponen, C. Körner, P.W. Leadley, M. Loreau. A. Minns, C.P.H. Mulder, G. O'Donovan, S.J. Otway, J.S. Pereira, A. Prinz, D.J. Read, M. Scherer-Lorenzen, E.-D. Schulze, A.-S. D. Siamantziouras, E.M. Spehn, A.C. Terry, A.Y. Troumbis, F.I. Woodward, S. Yachi, and J.H. Lawton. 1999. Plant Diversity and Productivity Experiments in European Grasslands. *Science* 286: 1123–1127.

Henry, C. 1974. Investment Decisions under Uncertainty: The "Irreversibility Effect." *American Economic Review* 64(6): 1006–1012.

Howarth, R.W., G. Billen, D. Swaney, A. Townsend, N. Jaworski, K. Lajtha, J.A. Downing, R. Elmgren, N. Caraco, T. Jordan, F. Berendse, J. Freney, V. Kudeyarov, P. Murdoch, and Z.-L. Zhu. 1996. Riverine Inputs of Nitrogen to the North Atlantic Ocean: Fluxes and Human Influences. *Biogeochemistry* 35: 75–139.

Hubbell, S.P. 2001. *The Unified Neutral Theory of Biodiversity and Biogeography*. Monographs in Population Biology, no. 32. Princeton: Princeton University Press.

Huston, M.A. 1997. Hidden Treatments in Ecological Experiments: Reevaluating the Ecosystem Function of Biodiversity. *Oecologia* 110: 449–460.

IPCC (Intergovernmental Panel on Climate Change). 2001. *Climate Change 2001: The Scientific Basis. Contribution of Working Group I to the Third Assessment Report of the Intergovernmental Panel on Climate Change*, edited by J.T. Houghton, Y. Ding, D.J. Griggs, M. Noguer, P.J. van der Linden, X. Da, K. Maskell, and C.A. Johnson. Cambridge: Cambridge University Press.

Johnson, D.G. 2000. Population, Food, and Knowledge. *American Economic Review* 90: 1–14.

Krutilla, J.V. 1967. Conservation Reconsidered. *American Economic Review* 57: 777–786.

Krutilla, J.V., and A.C. Fisher. 1975. *The Economics of Natural Environments: Studies in the Valuation of Commodity and Amenity Resources*. Baltimore: Johns Hopkins University Press for Resources for the Future.

Lehman, C.L., and D. Tilman. 2000. Biodiversity, Stability, and Productivity in Competitive Communities. *American Naturalist* 156: 534–552.

Likens, G.E., C.T. Driscoll, D.C. Buso, T.G. Siccama, C.E. Johnson, G.M. Lovett, T.J. Fahey, W.A. Reiners, D.F. Ryan, C.W. Martin, and S.W. Bailey. 1998. The Biogeochemistry of Calcium at Hubbard Brook. *Biogeochemistry* 41: 89–173.

Loreau, M. 1998a. Biodiversity and Ecosystem Functioning: A Mechanistic Model. *Proceedings of the National Academy of Science* 95: 5632–5636.

———. 1998b. Separating Sampling and Other Effects in Biodiversity Experiments. *Oikos* 82: 600–602.

———. 2000. Biodiversisty and Ecosystem Functioning: Recent Theoretical Advances. *Oikos* 91: 3–17.

Loreau, M., and A. Hector. 2001. Partitioning Selection and Complementarity in Biodiversity Experiments. *Nature* 412: 72–76.

Loreau, M., S. Naeem, P. Inchausti, J. Bengtsson, J.P. Grime, A. Hector, D.U. Hooper, M.A. Huston, D. Raffaelli, B. Schmid, D. Tilman, and D. A. Wardle. 2001. Biodiversity and Ecosystem Functioning: Current Knowledge and Future Challenges. *Science* 294: 804–808.

MacArthur, R.H. 1955. Fluctuations of Animal Populations and a Measure of Community Stability. *Ecology* 36: 533–536.

May, R.M. 1972. Will a Large Complex System Be Stable? *Nature* 238: 413–414.

McGrady-Steed, J., P.M. Harris, and P.J. Morin. 1997. Biodiversity Regulates Ecosystem Predictability. *Nature* 390: 162–165.

McNaughton, S.J. 1977. Diversity and Stability of Ecological Communities: A Comment on the Role of Empiricism in Ecology. *American Naturalist* 111: 515–525.

Naeem, S., and S. Li. 1997. Biodiversity Enhances Ecosystem Reliability. *Nature* 390: 507–509.

Naeem, S., L.J. Thompson, S.P. Lawler, J.H. Lawton, and R.M. Woodfin. 1994. Declining Biodiversity Can Alter the Performance of Ecosystems. *Nature* 368: 734–737.

Nordhaus, W.D., and E.C. Kokkelenberg (eds.). 1999. *Nature's Numbers: Expanding the National Economic Accounts to Include the Environment*. Washington, DC: National Academy Press.

Odum, E.P. 1953. *Fundamentals of Ecology*. Philadelphia: Saunders.

OECD (Organisation of Economic Co-operation and Development). 2001. *Agricultural Practices in OECD Countries*. Paris: OECD.

Rosenzweig, M.L. 1995. *Species Diversity in Space and Time*. New York: Cambridge University Press.

Sandor, R.L., and J.R. Skees. 1999. Creating a Market for Carbon: Opportunities for U.S. Farmers. *Choices* (Spring).

Schlesinger, W.H. 1997. *Biogeochemistry: An Analysis of Global Change*. San Diego: Academic Press.

Schulze, E.D., and H.A. Mooney. 1993. *Biodiversity and Ecosystem Function*. Berlin: Springer Verlag.

Settle, W.H., H. Ariawan, E.D. Astuti, W. Cahyana, A.L. Hakim, D. Hindayana, A.S. Lestari, Pajarningsih, and Sartanto. 1996. Managing Tropical Rice Pests through Conservation of Generalist Natural Enemies and Alternative Prey. *Ecology* 77: 1975–1988.

Smith, V.K. 1997. Pricing What Is Priceless: A Status Report on Nonmarket Valuation of Environmental Resources. In *International Yearbook of Environmental and Resource Economics 1997/1998,* edited by H. Folmer and T. Tietenberg. Cheltenham, UK: Edward Elgar.

Tilman, D. 1988. *Plant Strategies and the Dynamics and Structure of Plant Communities.* Princeton: Princeton University Press.

———. 1990. Constraints and Tradeoffs: Toward a Predictive Theory of Competition and Succession. *Oikos* 58: 3–15.

———. 1996. Biodiversity: Population versus Ecosystem Stability. *Ecology* 77(3): 350–363.

———. 1999. Diversity and Production in European Grasslands. *Science* 286: 1099–1100.

Tilman, D., and J.A. Downing. 1994. Biodiversity and Stability in Grasslands. *Nature* 367: 363–365.

Tilman, D., D. Wedin, and J. Knops. 1996. Productivity and Sustainability Influenced by Biodiversity in Grassland Ecosystems. *Nature* 379: 718–720.

Tilman, D., J. Knops, D. Wedin, P. Reich, M. Ritchie, and E. Siemann. 1997a. The Influence of Functional Diversity and Composition on Ecosystem Processes. *Science* 277: 1300–1302.

Tilman, D., C.L. Lehman, and K.T. Thomson. 1997b. Plant Diversity and Ecosystem Productivity: Theoretical Considerations. *Proceedings of the National Academy of Science* 94: 1857–1861.

Tilman, D., P.B. Reich, J. Knops, D. Wedin, T. Mielke, and C. Lehman. 2001. Diversity and Productivity in a Long-Term Grassland Experiment. *Science* 294: 843–845.

Toman, M. 1998. Why Not to Calculate the Value of the World's Ecosystem Services and Natural Capital. *Ecological Economics* 25(1): 57–60.

Tschirhart, J. 2000. General Equilibrium in an Ecosystem. *Journal of Theoretical Biology* 203: 13–32.

U.S. Environmental Protection Agency. 1997. *The Benefits and Costs of the Clean Air Act 1970–1990.* Washington, DC: EPA.

Vitousek, P.M., D.R. Turner, W.J. Parton, and R.L. Sanford. 1994. Litter Decomposition on the Mauna Loa Environmental Matrix, Hawaii: Patterns, Mechanisms, and Models. *Ecology* 75: 418–429.

Vitousek, P.M., J.D. Aber, R.W. Howarth, G.E. Likens, P.A. Matson, D.W. Schindler, W.H. Schlesinger, and D. Tilman. 1997a. Human Alteration of the Global Nitrogen Cycle: Sources and Consequences. *Ecological Applications* 7: 737–750.

Vitousek, P.M., H.A. Mooney, J. Lubchenco, and J.M. Melillo. 1997b. Human Domination of Earth's Ecosystems. *Science* 277: 494–499.

Wilson, E.O. 1988. The Current State of Biological Diversity. In *Biodiversity,* edited by E.O. Wilson. Washington, DC: National Academy Press, 3–18.

———. 1992. *The Diversity of Life.* Cambridge, MA: Belknap Press of Harvard University Press.

CHAPTER 5

Emerging Scarcities

Bioenergy–Food Competition in a Carbon Constrained World

Christian Azar

*E*NERGY IS *NOT* SCARCE. There is enough coal in the world to meet any plausible energy demand projections for centuries to come.[1] If coal is unacceptable, the global society may resort to the plentiful supplies of solar energy. The solar influx to the earth carries some 10,000 times more energy per year than the current annual global anthropogenic use of fossil fuels, nuclear power, and hydroelectric power combined (WEA 2000, *167*). Solar energy can be converted into clean energy carriers, such as electricity and hydrogen, and make our energy and transportation systems virtually emissions free.

Thus, the "energy problem" has nothing to do with physical scarcities of energy per se. Rather the energy problem, or problems to be correct, is related to economic, institutional, geopolitical, and environmental factors.

Poverty and various institutional factors, not physical scarcity of energy, are the key reasons why two billion people lack access to modern energy carriers in developing countries. This has tremendous health and environmental consequences. Moreover, *geopolitical scarcities* exist primarily in the context of oil—conventional oil reserves will likely become scarcer over the next decades, and this brings risks for new oil crises and military conflicts. Oil is cheaper than its alternatives—for example, liquefied coal or hydrogen from solar energy—and this is what creates the great interest in oil. Finally, *environmental scarcities* arise because the burning of fossil fuels emits carbon dioxide (CO_2). Local and regional pollutants, such as sulfur dioxide (SO_2), nitrogen oxides (NO_x), and particulate matter, are causing health and environmental hazards for billions of people around the planet.[2]

This chapter will focus on climate change and how policies to deal with climate change might increase competition for land between biomass and food. Four hypotheses are put forward:

- It is technically and economically feasible to meet stringent climate targets.

- There is a risk that new land scarcities will emerge in response to climate abatement policies. Bioenergy can be expected to play a key role in the transition to low CO_2 emissions because it is a relatively low-cost renewable fuel. But the potential supply is low compared to the required levels of carbon-free energy, almost regardless of whether one is optimistic or pessimistic about the potential biomass supply. Thus, more costly carbon-free energy sources will have to be used if low CO_2 targets are reached.

- The more costly carbon-free energy could raise energy prices to a level that would mean higher profits for the bioenergy sector. With these higher profits, farmers would have greater economic incentives to turn to bioenergy, unless food prices rose to the point where profits matched the energy sector. Thus, land and food prices are likely to be pushed upwards. In this chapter, it is estimated that land values might increase by an order of magnitude, and food prices by a factor of two to five.

- The socioeconomic impacts of higher land and food prices on the poor and malnourished of the world are complex and could be either positive or negative.

Arguments supporting the first three hypotheses will be laid out using back-of-the-envelope calculations to give a feel for the results, and more detailed energy-economy modeling will follow. See Walsh et al. (1996), Azar and Berndes (1999), Gielen et al. (2001), McCarl and Schneider (2001), and Johansson and Azar (2003) for other modeling studies. I then provide some empirical evidence for the land competition hypothesis and discuss how all this might affect global hunger patterns. Policy conclusions are offered in the final section.

Meeting Stringent Climate Targets

Atmospheric CO_2 concentration is currently around 375 parts per million (ppm); this is some 30 percent higher than the pre-industrial concentration, and it is expected to more than double by the end of this century unless carbon abatement policies are introduced. For that reason, the UN Framework Convention on Climate Change (FCCC) (United Nations 1992) calls for a stabilization of greenhouse gases at a level that prevents "dangerous anthropogenic interference with the climate system."

The FCCC does not attempt to define the concept of dangerous interference with the climate system, but governments (e.g., the European Union) and scientists—including Rijsberman and Swart (1990), the Scientific Advisory Council on Global Change to the Federal Government of Germany (WBGU 1995), Alcamo and Kreileman (1996), and Azar and Rodhe (1997)—have argued for setting an upper limit on the increase in the global annual average surface temperature at or around 2°C above pre-industrial temperature levels. Azar and Rodhe (1997) show that a 2°C target may require that concentrations be kept below 400 ppm, but also that there is some probability that higher concentrations, even 550 ppm, may be acceptable.[3]

From 1990 to 1999, the global society emitted an average of 6.3 GtC/yr from the combustion of fossil fuels and cement production (1 GtC is equal to a billion, or 10^9, metric tons of carbon). Land use changes, primarily deforestation in the tropics, released an additional 1.7 GtC/yr, within a wide uncertainty range, in the 1980s (IPCC 2001a, *190*). By the year 2100, so-called reference scenarios suggest that the global society may emit some 20 GtC/yr within a wide range (IPCC 1992, 1999).

In order to meet a 400 ppm target by 2100, total CO_2 emissions would have to drop to around 2 GtC/yr. If a higher stabilization target had been chosen, the reductions would not need to be as large, but they would nevertheless require a radical transformation of our energy systems. A 450 ppm and a 550 ppm target would require annual emissions to drop to 4 and 7 GtC, respectively (Wigley et al. 1996). But it is important to recognize that emissions eventually would have to drop to levels below 2 GtC/yr even for these higher concentration targets (IPCC 2001a).

The challenge is daunting. A per capita perspective may make this even more apparent. With a population of 10 billion people by the year 2100, per capita emissions would have to drop to 0.2, 0.4, and 0.7 tC/yr, for the 400 ppm, 450 ppm, and 550 ppm targets, respectively. Compare this to current emissions:

- the United States with per capita emissions above 5 tC/yr,
- Japan, the European Union, and the countries of the former Soviet Union in the range 2-3 tC/yr,
- developing countries such as China and Latin America at around 0.7 tC/yr, and
- India and Africa around 0.3 tC/yr.

Thus, global per capita emissions would eventually have to drop to a level below that of India today.

Back-of-the-Envelope Global Energy Scenario

Energy systems models are generally more detailed than transparent. A back-of-the-envelope calculation hopefully provides a clearer picture of the assumptions underlying many low-carbon emission scenarios and enables readers to check the results themselves. Here a simple scenario is offered to demonstrate the technical feasibility of meeting stringent carbon targets by the year 2100. Later, a global energy scenario towards a 400 ppm target will illustrate the full dynamics of that transition.

Assume that income grows so that by 2100 the average global citizen can enjoy roughly the same material standard of living as we do in the developed world today. Assume that there will be 10 billion people and that each will use around 200 GJ/yr (primary energy), as much energy as a typical OECD citizen. This means that the total energy supply would be as high as 2,000 EJ/yr (1 EJ = 10^{18} J, approximately 1 quad). This is five times as much as the current global primary energy supply.

Lower CO_2 emissions can be achieved through a combination of

- reduced energy use via technical energy efficiency measures, lifestyle changes, and slowed population growth,
- reduction in the emissions of CO_2 per unit of primary energy supply via increased use of renewables, including biomass, wind, solar and hydro, nuclear, and natural gas at the expense of coal and oil, and
- carbon sequestration from fossil fuels and biomass.

It is beyond the scope of this chapter to make a detailed assessment of these technologies. The interested reader is referred to the *World Energy Assessment* for a

Table 5-1. *Primary Energy and Negawatts by the Year 2100[a]*

Source	Supply (EJ/yr)	Area requirements (million km²)
Negawatts: Energy efficiency improvements[b] –0.7%/year	1,000[b]	—
Conventional fossil	100	—
Fossil with carbon sequestration	200	—
Biomass energy	200	5–10
Solar and wind	500	1
Total	2,000	—

[a]Assuming that everybody in the world will use energy services to the same extent as the average OECD citizen today (primary energy supply in round numbers is 200 GJ (Gigajoules) /cap/year). Technical, organizational, and behavioral changes are assumed to reduce the demand to 100 GJ/cap/year without compromising the energy services currently provided. On average this would require an energy efficiency improvement by 0.7 percent/yr, throughout this century.

[b]Clearly, energy efficiency is not a source of energy but is included here to clearly display the enormous potential this option has when compared with supply options.

Source: Author calculations based on WEA (2000).

recent detailed and authoritative analysis of nuclear, fossil, and renewable energy technologies (WEA 2000).

Energy efficiency improvements may reduce this demand for energy by half, or down to 100 GJ/yr per capita, but only if policies to enhance efficiency are introduced (i.e., higher energy prices or standards). This would correspond to an increase in energy efficiency by 0.7 percent/year. Technical measures include fuel cells for electricity generation that could raise the efficiency to 50–80 percent (current efficiency is 30–40 percent for coal and about 50 percent for modern natural gas–fired combined-cycle plants); fuel cells for automobiles that could approximately double their efficiency; well-insulated houses in cold climates that do not require a heating system (such houses exist in Sweden, the United States, and elsewhere). For more on energy efficiency, see Ayres (1994) and WEA (2000).

Thus, 100 GJ of primary energy supply could offer the same energy services as a typical OECD citizen enjoys today while expending 200 GJ/year. Our task is to provide the world with 100 GJ/yr per capita—1,000 EJ/year—by the year 2100 (see Table 5-1).

With a carbon emission target of 2 GtC/yr, the world would get some 100 EJ/yr from conventional uses of fossil fuels.[4] This means that the world needs to find 900 EJ/yr from other sources.

Hydroelectric power supplies some 10 EJ/yr. The potential for large-scale expansion is limited, and even if the supply doubled over the next hundred years, it would be less than 10 percent of the total demand for carbon-free energy. Often it is of interest as a low-cost option, but we need not dwell on it here because changes in its supply by a factor of two or more will not affect the extent to which competition for land may arise.

Biomass—organic materials from agriculture and forestry—is a promising low-cost renewable energy source that could supply perhaps 200 ± 100 EJ/yr. It should be noted that 200 EJ/yr is a very impressive amount of enery;[5] it corresponds, in rough terms, to the size of the current global food system, or ten times current commercial roundwood (e.g., paper, furniture) production. Some suggest that even higher yields, perhaps twice as high, are achievable (see WEA 2000 and Berndes et al. 2003 for reviews of other studies), but these estimates tend to be optimistic.

The question is not whether it is technically possible to supply 200 EJ/yr of biomass energy, or even twice this amount; it is possible. The point is that even this level would begin to stretch what can be achieved in a socially and environmentally acceptable way.

CO_2 emissions from the use of fossil fuels—and bioenergy—can be captured and stored.[6] Technically, capture and storage technologies are commercially available and proven, but increases in scale are needed, as are improvements in performance and reductions in costs. The cost of capturing and storing carbon is expected to stay below a few hundred US$/ton C (see WEA 2002; Azar et al. 2004). One critical issue is the storage potential. A technically feasible but politically contentious option is storage in geological reservoirs, including in depleted oil and natural gas fields, deep coal beds and saline aquifers, and the oceans.[7] The potential for CO_2 storage clearly determines the role fossil fuels can play under stringent climate policies, but how much can be stored is uncertain. If ocean sequestration is ruled out, depleted oil and gas fields and saline aquifers with structural traps can store several hundred GtC (WEA 2000; Grimston et al. 2001). If structural traps are not deemed necessary, saline aquifers could sequester as much as several thousand GtC.

Finally, a word on acceptable leakage rates. If 2000 GtC were sequestered, a leakage rate of 0.1 percent/yr would lead to releases of around 2 GtC/yr for several centuries. This would not be compatible with atmospheric CO_2 stabilization at the levels discussed in this chapter. Thus, if thousands of GtC are stored, then leakage rates must be significantly lower than 0.1 percent.[8]

For the purpose of this exercise, let us assume that 500 GtC can ultimately be stored safely. This would mean that around 200 EJ/yr of energy could be delivered annually over a period of 100 years without any carbon emissions.[9] But after that, renewables would need to replace this contribution from fossil fuels with carbon sequestration.

Nuclear power is currently in a stalemate, but it should be recognized that the introduction of stringent carbon abatement policies will most likely increase electricity prices and therefore enhance the economic competitiveness of nuclear power. Some politicians, industry representatives, and researchers argue that we cannot reduce CO_2 substantially without resorting to nuclear energy (see Sailor et al. 2000; Lake et al. 2002).

It is not possible in this chapter to study the details of the role of nuclear energy in a carbon-constrained world, but for nuclear energy to make a substantial contribution, around an order of magnitude expansion would be required: a supply of 100 EJ of electricity or hydrogen (heat losses not counted) would require 4,000 reactors (each at 1,000 MW_e and with a load factor of 0.8). This expansion would correspond to an increase of the number of reactors in the world by a factor of 10. Such an expansion would substantially increase the risk of accidents and nuclear proliferation, in particular if it involved a transition to breeder reactors.[10] For critical arguments about a large-scale expansion of nuclear power, see Abrahamson and Swahn (2000).

Wind and *solar energy* could both offer large amounts of electricity, hydrogen, and heat. The global annual wind resource is estimated at 5,800 EJ/yr (without any land constraints) and 231 EJ/yr (if less than 4 percent of the land area is used).

The solar influx to the earth is large enough to supply any reasonable energy demand level. The influx is typically 7 $GJ/m^2 \cdot yr$ in temperate countries, and half of that in northern countries. Assuming a conversion efficiency of 10 percent into hydrogen or electricity, and a land use efficiency of 50 percent, each square meter of land allocated to solar energy would offer 350 MJ/yr. Thus, 1 million km^2 would be needed to supply 350 EJ/yr, and this corresponds roughly to 3.6 percent of the area of the world's deserts, polar regions excluded (IPCC 2001a, *192*).

This is of course a large area, but not nearly as large as that envisioned for biomass energy plantations in most global energy scenarios; more importantly, green bioproductive areas would not need to be targeted. WEA (2000, *163*) estimates a minimum and maximum technical solar energy potential at 1,500–50,000 EJ/yr.

In conclusion, a prescribed primary energy supply of 1,000 EJ/yr by the year 2100 could be met by 100 EJ/yr conventional fossil fuel use, 200 EJ/yr fossil fuel use with carbon capture and storage, 200 EJ/yr of biomass energy, and some 500 EJ/yr from solar and wind energy. Significant contributions could also be obtained from hydro, geothermal, and nuclear energy and ocean energy flows, but these are omitted from this analysis. Nuclear energy has the potential to be important, but concerns about weapons proliferation, radioactive waste disposal, and accidents may virtually eliminate it.

Obviously these numbers are not exact and should only be thought of as indications of potential orders of magnitudes. Perhaps biomass could contribute some 100 EJ/yr more or less, and fossil fuels with carbon sequestration could be much more important or possibly much less depending on storage options and political feasibility. The key point to understand is that the contribution from bioenergy is likely to be minor compared to the overall demand for carbon-free energy sources, even if optimistic numbers are assumed.

Emerging Scarcities—Food versus Fuel

Policies intended to reduce CO_2 emissions will increase the demand for other technologies and materials, which might induce new scarcities in, say, CO_2 storage capacity, uranium, and metals for solar cells, fuel cells, or batteries (see Andersson and Råde (2002) for the metals case). In the process of solving one problem, we must be careful not to create others.

Carbon abatement policies may increase the demand for bioenergy and, therefore, lead to competition with other sectors that use land, in particular the food, paper, and materials sectors. This chapter focuses on food, but climate policies would also affect the economics of other land uses, such as the conversion of forests into cropland or pastures,[11] or the profitability of using biomass for the production of woody materials or paper. Such competition has already started in Sweden, and there have been requests that the government tax the use of biomass for energy purposes.

The Potential Supply of Biomass Energy

Biomass for energy can be obtained from dedicated plantations and from residues (e.g., forestry, agriculture, and households). The supply from plantations includes eucalyptus, poplar, or herbaceous grasses and depends on the yield and the amount of land that can be made available. Plantations can be established on agricultural lands, pastures, and forestlands; the extent will be determined by a combination of social, ecological (precipitation, soil quality, etc.), and economic factors, as well as environmental concerns (conservation and biodiversity protection).

Currently, global agricultural cropland covers some 1.5 billion hectares; tropical savannahs and grasslands, and temperate grasslands and shrublands, largely used for grazing, cover some 3.5 billion to 4.5 billion hectares, and forests—tropical, temperate, and boreal—cover 4.2 billion hectares (IPCC 2001a, *192*).

Analysis of global food demand suggests that most of the additional supply over the next 50 years will come from increased yields (Dyson 1996), but some studies suggest that several hundred million hectares of cropland expansion will be required (e.g., Tilman et al. 2001). Establishing short rotation plantations on previously forested lands is objectionable for several reasons.[12] Thus, marginal croplands and pastures, in particular degraded areas in developing countries, are seen by many as prime candidates for plantations.

However, Berndes et al. (2003), having reviewed the literature on global bioenergy potentials, conclude that "none of the reviewed studies presents an autonomous assessment of degraded land that is suitable and available for plantation establishment. Instead reference is made to other studies.... However, the studies referred to did not focus on availability of degraded land for plantation establishment." It should also be kept in mind that "degraded lands" is an ambiguous concept that includes all lands that have lost some quality and that these lands are most often inhabited and cultivated by poor subsistence farmers.

It is clear that if the demand for CO_2-free or -neutral energy is extremely high and the price of energy is sufficiently high, large tracts of lands could technically and economically be converted to plantations. The amount of land that will actually be converted depends on the social and environmental acceptability of these plantations. In order to estimate the acceptable potential supply of biomass, one is forced to make assumptions about social and environmental reactions, and this is largely guesswork that reflects the analyst's subjective views.

Some argue strongly in favor plantations. For instance, Hall et al. (1993, *644*) argue that large areas of lands are suitable for plantations in the United States, Europe, Latin America, and Africa, but point out that competition with food production could become significant in Asia. They write that "in light of the prospective favorable economics and the environmental and social benefits that can be derived from plantations' biomass energy, intensive efforts to develop such plantations around the globe are warranted."

However, one can expect that social resistance against plantations will increase once they cover large areas of land. This critique has already started even though the current area of short rotation forests only amounts to a small share of what is hoped for in more bioenergy-oriented scenarios (Carrere and Lohmann 1996).

For the purpose of this chapter, I will assume that 500 million hectares can be freed for plantations under socially and environmentally acceptable conditions. This is consistent with a renewable energy–intensive scenario (Johansson et al. 1993; Hall et al. 1993). But it is significantly less than the 1.2 billion hectare potential assumed in WEA (2000, *158*). Others would argue that 500 million is far too much.

My assumption that 500 million hectares can be freed for plantations does not necessarily mean that I believe that this much should be used. It is too early for strong views. Rather it should be seen as a starting point for discussions. The world should experiment with plantations with multifunctional benefits (Börjesson et al. 2002), in particular on degraded lands, and carefully observe the social and environmental implications. Local decisions will then determine how much bioenergy plantations will eventually produce.

Annual yields of short rotation crops range from 5 to 20 metric tons dry matter per hectare per year(t/ha·yr), depending on production system, precipitation, etc. The higher estimate represents very good conditions (precipitation of 2,000 mm/yr on well-managed experimental plantations in Brazil). Over the next hundred years, improvements might be expected as a result of research and experience, as corn, wheat, and rice yields have improved over the past hundred years. But this aspect should not be overstated: we already have experience with managed plantations for pulp and timber production (there are millions of hectares of eucalyptus and pine in Brazil, India, Chile, and several other countries). Further, average cereal yield today is slightly below 3 ton grain/ha·yr on somewhat less than 800 Mha of lands (FAO 2002; Dyson 1996). Thus, it might be reasonable to assume an average yield of 10 t/ha·yr if very large areas of land are used, some of which would probably be short of rain.

It is of interest to note that 10 t/ha·yr corresponds to 200 GJ/ha·yr, which is roughly equal to the per capita primary energy supply in the OECD countries. The populations of North America and Europe combined total 876 million, and lands used for crops occupy 400 Mha, that is, around half a hectare of cropland per person. If all cropland in these regions were used for bioenergy plantations (a completely unrealistic assumption just for the sake of illustration), the yield would only be equivalent to half of their current primary energy supply.

The amount of residues is determined by the demand for food and industrial roundwood. This makes estimates of the potential from residues uncertain, but it might be 20–100 EJ/yr (see Berndes et al. 2003 for a survey of different studies, and Wirsenius 2000 for a detailed assessment of biomass flows in the global food system).[13]

Thus, assuming an average yield of 10 t/ha·yr and an area of dedicated plantations covering 500 Mha, we would get 5,000 Mt of biomass per year globally. This corresponds to 100 EJ/yr. In combination with up to 100 EJ/year from residues, the potential supply totals 200 EJ/yr in round numbers, give or take 100 EJ/yr to account for all the uncertainties.

Impact on Food and Land Prices

Climatic change and countervailing policies might affect food production in several different ways. Climatic changes and changes in atmospheric CO_2 concentration will affect food production as well as the productivity of biomass in general (IPCC 2001b, chapter 5), and greenhouse gas taxes will increase the cost of agricultural activities (inputs like fuel, electricity, and fertilizers and the cost of emissions, such as methane from cattle and rice paddies). But a CO_2 tax will also increase the price of energy, and thus the profitability of biomass energy plantations, thereby affecting land prices and the economic conditions for agriculture.

In the preceding section we concluded that the demand for biomass is likely to be higher than the potential supply. A supply of 200 EJ/yr is a very impressive amount—larger than current global oil use—but it would nevertheless be a third or less of the global need for carbon-free energy by the end of the century. The inevitable conclusion is that other more costly energy sources must enter the picture. They will eventually set the marginal price of energy.

If a transition from CO_2 emissions were to materialize, substantial profits in the bioenergy sector could be expected, initially as a result of carbon taxes (or a cap-and-trade system) driving up the price of fossil energy. Prices would eventually level out at the cost of the back-stop technology (here assumed to be solar energy for heat, electricity, and hydrogen production). The higher profits in the bioenergy sector would translate into higher land prices and consequently higher food prices. This mechanism would operate under free market conditions, but could of course be modified by specific policies if that were considered desirable.

Studies of food and bioenergy interactions along these lines have been carried out by Walsh et al. (1998), Azar and Berndes (1999), Gielen et al. (2001), McCarl and Schneider (2001), and Johansson and Azar (2003). Edmonds et al. (1996) have also analysed bioenergy and food in an integrated model, but without the introduction of any carbon abatement policies.

Back-of-the-Envelope Calculation of the Willingness to Pay for Biomass

Biomass energy from dedicated plantations, potentially the largest source of biomass, costs around 2 US$/GJ in the tropics and around 3–4 US$/GJ in Europe (WEA 2000, *226*; Azar and Larson 2000; Börjesson et al. 2002). Biomass can be converted into process heat, electricity, hydrogen, or essentially any liquid hydrocarbon (including methanol, ethanol, etc.). The cost of producing hydrogen from biomass (assuming 2 US$/GJ feedstock cost) can be expected to fall to 8 US$/GJ assuming a conversion efficiency of 60 percent (IPCC 1996, chapter 19; WEA 2000, *226*). Currently, these cost levels cannot be met without technical improvements, for instance in biomass gasification.

In order to estimate the willingness to pay for biomass, it is illustrative to compare the costs of hydrogen from biomass and from solar energy. With experience, further technical improvements, and mass production, the latter cost might fall to 20 US$/GJ (IPCC 1996, chapter 19), but this could be optimistic.[14]

If we treat solar hydrogen as the back-stop energy technology that sets the marginal energy price, then biomass can be sold to hydrogen manufacturers at a price of 9 US$/GJ biomass. At this price and a conversion efficiency of 60 percent, the feedstock cost for hydrogen production would be 15 US$/GJ hydrogen. Capital and conversion costs are estimated at around 5 US$/GJ, which brings the total cost to 20 US$/GJ, which is in line with the cost of solar hydrogen. If the price of biomass ends up at 9 US$/GJ, and the biomass production cost is 2 US$/GJ, the profit would be 7 US$/GJ. Given yield levels of 200 GJ/ha·yr, the profit per hectare comes to 1,400 US$/ha·yr.

It is of interest to compare these rents with current land values.[15] Land rents in the United States range from 50 to 300 US$/ha·yr; the average value is 170 US$/ha·yr (USDA 2002). Northern European land rents typically fall in the range 100–300 US$/ha·yr (Jordbruksverket 2001). Thus, land rents would increase by a factor of 5 to 10 if these scarcity costs were to materialize.

These estimates strongly suggest that bioenergy might come to compete with food production for bioproductive lands. Consider wheat production: average yields under good conditions in mature production systems are close to 6 ton grain (dry matter)/year/hectare in the European Union and around 3 ton/year/hectare in the United States. The price of wheat is roughly 100–150 US$/ton grain, so total revenues equal 300–450 US$/ha·yr for U.S. farmers and 600–900 US$/ha·yr in the European Union. To this one should add subsidies, which account for a substantial contribution to farmers' income. But subsidies vary substantially from country to country, and that factor is omitted here.

Consider a farm with an annual wheat yield of 5 t/ha·yr. If farmers under market conditions produced wheat rather than energy crops, the wheat farming profits would have to increase to 1,400 US$/ha·yr or around 280 US$/ton grain (still assuming a willingness to pay 9 US$/GJ for bioenergy). The price of wheat would have to rise to around 400 US$/ton, an increase by as much as a factor of three. The social and political implications of this could be drastic, but are not yet well understood.

Sensitivity Analysis

These numbers should be understood as no more than indications of the increase in land values that could follow from a concerted effort to reduce global CO_2 emissions. Below, we evaluate the sensitivity of these results with respect to some critical assumptions. Assume in all cases that biomass supplies are constrained and that solar energy is the back-stop technology, which can supply large amounts of energy at a constant marginal price, assumed here at 20 US$/GJ unless otherwise stated.

- *The cost of back-stop carbon-free energy.* A lower cost for the back-stop energy source, say solar hydrogen, would reduce the profits in the bioenergy sector, and thus the increase in land values would be less dramatic. Conversely, a higher cost would increase the biomass scarcity rents. The willingness to pay for biomass increases by roughly 0.6 US$/GJ for every 1 US$/GJ increase in the back-stop energy cost.

- *The potential supply of biomass.* Let us assume that average yields are 50 percent higher than 10 t/ha·yr and that the land area covered with dedicated plantations is 100 percent larger (1,000 Mha rather than 500 Mha). The world would then enjoy a supply of 300 EJ/yr of bioenergy from dedicated plantations. This would still amount to a minor share of the total global demand for carbon-free energy, and more costly sources of energy would nevertheless be needed. Assuming higher yields or larger areas of dedicated plantations will thus not change the basic picture that the potential demand for biomass energy is larger than the acceptable supply.

- *Higher yields of crops or bioenergy or both.* Higher profits in the food and bioenergy sectors would offer incentives to increase yields. Higher grain yields would reduce the demand for land for food production, and this would free lands for biomass cultivation. Furthermore, it would lead to smaller food price increases because the higher land rents would be based on a larger crop yield. Assume, for instance, that land rents are 1,400 US$/ha·yr as in the original example. A higher wheat yield, say 10 t/ha·yr would then lead to a price of 140 US$/ton wheat, instead of the 280 US$/ton estimated earlier. On the other hand, if these high crop yields were attainable, then higher yields could also be expected for bioenergy production systems, and this would increase land values even further. For instance, 400 GJ/ha·yr yield would increase land rents to 2,800 US$/ha.
- *Lower bioenergy yield on degraded lands.* This would not have any significant impact on food prices. It would only lower the value of degraded lands and not affect the opportunity cost of growing bioenergy on prime agricultural lands.
- *Selling crop residues for energy purposes.* There is a substantial amount of energy in agricultural harvest residues (typically 50 percent of the dry matter of the crop can be eaten by humans). Part of the residues should be left on the soil for soil quality reasons, but part could be sold for its energy content if energy prices escalated enough. These sales would make food production more economical, and grain price increases would be less pronounced. Let us assume that two tons of residues are sold per hectare per year, with an energy content of 20 GJ/ton, that the value of the residues is 9 US$/GJ, and that the cost of collecting them is 1 US$/GJ. We would then get 320 US$/ha·yr from the sale of residues, or around 60 US$/ton wheat. The grain price would then be 220 US$/ton grain instead of 280 US$/ton (in order to achieve profit parity with bioenergy production). More optimistic assumptions about the recoverability of such residues would further reduce the price increase on wheat (Johansson and Azar 2003).
- *Where the value of biomass is highest.* Earlier we compared biomass to solar hydrogen. The difference in cost is even higher when it comes to process heat and steam production. If solar hydrogen costing 20 US$/GJ is replaced by biomass (burnt directly) at a cost of 2 US$/GJ, then the willingness to pay for biomass might reach 20 US$/GJ (assuming the same cost for the heat plant). This is behind the higher bioenergy values in the more detailed models presented below.
- *Carbon capture and storage from fossil fuels as the back-stop technology.* This would imply that the cost of the carbon-free energy carrier would be much lower, but if this option became available on a large scale (if storage capacity were large, and capture and storage technologies became competitive), then this technology could also be used in biomass conversion facilities. Biomass energy with carbon capture and storage could make it possible to produce carbon-free energy carriers—heat, electricity, and hydrogen—and at the same time remove carbon from the atmosphere (Obersteiner et al. 2001). This would add to the competitiveness of biomass. In Figure 5-1, the cost of electricity from coal, natural gas, and biomass power plants with and without carbon capture is depicted. The profit margin for biomass increases linearly with the carbon tax in both

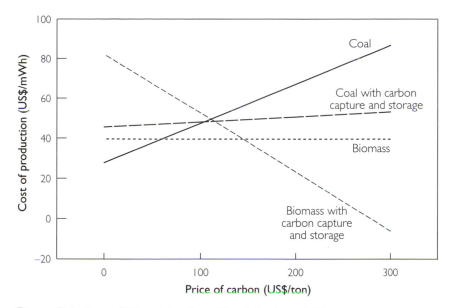

Figure 5-1. *Cost of Electricity from Fossil Fuels and Biomass*

Note: The cost of electricity from fossil fuels and biomass with and without carbon capture is shown as functions of the carbon tax. Transportation and storage costs are included.

Source: Azar et al. 2004.

cases (Azar et al. 2004). It is this profit margin that translates into a willingness to pay more for biomass than the actual production cost.

A More Detailed Model of Food–Fuel Competition

A back-of-the-envelope calculation of global energy supply and demand, and of the biomass contention between food and fuel, may give a transparent overview of the issues at stake; it says very little about progress toward low CO_2 targets and how energy, food, and land prices evolve over time. To evaluate such issues, a more detailed, and therefore less transparent, model is required.

Figures 5-2 and 5-3 present the results of such modeling efforts using the GET model, developed by Azar and Lindgren (Azar et al. 2003, 2004). Other more detailed energy scenarios meeting stringent climate targets include those of IPCC (1996, chapter 19), the C1 scenario in IIASA/WEC (1995), and Azar et al. (2003). The economic cost of achieving such low emissions is discussed in Azar and Schneider (2002).[16]

The GET model seeks to minimize the energy and transport costs for any given CO_2 constraint (tax, emission constraint, or atmospheric stabilization target). Energy demand levels for electricity, transportation fuels, and stationary fuel use are roughly in line with IIASA/WEC so-called ecologically driven scenarios. These scenarios imply the need for substantial efforts to improve energy efficiency.

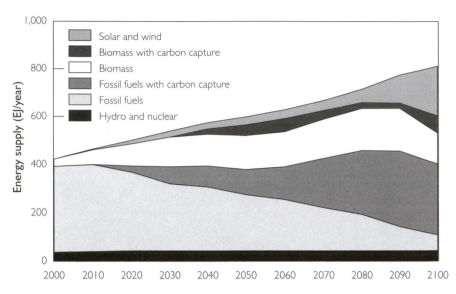

Figure 5-2. *World Primary Energy Supply*

Note: World primary energy supply meeting a 400 ppm concentration target by the year 2100. Generated with the GET model (for details about this model, see Azar et al. 2005). 1 EJ (one Exajoule) is equal to 10^{18} J, which corresponds to 277 TWh.

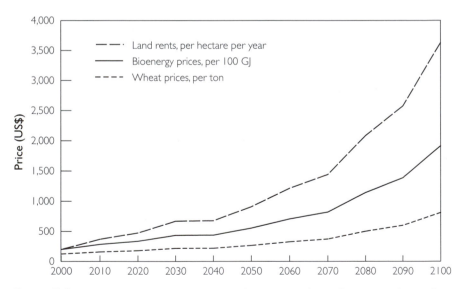

Figure 5-3. *Biomass Scarcity Prices, Land Rents, and Food Prices Obtained with the GET Model (for details about this model, see Azar et al. 2005)*

Energy supply options include coal, oil, natural gas, solar, biomass, wind, nuclear, and hydro. Solar energy may be converted into electricity, hydrogen, and heat, and its potential is many times larger than the current global primary energy supply. Carbon capture and storage is assumed technically feasible on coal, oil, natural gas, and biomass. Biomass energy with carbon capture and storage opens up the possibility of turning the entire energy system into a continuous carbon sink. For assumptions on energy availability, technology cost parameters, conversion efficiencies, and storage options for captured CO_2, see Azar et al. 2005.

The global carbon tax is set at 75 US$/tC in 2010 and then increased by 2.5 percent/yr. The maximum supply of biomass energy is set at 200 EJ/yr, and the full potential can be reached by 2060. Nuclear and hydro power are exogenously set at roughly the current level throughout the next century. Ultimately recoverable oil and natural gas resources are assumed to be twice current reserves, and these will be exhausted before the end of the century.

Figure 5-2 shows the least-cost scenario under the increasingly higher CO_2 tax. Biomass, then fossil fuels and biomass with carbon capture, and ultimately solar and wind energy used to produce heat, electricity, and hydrogen grow in importance over time. Carbon emissions drop to less than 1 GtC/yr toward the end of the century; atmospheric concentrations of CO_2 are kept below 425 ppm throughout the century and drop to 400 ppm by the year 2100.

The higher cost of solar and fossil fuels with carbon capture implies that the price of bioenergy, hence profits in this sector, will increase. Shadow prices for biomass are generated by the model (see Figure 5-3). The higher biomass price implies that land values will increase as consequently will the price of food. Land prices are calculated as the price of biomass times the yield per hectare minus production costs (assumed to be 1 US$/GJ for nonland inputs).

The wheat price is calculated as the price required to make wheat production as profitable as biomass production (a yield of five t/ha·yr and a nonland-related production cost equal to 425 US$/ha). Current wheat prices are set at 125 US$/ton and biomass prices at 2 US$/GJ, both of which give a land rent of 200 US$/ha·yr.

In a modeling study of the greenhouse gas mitigation options in U.S. agriculture and forestry, McCarl and Schneider (2001) specifically analyse the competition for land between sinks, food, and bioenergy and conclude that a carbon tax of 100 US$/ton C would cause crop prices to increase by 30 percent, whereas 500 US$/ton C leads to crop price increases by almost a factor of three. (This carbon tax level is probably high enough to basically phase out all carbon emissions and therefore equivalent to our solar hydrogen future described above.) Their results are somewhat lower than the estimates above, but comparable to Johansson and Azar (2003). Walsh et al. (1996) also find that crop prices increase as a function of higher biomass prices, but they look at more modest carbon abatement policies (modeled as willingness to pay 3 US$/GJ for biomass).

Empirical Evidence of Food–Fuel Competition

Currently, very little carbon abatement takes place in the world, so this analysis might seem purely theoretical. However, there are a few cases where the competition between bioenergy, pulpwood, and food can be observed in the real world.

Sweden is one of the few countries that has actually implemented a carbon tax. The tax is levied at around 400 US$/ton C for households, district heating, and the transport sector, and it has already induced some competition between different end users of the raw materials. Farmers commonly use cereals for residential heating (personal interviews, October 2002), and there are plans to base biogas production on cereals (Köhler 2002). Grain is currently converted into ethanol, but this is primarily driven by large tax reductions for ethanol users and less by the carbon tax.

Producers of furniture based on particle board face bankruptcy because of increasing competition for their raw materials from the energy sector, and Swedish pulp manufacturers are complaining that the high carbon taxes make it profitable for energy companies to buy the wood needed for paper production and that this leads to higher pulpwood prices.[17] The Swedish government has announced that it will investigate this situation (*Dagens Nyheter* 2003).

European forest-based industries are worried that the EU targets for bioenergy will cause substantial losses in wood industries, that many mills will be closed, and so on, because of competition for raw materials (Dielen et al. 2000). The paper industry has expressed similar fears about competition with carbon sinks (in the context of the Kyoto Protocol). Maximizing sinks means harvesting less, thereby increasing wood prices.

During a meeting organized by the Swedish Environmental Protection Agency on corporate social responsibility, the head of environmental affairs of a large pulp and paper manufacturer pointed out that their products were renewable and could eventually be used for energy purposes without carbon emissions. Being the chair of the meeting, I asked him whether that meant that he was in favor of higher carbon taxes. He answered that his company is concerned that this would lead to higher costs for their raw materials because of competition from the energy sector.

On the other hand, he noted that if all countries levied the same tax, this would be less of a problem because the price of paper would rise (in the same way as food prices can be expected to rise). If the tax were applied exclusively in Sweden, it would put Swedish paper producers at a disadvantage on the international market.

In Brazil, sugar is converted into ethanol. Although this is not driven by climate considerations, it shows that a subsidy for ethanol production from sugar cane can induce a large-scale transformation of agricultural products' use. It is now being argued that ethanol from Brazilian sugar cane is nearly competitive with gasoline (Moreira and Goldemberg 1999), and several Annex-1 countries with carbon targets, for example, Holland and Japan, are investigating the possibility of importing ethanol or wood from short rotation forests in tropical countries (Faaij et al. 2002). In Sweden, ethanol manufacturers claim that they would have to close down if ethanol from Brazil were taxed in the same way as domestic ethanol; Brazilian ethanol is seen by many as the most reasonable way to meet the EU biofuels directive, which stipulates an increasing share of biofuels in the transportation sector.

Large-scale biomass trade already takes place in the pulp sector, where even thickly forested Finland imports eucalyptus logs from Uruguay. The eucalyptus is planted on lands previously used for grazing.

Socioeconomic Consequences for Rural and Urban Poor

The socioeconomic consequences of higher land values and food prices are complex. In the optimistic perspective, higher food prices and demand for bioenergy from marginal lands will generate additional income for rural communities around the world. Higher food prices will also offer incentives to increase food output per hectare. The higher income that this may generate will help farmers buy fertilizers and other productivity aids, which will lead to higher yields and revenues in a positive feedback loop. Most of the world's poor live in rural areas, so higher food and land prices will benefit the rural poor.

This argument echoes current discussion of phasing out agricultural subsidies in the rich world. Such a phase-out would raise the price of agricultural crops in poor countries, a result most see as beneficial. In fact, an increased demand for bioenergy could actually make the phase-out of subsidies in developed countries politically feasible, as a large demand for bioenergy could save OECD farmers from losing income.

But this appealing prospect should not blind us to the potentially negative consequences. Today some 800 million people are chronically hungry and undernourished, although the world produces enough food for everyone.[18] The primary reason is the vastly skewed distribution of income (and entitlements in general).[19] Higher food prices could make more people hungry, in particular in urban areas but also among the landless in rural areas. The impact of higher land values would depend on whether you are rich or poor, whether you live in urban or rural areas, and how effectively social security and legal systems work. Here I will concentrate on the impact on the rural poor.

At first one might expect that *rural poor landowners* would benefit, but the situation is complex. Higher land values would make their land more attractive to large landholders, and the poor might lose their land through a variety of mechanisms.

- It might be bought, but uninformed farmers might sell at too low a price.
- Farmers heavily in debt would be in a weak bargaining position.
- Poor farmers might not possess legally binding contracts to their land—in many countries this is the norm rather than the exception (see de Soto 2000 for a detailed account of property rights in developing countries)—and might simply be expelled.

For the *rural landless poor*, higher land values and profits in crop production might mean higher salaries, but modernization of agriculture or an expansion of tree plantations on lands previously under subsidence farming will mean a reduction in the labor demand per hectare. Tree plantations are generally much less labor intensive than agriculture.

The ethanol program in Brazil is worth examining. Proponents argue that it has created jobs for the rural poor, whereas others refer to it as an "exodus" because it has forced thousands of rural subsistence farmers off their lands. I have traveled in the sugar cane regions of northeastern Brazil and witnessed extreme hunger among rural poor and their families. According to rural workers I interviewed, the price of food was higher in the sugar belt than outside it, since the sugar barons controlled not only sugar production but virtually every economic transaction in the region.

A similar example is the growing soybean production in Brazil (in the Cerrado, one of the most biodiversity-rich ecosystems in the world), which has forced poor farmers to move into the Amazon region (Fearnside 2001). Cases of contemporary slavery (heavily indebted rural workers forced to work for local landowners) continue to flourish in the Amazon region according to the Brazilian government (Rohter 2002).

The expansion of eucalyptus plantations for pulp and paper production throughout the tropics offers empirical evidence of the social and environmental consequences that a large-scale expansion of eucalyptus for energy might have. Although the total world area planted in eucalyptus is small, extensive local opposition to plantations in Asia and Latin America is documented in Carrere and Lohman (1996). In Brazil, the state of Espirito Santo has prohibited the Brazilian pulp giant Aracruz from establishing more plantations.

The implications of heightened competition for land are complex and difficult to assess, in particular since it is closely related to the overall development process. If legal systems work and economic development benefits everybody, the competition for land is less troubling.

Some Policy Conclusions

This chapter has argued that climate change, local and regional environmental pollution, and eventual resource scarcities associated with the use of exhaustible resources can be dealt with. I have attempted to show that new energy supply technologies, in particular solar energy, and a wide array of substitutions and efficiency improvements at different levels in society can ease the transition to very low CO_2 emissions. However, this transition will not take place on its own. New policies are required, and even if I feel optimistic that technological solutions are available, I am less optimistic about the prospects for effecting the necessary policies, at least at a sufficiently rapid pace.

The government policies required in order to pave the way for the new technologies include the introduction and continuous increase of the cost of emitting CO_2, for instance, through a carbon tax or a cap-and-trade system. However, the tax or permit fees required to meet near-term targets (such as the Kyoto Protocol) are too low to offer economic incentives to develop and commercialize the more advanced technologies (solar cells, fuel cells, etc.) required to meet long-term targets. This means that policies such as public R&D and the creation of niche markets complementary to the carbon fees are needed today in order to decrease uncertainty, improve performance, and reduce costs of future technologies (Sandén and Azar 2005).

A second key observation is that when, or if, governments intervene and raise the price of CO_2, the value of bioenergy and land can be expected to increase, and this may lead to grain price increases by a factor of two to three during this century. This competition for land might have both positive and negative social and environmental consequences. Higher land vales will offer incentives to improve the quality of degraded lands and might generate income for the rural poor. On the other hand, human demand for domesticating more bioproductive lands might lead to the conversion of biodiversity-rich ecosystems into monocultures, and poor people might be evicted from their lands. Policies are required to foster the positive impacts and discourage the negative. Governments should at least

- strengthen the legal rights of the rural poor to lands that they cultivate;
- introduce rules for the use of residues in agriculture and forestry, because too rapid extraction could threaten soil quality and long-term productivity;
- protect the integrity and biodiversity of valuable ecosystems;
- value the carbon content of biota;
- encourage multifunctional bioenergy systems and agroforestry where appropriate (e.g., to counteract erosion on sensitive soils);
- monitor carefully social impacts of higher land and food prices;
- consider countervailing policies to reduce the impact on urban and rural poor in developing countries that might face higher staple food prices; and
- be prepared to tax biomass energy if the demand becomes very high.

Acknowledgments

This paper builds on invaluable collaboration with Göran Berndes, Daniel Johansson, Eric Larson, and Kristian Lindgren. Furthermore, I would like to thank Dean Abrahamson, Faye Duchin, Dolf Gielen, and Bruce McCarl for detailed comments on the manuscript, and Stefan Wirsenius and Janne Wallenius for discussions on specific parts of it. I would also like to thank the Swedish Energy Agency, Formas, and Vinnova for financial support.

Notes

1. Coal reserves amount to 20,000 EJ, and the resource base is estimated at 200,000 EJ. This can be compared to the annual use of 90 EJ, and the global primary energy supply of around 400 EJ/yr (WEA 2000, *166*).

2. Local air pollution could be significantly reduced. This has already happened in many, if not most, OECD countries. Improvements can also be seen in many developing countries. Improved combustion technologies, catalytic converters, removal of lead as an additive to gasoline, and improved fuel qualities (e.g., sulfur removal) in response to policy regulations have brought about these improvements despite increases in energy demand and transportation volumes.

3. The European Union also adopted a maximum of 550 ppm CO_2 equivalent target, but a 550 ppm concentration might mean that the temperature increase could be as high as 5–6°C. Thus, the EU negotiating positions and their energy policies need to recognize the possibility that a stringent concentration target as low as 400 ppm may be required (pending the contribution from other gases and climate sensitivity).

4. The emission factors for coal, oil, and natural gas are 25, 20, and 14 gC/MJ, respectively. Thus a balanced mixture of these fossil sources would release 2 GtC for every 100 EJ of energy supplied.

5. Global food intake equals 17 EJ/yr (metabolizable energy), and gross energy biomass production aimed at food production (including residues on crop lands and grass on pastures not consumed by animals) is roughly equal to 220 EJ/yr (Wirsenius 2000). One hundred fifty EJ/yr

of biomass from dedicated plantations would correspond to 15 billion m^3/yr of wood. Current commercial roundwood production is roughly equal to 1.5 billion m^3/yr.

6. If CO_2 released from burning sustainably grown biomass were captured and stored, then this system would deliver carbon-free energy carriers (e.g., electricity) and at the same time remove CO_2 from the atmosphere (Obersteiner et al. 2001).

7. Storage in the oceans is politically sensitive, and one research project where only 20 tons of liquefied CO_2 would be injected at the bottom of the ocean floor just outside Hawaii was recently canceled before it started because of local opposition (GECR 2002). The project was then invited to Norway, but opposition from Greenpeace and the World Wide Fund for Nature led the government to withdraw an approval granted by the Norwegian Pollution Control Authority (Giles 2002).

8. This leakage rate might be acceptable if biomass energy with carbon capture and storage were applied as a counteracting measure. The leakage of 2 GtC/yr could be compensated by sequestering 2 GtC/yr from bioenergy. On the other hand, this would require that societies continue to sequester carbon from bioenergy and store it (to compensate for the leaking reservoirs) for hundreds or thousands of years, and one might be more than skeptical about committing future generations to such a large-scale geoengineering project, particularly in light of the fact that 2 GtC/yr corresponds to the carbon flow in 400 Mha of bioenergy plantations with an average yield of 10 t/ha·yr.

9. Here an average carbon "emission" factor corresponding to 25 gC/MJ has been used. This corresponds to the emission factor of coal. In reality, the emission factor per unit of useful energy could be either higher as a result of energy efficiency losses (due to the energy cost of carbon sequestration) or lower because less carbon intense fossil fuels could also be expected to play a role.

10. A large-scale expansion of nuclear energy might lead to uranium scarcities. World reserves of natural uranium are currently estimated at 3.9 million tons (recoverable resources at a cost less than 130 US$/kg: NEA 2002) and would be consumed over the next hundred years at current mining rates (36,000 ton U/yr supplies 56 percent of the reactor requirements, and the remainder is from secondary sources, such as reprocessed spent fuel). This would not suffice for the large-scale expansion envisioned here. However, current reactors only use a fraction of the available energy in natural uranium (natural uranium consists of 0.7 percent U-235 and 99.3 percent U-238, and current reactors rely almost exclusively on U-235). Breeder reactors, however, make use of the U-238 in the natural uranium. They produce as much fuel as they consume, through the production of plutonium from neutron capture by U-238 (followed by two beta decays). The problem is that this plutonium would have to be recovered from the reactor—taken out of the reactor and separated into new fuel. This would greatly expand risks of civilian plutonium being spread to nation-states or subnational groups. A light-water reactor of 1 GW electric output typically produces 200–300 kg of plutonium per year, and a breeder reactor around one ton. Four thousand 1 GW breeder reactors would put 4,000 tons of plutonium in the global economy every year. The construction of a nuclear bomb would not require more than about 6 kg. A comprehensive analysis of plutonium management options can be found in NAS (1994). One way to avoid this plutonium intensive scenario, while still expanding nuclear energy, would be to rely on uranium from seawater. It is currently considered too expensive to "mine" uranium from seawater, but the resource is huge—4.5 billion tons—and technological improvements might reduce costs to 300 US$/ton, which corresponds to 0.004 US$/kWh electricity (WEA 2000, 316).

11. The carbon prices required to meet stringent climate targets (most likely several hundred US dollars) would have a significant impact on the profitability of establishing forests as carbon sinks or protecting standing forests to avoid emissions related to deforestation. For instance, a 200 US$/ton carbon price would mean that deforestation in the tropics (which releases some 200 ton C/ha) would cost 40,000 US$/ha. This is an order of magnitude larger than the value of high-

quality cropland in industrialized countries, and it would be uneconomical to convert forestland to pasture or cropland. Thus, if a global climate regime came to include carbon emissions from deforestation and carbon sinks in forests, then this would have an enormous effect on the way countries govern their forests. Of course, the issue of biospheric carbon stocks is closely linked to the issue of bioenergy plantations, but in order to avoid the additional complexities of carbon sinks, and for reasons of space, it is addressed in a separate paper (Hedenus and Azar 2003a).

12. It would imply a significant loss of biodiversity because plantations generally are monocultures. But more intensively managed forests could become of interest. Börjesson et al. (1997) have argued in favor of fertilizing certain managed forests in Sweden to increase yields from 2–5 m^3/hectare per year (ha·yr) to perhaps 30 m^3/ha·yr.

13. To illustrate these numbers, assume that the average citizen of the world consumes 500 kg of cereals per year (direct and indirect consumption). North America and Europe consume 600–800 kg/yr per capita, whereas Africa consumes 150 kg, South Asia 237 kg, and the current world average is 363 kg/yr (Dyson 1996). Each kilogram of grain yields somewhat less than a kilogram of residues (e.g., straw). Assume that a third of this residue stream is recoverable for energy purposes. Further assume that every person consumes 400 kg of paper and woody products per year. Recycling will lower the per capita input demand to, say, 200 kg/year. All of this cannot be burned, but there will also be losses in the production, harvest, and processing of the raw material, so assume that 200 kg/year per capita will be made available for energy purposes. (Per capita paper consumption by the 20 percent in the world that consume most paper is around 200 kg/yr; the poorest 20 percent consume around 1 kg/yr; see Hedenus and Azar 2003b). In total we have around 350 kg or 7 GJ/yr per capita; with a global population of 10 billion people, we have 70 EJ/year. The number can be compared with the current noncommercial use of biomass in developing countries, which amounts to some 33–55 EJ/year (WEA 2000, 156).

14. Assume that the hydrogen is produced through electrolysis, with the electricity being produced from solar energy. An electricity production cost of .05 US$/kWh, corresponds to 16 US$/GJ in energy costs (assuming an 85 percent conversion efficiency from electricity to hydrogen). On top of that, capital, OM, and transportation costs would have to be added. Currently the production cost for solar electricity is at least three times as high.

15. Comparisons with cropland values are difficult because these values are distorted by subsidies that may amount to several hundred US$/ha·yr (at least in OECD countries).

16. They show that even pessimistic studies of the cost to stabilize the atmosphere acknowledge that the cost "only" amounts to a few years' delay in achieving an impressive *increase* in global income levels. For instance, a 350 ppm target implies that the world GDP is 10 times larger in April 2102 rather than in January 2100. This is not to say that the costs are insignificant. The estimates do point toward trillion US$ levels net present value cost, but this alternative way of expressing the costs put them in a perspective worth highlighting for policymakers and the general public.

17. On a US$/GJ basis, pulpwood is cheaper than food, and therefore competition can be expected in this sector earlier than in the food sector. For that reason I have included this aspect here.

18. On croplands, 15,000 kcal/day per capita of biomass is being produced, 3,000 kcal/day of which is pasture biomass and animal forage crops. Out of the remaining 12,000 kcal/day, 5,000 kcal/day is edible products, and this is twice as much food as is needed (Wirsenius 2000). However, conversion losses in food processing and, even more importantly, the conversion of cereals into animal foodstuffs reduce this amount to less than 2,000 kcal/day. Animal products add some 270 kcal/day. Thus, there would be enough food for everybody were it not for the choice to use cereals to produce animal products. Note that a decision not to use cereals as feed would not necessarily solve this situation, because it might only result in a lower production of cereals. (Statistics on food availability generally suggest that 2,700 kcal/day per capita is available;

see FAO 2002 or Dyson 1996. But this number refers to what is available at the wholesale level, not to actual intake). The number of people suffering from starvation is much lower and cases of starvation are much more rare in democratic countries than in dictatorial regimes, primarily because democratic governments need the support of their people to be reelected (Sen 1981). On the other hand, even democratic governments have not been able to eradicate hunger, as is evident in countries such as Brazil and India.

19. Sometimes this observation has been perceived as an argument that a redistribution of income (or land, capital, work opportunities) is the key step toward achieving an end to hunger. But a more effective approach would be to *also* increase the production of food. It is difficult to envision a redistribution of income, land, and capital that does not lead to increases in overall food production.

References

Abrahamson, D., and J. Swahn. 2000. The Political Atom. *Bulletin of the Atomic Scientists* 56: 39–44.

Alcamo, J., and E. Kreileman. 1996. Emission Scenarios and Global Climate Protection. *Global Environmental Change* 6: 305–334.

Andersson, B.A., and I. Råde. 2002. Material Constraints on Technology Evolution—The Case of Scarce Metals and Emerging Energy Technologies. In *Handbook of Industrial Ecology*, edited by R. Ayres and L. Ayres. Cheltenham, UK: Edward Elgar, 391–404.

Ayres, R.U. 1994. On Economic Disequilibrium and Free Lunch. *Environmental and Resource Economics* 4: 434–454.

Azar, C., and E. Larson. 2000. Bioenergy and Land-Use Competition in the Northeast of Brazil: A Case Study in the Northeast of Brazil. *Energy for Sustainable Development* IV(3): 64–72.

Azar, C., and G. Berndes. 1999. The Implication of CO_2-Abatement Policies on Food Prices. In *Sustainable Agriculture and Environment: Globalisation and Trade Liberalisation Impacts,* edited by A. Dragun and C. Tisdell. Cheltenham, UK: Edward Elgar.

Azar, C., and H. Rodhe. 1997. Targets for Stabilization of Atmospheric CO_2. *Science* 276: 1818–1819.

Azar, C., and S.H. Schneider. 2002. Are the Economic Costs of Stabilizing the Atmosphere Prohibitive? *Ecological Economics* 42:73–80.

Azar, C., K. Lindgren, and B. Andersson. 2003. Global Energy Scenarios Meeting Stringent CO_2 Constraints—Cost-Effective Fuel Choices in the Transportation Sector. *Energy Policy* 31: 961–976.

Azar, C., K. Lindgren, E. Larson, and K. Möllersten. 2005. Carbon Capture and Storage from Fossil Fuels and Biomass: Costs and Potential Role in Stabilizing the Atmosphere. *Climatic Change*. In press.

Berndes, G. 2002. Bioenergy and Water—the Implications of Large Scale Bioenergy Production for Water Use and Supply. *Global Environmental Change* 12: 253–271.

Börjesson, P., L. Gustavsson, L. Christersson, and S. Linder. 1997. Future Production and Utilisation of Biomass in Sweden: Potentials and CO_2 Mitigation. *Biomass and Bioenergy* 13: 399–412.

Börjesson, P., G. Berndes, F. Fredriksson, and T. Kåberger. 2002. *Multifunctional Bioenergy Plantations—Final Report to the Swedish Energy Agency* (Multifunktionella bioenergiodlingar. Slutrapport till Energimyndigheten). Eskilstuna: Swedish Energy Agency.

Carrere, R., and L. Lohmann. 1996. *Pulping the South*. London: Zed Books.

Dagens Nyheter. 2003. Paper Industry Battles with Politicians (Pappersindustrin i kamp med politiker). June 12.

de Soto, H. 2000. *The Mystery of Capital—Why Capitalism Triumphs in the West and Fails Everywhere*. London: Black Swan.

Dielen, L.M.M., Guegan, Lacour, Mäki, Rytkönen, and Stolp. 2000. *EU Energy Policy Impacts on the Forest-Based Industries: Modeling Analysis of the Influence of the EC White Paper on Renewable*

Energy Sources on the Wood Supply to the European Forest-Based Industries. Summary Report. Wageningen: Afocel. Nangis.

Dyson, T. 1996. *Population and Food—Global Trends and Future Prospects.* London: Routledge.

Edmonds, J.A., M.A. Wise, R.D. Sands, R.A. Brown, and H. Khesghi. 1996. *Agriculture, Land Use and Commercial Biomass Energy.* Washington, DC: Pacific Northwest National Laboratory.

Faaij, A., B. Schlamadinger, Y. Solantausta, and M. Wagner. 2002. Large-Scale International Bioenergy Trade. Paper presented for the 12th European Conference and Technology Exhibition on Biomass for Energy, Industry, and Climate Protection. June 17–21, Amsterdam, the Netherlands.

FAO (Food and Agriculture Organization of the United Nations). 2002. FAO Database. http://www.fao.org.

Fearnside, P.M. 2001. Soybean Cultivation as a Threat to the Environment in Brazil. *Environmental Conservation* 28: 23–38.

GECR (Greenpeace and the World Wide Fund for Nature). 2002. *Global Environmental Change Report.* July 26.

Gielen, D.J., M.A.P.C. de Feber, A.J.M. Bos, and T. Gerlagh. 2001. Biomass for Energy or Materials? A Western European Systems Engineering Perspective. *Energy Policy* 29: 291–302.

Giles, J. 2002. Norway Sinks Ocean Carbon Study. *Nature* 419: 6.

Grimston, M.C., V. Karakoussis, R. Fouquet, P. van der Vorst, M. Pearson, and Leach. 2001. The European and Global Potential of Carbon Dioxide Sequestration in Tackling Climate Change. *Climate Policy* 1: 155–177.

Hall, D., F. Rosillo-Calle, R. Williams, and J. Woods. 1993. Biomass for Energy: Supply Prospects. In *Renewable Energy: Sources for Fuel and Electricity*, edited by T.B. Johansson et al. Washington, DC: Island Press.

Hedenus, F., and C. Azar. 2003a. Carbon Sinks versus Bioenergy—Physical and Economic Perspectives. Work in progress.

———. 2003b. World Income and Resource Inequality Trends. *Ecological Economics.* In press.

IIASA/WEC, 1995. *Global Energy Perspectives to 2050 and Beyond,* London: World Energy Council.

IPCC (Intergovernmental Panel on Climatic Change). 1992. *Climate Change 1992: The Supplementary Report to the IPCC Scientific Assessment,* edited by J.T. Houghton, B.A. Callander, and S.K. Varney. Cambridge: Cambridge University Press.

———. 1996. *Impacts, Adaptation and Mitigation Options.* IPCC Working Group II. Cambridge: Cambridge University Press.

———. 1999. *Special Report on Emission Scenarios,* edited by N. Nakicenovic. Cambridge: Cambridge University Press.

———. 2001a. *Climate Change 2001: The Scientific Basis.* Contribution of Working Group I to the Second Assessment Report of the Intergovernmental Panel on Climate Change, edited by Houghton et al. Cambridge: Cambridge University Press.

———. 2001b. *Climate Change 2001: Impacts, Adaptation, and Vulnerability.* Contribution of Working Group II to the Second Assessment Report of the Intergovernmental Panel on Climate Change, edited by J.J. McCarthy, O.F. Canziani, N.A. Leary, D. Dokken, and K.S. White. Cambridge: Cambridge University Press.

Johansson, D.J.A., and C. Azar. 2003. Analysis of Land Competition between Food and Bioenergy. *World Resources Review* 15: 165–175.

Johansson, T.B., H. Kelly, A.K.N. Reddy, and R.H. Williams. 1993. *Renewable Energy: Sources for Fuel and Electricity.* Washington, DC: Island Press.

Jordbruksverket. 2001. Utveckling av arrende, mark och fastighetspriser i Sverige (Trends in agricultural rents, land prices and real estate prices, report in Swedish with English summary). Swedish Board of Agriculture Report no. 2001: 8. http://www.jordbruksverket.se.

Köhler, N. 2002. Fermented Cereal Grain Can Produce Inexpensive Biogas without Government Subsidies (in Swedish). *Ny Teknik* 35: 5.

Lake, J.A., R.G. Bennett, and J.F. Kotek. 2002. Next Generation Nuclear Power. *Scientific American* 286(1): 70–79.

McCarl, B.A., and U.A. Schneider. 2001. Greenhouse Gas Mitigation in US Agriculture and Forestry. *Science* 294: 2481–2482.

Moreira, J.R., and J. Goldemberg. 1999. The Alcohol Program. *Energy Policy* 27: 229–245.

NAS (National Academy of Sciences). 1994. *Management and Disposition of Excess Weapons Plutonium.* Washington, DC: National Academy Press.

NEA (Nuclear Energy Agency). 2002. *Uranium 2001: Resources, Production, and Demand.* Paris: Nuclear Energy Agency.

Obersteiner, M., C. Azar, P. Kauppi, K. Mollersten, J. Moreira, S. Nilsson, P. Read, K. Riahi, B. Schlamadinger, Y. Yamagata, J. Yan, and J.P. van Ypersele. 2001. Managing Climate Risk. *Science* 294: 786–787.

Rijsberman, F.R., and R.J. Swart (eds.). 1990. *Targets and Indicators of Climatic Change.* Stockholm: Stockholm Environment Institute.

Rohter, L. 2002. Trapped Like Slaves on Brazilian Ranches—Forced Labor Clears Forest for Cattle. *International Herald Tribune,* March 26.

Sailor, W.C., D. Bodansky, C. Braun, S. Fetter, and B. van der Zwaan. 2000. Nuclear Power—A Nuclear Solution to Climate Change? *Science* 288: 1177–1178.

Sandén, B.A., and C. Azar. 2005. Near-Term Technology Policies for Long-Term Climate Targets. *Energy Policy* 33: 1557–1576.

Sen, A. 1981. *Poverty and Famines—An Essay on Entitlement and Deprivation.* Oxford: Clarendon Press.

STEM, 2001. Energiläget 2001 (Energy situation in Sweden 2001). Eskilstuna: Swedish Energy Agency. www.stem.se.

Tilman, D., J. Fargione, and B. Wolff. 2001. Forecasting Agriculturally Driven Global Environmental Change. *Science* 292: 281–284.

USDA (U.S. Department of Agriculture). 2002. Pasture Rents Up Slightly in Wyoming. http://www.nass.usda.gov/wy/cashrent.htm.

United Nations. 1992. Framework Convention on Climate Change. http://www.unfccc.ch.

Walsh, M., R.L. Graham, D. de la Torre Uguarte, S. Slinsky, D. Ray, and H. Shapouri. 1998, Economic Analysis of Energy Crop Production in the US—Location, Quantities, Price, and Impacts on Traditional Agricultural Crops. Paper presented at BioEnergy 1998: Expanding Bioenergy Partnership, October 4–8, 1998. http://bioenergy.ornl.gov/papers/bioen98/walsh.html.

WBGU. 1995. *Scenarios for Derivation of Global CO_2 Reduction Targets and Implementation Strategies.* Bremerhaven, Germany.

WEA (*World Energy Assessment*). 2000. United Nations Development Programme and World Energy Council.

Wigley, T., R. Richels, and J. Edmonds. 1996. Economics and Environmental Choices in the Stabilization of Atmospheric CO_2 Concentrations. *Nature* 379: 240–243.

Wirsenius, S. 2000. Human Use of Land and Organic Materials: Modeling the Turnover of Biomass in the Global Food System. Ph.D. diss., Department of Physical Resource Theory, Chalmers University of Technology.

CHAPTER 6

Sustainability and Its Economic Interpretations

John C.V. Pezzey and Michael A. Toman

WELL BEFORE *Our Common Future* (WCED 1987) popularized the concept of sustainability, the idea (if not the precise term) figured prominently in debates about scarcity and growth. The original *Scarcity and Growth* volume observed,

> The classical economists—particularly Malthus, Ricardo and Mill—predicted that scarcity of natural resources would lead to … retardation and eventual cessation of economic growth. (Barnett and Morse 1963, *2*)

One of the seminal essays on economic growth and nonrenewable resources defined its topic thus:

> This study is an attempt to determine more precisely under what conditions a sustainable level of per capita consumption is feasible. (Stiglitz 1974, *123*)

The foreword of *Scarcity and Growth Revisited* noted that

> the public's recent questioning of whether or not it will be possible to provide and maintain a reasonably high material standard of living for all while ensuring that the overall quality of life remains unchanged. (Spofford 1979, *xi*)

The introduction to the same book ended its first paragraph by stating,

> not until the 1900s was any general concern expressed regarding the adequacy of supplies of raw materials to sustain continuous economic growth. (Smith and Krutilla 1979, *1*)

The topic of sustainability is thus central and well established, if not always explicit, within the concept of scarcity and growth.

Scarcity and Growth listed four factors sufficient to overcome the problem of *marketed* natural resource depletion and allow economic growth to be sustained indefinitely:

* use of lower-grade natural resources,
* substitution by other resources or human-made capital,
* increased exploration and recycling, and
* technical progress in resource extraction.

So far these factors appear to have been effective (see Krautkraemer in this volume), though concerns about the physical availability of some key raw materials continue to be raised (see Menzie et al. in this volume).

As emphasized in many other parts of this volume, the focus of scarcity and growth analysis has shifted considerably in the past two decades toward new concerns about depletion of *nonmarketed* environmental resources such as biodiversity and the ability of the atmosphere to hold greenhouse gases without "dangerous" impacts. One analogue to the four factors listed in *Scarcity and Growth* is adapting patterns of consumption and production to less available or more degraded flows of ecological services. However, the feasibility of perpetual economic growth in the face of the "New Scarcity" is more hotly debated. If perpetual economic growth is not feasible, moreover, an old question from the earlier scarcity and growth literature assumes new prominence: Is there an ethical case for deliberately curbing growth now, so that future generations will not face a shrinking economy and declining well-being?

We have not attempted here an up-to-date, comprehensive review of even the economics of sustainability, let alone discussions of sustainability from other disciplinary perspectives. That is impossible given the sheer volume of recent publications. We aim instead to provide a nontechnical overview of what the recent economic literature on sustainability can offer to the debate on scarcity and growth. A more complete and somewhat more technical treatment of sustainability can be found in Pezzey and Toman 2002. To keep our scope manageable, we also focus on overall sustainability in the context of *equity across several generations*. We give no attention to "sectoral sustainability" (sustainable agriculture, sustainable transport, etc.) as such. We have also excluded intragenerational equity (a key feature of what some call "social sustainability") from our definition of sustainability and left out any discussion of population growth. On both these last counts, sustainability in developing countries simply does not receive the attention here that it merits.

First we review the most useful approaches to defining sustainability in an economic context. We first list issues to be resolved: what is to be sustained and why, when and where it is to be sustained, and basic "weak" and "strong" views on how it can be sustained. Then we give a simplified synthesis of what has been derived from "neoclassical economic theory"[1] about threats to, opportunities for, measurement of, and policies to achieve sustainability (mainly at a national level). We also touch on some economic views that include nonneoclassical elements and some physical measures of sustainability. Last we offer some conclusions.

Defining Sustainability

Basic Economic Concepts and Assumptions

Before setting out our approach to sustainability, it may be helpful to the non-economist reader if we explain some basic economic concepts. Economics is the study of (a) how scarce inputs of human labor, accumulated capital stocks of different types, and environmental resources are allocated to different parts of society over time in order to produce outputs of goods and services that people want; and (b) how the outputs are distributed. In doing this, economists typically make several assumptions about human behavior that are often collectively labeled as the concept of "*Homo economicus.*" They assume that people or households rationally (i.e., consistently) try to maximize their *welfare*, which is some aggregate over time of the instantaneous well-being or *utility* that people receive from different forms of consumption at each moment in time, including marketed and nonmarketed goods and services.[2] Welfare is usually taken to be the *present value* (PV), or discounted sum, of utility from some time onwards, using a constant discount rate. This reflects the observation that people usually prefer benefits sooner rather than later (and costs later rather than sooner); they are "impatient." As discussed below, when thinking about intergenerational resource allocations, the issues of whether and at what rate to discount future flows of well-being (utility) are more complex. The economic model of individual behavior also typically assumes that people are well (though imperfectly) informed, so that the long-term as well as short-term consequences of decisions can be reasonably well assessed.

It is also usually assumed that smooth trade-offs, or *substitutions*, are always possible among different inputs used to produce goods and services and among goods and services that yield well-being or utility. In particular, typically unlimited (though not necessarily perfect) substitutability between human-made and natural inputs to production and utility is assumed.[3] Economists typically further assume that an economy can in principle attain an *equilibrium* where all supplies and demands of inputs and outputs are in balance. In practice the economy is always changing because of evolving circumstances, but this can be reflected in an economic model of dynamic equilibrium incorporating markets for savings and investment, for potential future consumption ("forward markets"), and for reallocating risk (like insurance). A competitive equilibrium is one in which producers and consumers act independently in response to various price signals. Economic analysis also can include equilibrium with noncompetitive markets (e.g., price-setting monopolies).

Given all these assumptions and another crucial assumption, that "markets are complete," it is possible to prove that the economy's equilibrium will be *efficient*. In economics this means that it would be impossible to make anyone better off without first making someone else (perhaps at a different time) worse off (Arrow and Debreu 1954).[4] The complete markets assumption requires that every good and service that affects well-being has a market-determined price; there are no goods and services outside the market system. This establishes the demanding circumstances under which Adam Smith's (1776) intuition holds, that markets act as an "invisible hand" that guides the economy to a benevolent outcome. It can be seen both as a cautious prescription in favor of markets for allocating resources and as a warning of the many reasons why markets alone will *not* produce a desirable outcome when markets are incomplete. For our purposes, one of the most important examples of a missing market is the "external costs" imposed by some on others via uncompensated degradation of the natural environment.

Economic Definitions of Sustainability

Importantly, economic efficiency (as defined above) says nothing about sustainability, at least as we choose to define the latter. As noted, we focus on sustainability in a context of concern for intergenerational equity over several generations and on the potential role of finite environmental resources in impeding intergenerational equity, given the importance of this subject in both sustainability and scarcity-and-growth debates. Centering on intergenerational equity differs from the now everyday use of sustainable to mean just "environmentally desirable or responsible." The latter usage makes no distinction between momentary environmental problems (such as foul smells that cause no lasting health effects) and permanent or cumulative problems (such as species loss or climate change). Correction of the former type of external cost increases the efficiency of resource allocation. On the other hand, it is possible that "correcting" the latter type of externality will lead to an efficient, yet not sustainable, allocation of resources over time. We develop this point in more detail below.

Most economic definitions of sustainability ignore any age, geographical, and especially economic inequities within a society. Inequities can perhaps influence sustainability, either through production (because of disruption and social breakdown) or consumption (a future with rising consumption but sharply rising inequity may be seen as one in which "society's" quality of life is not sustained), but we do not discuss any such influences here.

One of the most common ways of approaching sustainability, and the one on which we will rely, is a heuristic *constraint on utility over time*. One of the first variants considered in the economics literature was forever constant utility (as in Solow 1974), drawing on the philosophical theory of justice developed by Rawls (1971). Subsequent theorists extended this to forever nondeclining utility (Pezzey 1989), which allows for the possibility of rising well-being over time, and forever nondeclining intertemporal welfare (Riley 1980; Dasgupta and Maler 2000). The specific formulation we use below states that sustainability means that the current *actual* utility must not exceed the current *maximum sustainable* utility (Pezzey 1997); if it does, some future decline in actual utility is inevitable.

Sustainability can instead be defined as a *constraint on changes in opportunities*, rather than changes in outcomes. This represents quite a shift in political philosophy, from concern over what future generations will enjoy (they can have a good inheritance, but still squander it if they choose) to concern over what they will inherit from us. The constraint most frequently suggested is nondeclining wealth or aggregate capital, instead of nondeclining utility. The two approaches can be linked with each other, though by no means equated.

Given this heuristic but seemingly plausible definition of sustainability as nondeclining well-being, what can economics tell us about the possibility of growth being sustainable or unsustainable? How can we make informed judgments in practice about whether a particular path is sustainable? The ability to empirically gauge sustainability we defer to the next section. First, three classic papers by Dasgupta and Heal (1974), Solow (1974), and Stiglitz (1974) provide intuitive insights that are still very useful in establishing a conceptual base, and in highlighting the challenges in developing a useful operational theory of sustainability.

Dasgupta and Heal consider a mathematical model of a simple economy that produces a single good, usable for consumption and investment, by combining the efforts of labor and capital with a depletable natural resource. Ignoring for simplicity all issues of external costs and uncertainty, they show that along an inter-temporally efficient path (as defined above) that maximizes the discounted present value of utility from consumption, the rate of consumption and utility must eventually decline. In other words, what we will call for brevity the *PV-optimal* path of this simple stylized economy must be unsustainable as we have defined the latter term. Deeper analysis of behavior along the PV-optimal path shows that even if the economy is technically able to overcome rising natural resource scarcity by rapidly investing in built capital—a controversy we revisit below—an adequate diversion of current consumption into such investment will not take place given the discounting of future benefits from investment. Discounting imposes an inherently "presentist" tilt to consumption choices that undermines the ability of the economy to grow sustainably.

Solow's analysis roughly mirrors Dasgupta and Heal's, as he presumes that utility should be sustained over time and examines conditions under which this is technically feasible. He finds that such conditions exist, but in the context of his model they are restrictive. Solow's analysis also serves to further emphasize that any economic path that sustains utility over time, at least given the simple assumptions of the economic model with which he is working, is not PV-optimal. This leads to questions about why people would want to follow such a path if they do have impatient or presentist preferences. We return to this point below.

Stiglitz approaches the problem a different way, by asking how serious the scarcity problem is for growth. Continuing to work with a simple model like that of Solow, but adding the possibility of exogenous technical progress that increases the productivity of the natural resource and thereby offsets its increasing natural scarcity, Stiglitz shows that if technical progress is rapid enough and sustained, the economy can achieve a path of indefinitely expanding utility over time.

It seems easy at first to criticize this result as an absurd caricature of scarcity and growth. But, in fact, the modern economy does have a number of ways to work around natural resource scarcity—through substitution of less scarce resources, as well as through increased efficiency triggered by scarcity in the use of all scarce resources, including natural resources. On the other hand, as emphasized in other papers in this volume, the New Scarcity of nonmarket environmental and ecological resources is probably more significant for future economic growth than the scarcity of natural resources per se, and experience is a less robust guide to how well the economic system can respond to this form of scarcity.

Because of externalities, actual market prices and quantities may be appreciably different from what would be observed after policy intervention has shifted the economy to a PV-optimal path. There are thus two types of unsustainable economy possible, as noted by Pezzey (1989): those where policies that move to PV-optimality would also make the economy sustainable and those where even the PV-optimal economy will be unsustainable, for example, because of depletion of nonrenewable resources as in Dasgputa and Heal. Explicit sustainability promotion policies, beyond correction of environmental externalities and other conventional market failures, would then be needed in the latter case to achieve sustainability (Pezzey 2004).

The focus until now has been on quantities of consumption, investment, and resource depletion. In economic analysis, however, *prices* play a central role in conveying information about scarcity. In a well-functioning economy that efficiently allocates resources, prices (including prices on hypothetical markets for future consumption, as well as prices for allocating risk) effectively signal economic scarcity by characterizing both the values that users place on resources and the opportunity cost of increasing provision of the resources. When external costs impede the efficient allocation of resources, economic policy attempts in various ways to correct quantities *and* prices in order to improve efficiency. Pollution is a classic example: the government may intervene directly, by taxing pollution (or goods that are its sources in the economy), or indirectly, by regulating the quantities of production, consumption, and emissions.

Sustainability as intergenerational equity may be an additional constraint on the allocation of goods and services, as illustrated by the above comparison of Dasgupta and Heal (1974) and Solow (1974). Therefore, intervention in the economy not just to correct externalities, but also to modify a PV-optimal allocation of resources to meet a sustainability constraint, will have a further impact on prices. Pezzey (1989) provides a simple example of this, where in an economy with both external costs of pollution and an unsustainable PV-optimal consumption path from resource depletion, two separate policy instruments are needed: one to internalize the external costs of pollution, the other to increase investment and dampen consumption in order to avert declining utility over time (see also Howarth and Norgaard 1992; Mourmouras 1993; Howarth 1998; and Pezzey 2004).

Prices that would induce an economy to achieve a PV-optimal allocation of resources are normally referred to as *efficient prices*. By analogy, we will refer to prices that induce a sustainable time path of utility as *sustainability prices*. Sustainability prices figure prominently in what economics can and cannot say about the measurement of sustainability, which is discussed later.

Conundrums in Defining Neoclassical Sustainability

The discussion above emphasizes a heuristic criterion for sustainability. Critieria for sustainability—or for intergenerational equity more generally—can be justified by choosing a set of basic *axioms* that the agent's utility path over time must satisfy. The best-known example is the constant utility path with utility at its *maximum sustainable level* first explored by Solow 1974 (after Rawls' 1971 "maximin" criterion). A number of alternative criteria, and the axioms supporting them, are in Asheim et al. (2001) and works cited there, as well as in Chichilnisky (1996) and Pezzey (1997).

Many economists, however, defend the criterion of maximizing the PV of utility as an acceptable reflection of intergenerational equity, based on an axiomatic approach predating any sustainability ideas in Koopmans (1960). Others criticize in varying degrees the effect of constant discounting as very inequitable to what we call the *far future*. The ethics of intergenerational equity criteria can be debated in terms of the appeal of the underlying implicit or explicit axioms. The key question on the part of critics of a separate sustainability criterion would then be, why should the government be interested in sustainability if private agents wish to maximize PV? Maximizing PV generally has nothing to do with sustainability and gives a complete and unique prescription for the time paths of every decision that ever has to be made in the economy. This is why, for example, Beckerman (1994) regards any sustainability constraint on PV-maximization as morally repugnant. This view also implies that there is no apparent motive for measuring sustainability on a PV-optimal path. Likewise, if unsustainability is detected, there is no apparent justification for a policy intervention that may be needed to make the economy depart from its PV-optimal path to achieve sustainability.

The resolution of this paradox has to lie, we believe, in some kind of split between private and public concerns about the far future. We must assume that individuals choose their own actions to maximize some form of present value, but vote for a government that applies a sustainability concern, both by measuring sustainability and taking action to achieve it if necessary. People treat private economic decisions as the domain of "the Economic Man" and governmental decisions as the domain of "the Citizen" (see for example Marglin 1963, *98*).[5] One good reason for this is that individuals cannot provide personally for their distant descendants, because of the mixing of bequests that occurs over several generations (Daly and Cobb 1989, *39*; Pezzey 1995).

The philosophical basis of neoclassical sustainability economics is thus distinct from three other main philosophical approaches to social choice over time (see Pezzey and Toman 2002, and references therein, for details). It rejects classical utilitarianism, which prohibits any discounting, and it rejects any purely rights-based view that future generations' resource opportunities matter, not utility outcomes. But it also rejects neoclassical utilitarianism, which sees PV-optimality as a complete prescription for intertemporal equity.

The behavioral assumptions of "Homo economicus" set out above are challenged by other disciplines, particularly psychology (Rabin 1998). Real people are not perfectly informed and far-sighted, it is argued. Nor do they value only absolute levels of goods and services: the degree to which they are influenced by their position relative to their peer groups (or to their own past) means that materialistic economic growth is much less likely to increase well-being, thus increasing the importance of environmental protection (Howarth 1996). People often do not use constant discount rates (Frederick et al. 2002), and the assumption that economies would eventually converge to equilibrium if left alone is fundamentally untestable.

Critics of the economic approach to sustainability (from several disciplines) particularly highlight the apparent paradox of constructing theories and measurements to address a concern in which uncertainty, and thus fundamental problems of unmeasurability, seem pervasive. From this perspective, because of the scales of time and space involved the prospect of physical limits to growth that cannot be fully pushed back through innovation seem endemic. The scales involved give rise specifically to the possibility of catastrophes: unlikely but highly undesirable outcomes (see for example Roughgarden and Schneider 1999). Such outcomes are not just difficult by their nature to forecast; they raise questions about the extent to which the standard expected utility model of economics is applicable (Camerer and Kunreuther 1989). In any event, the inherently long time scales raise doubts for some about the ability of analysts to measure the values placed by future generations on different components of natural and other capital.

Although these points are important, the various alternatives offered for dealing with them have their own difficulties. Faucheux and Froger (1995) argue that pervasive uncertainty heightens the importance of "procedural rationality": the need to ensure, among other things, an open and transparent process for bringing together different values while tolerating more "satisficing" in terms of outcomes. Although we would agree that complex sustainability problems raise the importance of public awareness and agreement, procedural rationality alone does not avoid the need to weigh how risk-averse the current generation should be on behalf of its descendants—especially when those participating in the decision often are not well informed about the choices and consequences.

Sustainability and Views on Resource Substitutability

Even if one accepts in principle the importance of intergenerational equity, there remains disagreement about how difficult intergenerational equity will be to achieve in practice through economic growth. The disagreement stems from different opinions, often polarized and held with some passion, on the substitutability, and to some extent the relative importance, of human-made and natural inputs to both production and well-being (utility) in the far future. The disagreement is often presented in terms of a simple dichotomy of *weak sustainability* versus *strong sustainability*. This terminology, which admittedly can be confusing to those outside the field, refers to how serious a constraint "natural capital" depletion and degradation may be for sustainability as we have defined it (nondecreasing utility over time). Weak sustainability assumes significant possibilities for substitution between natural capital and other inputs, so that natural capital is only one of a number of inputs that can be carried over by future generations to sustain their well-being (the sustainability constraint on natural capital is thus "weak"). Strong sustainability sees certain kinds and quantities of natural capital as more inherently needed for sustaining economic activity, let alone growth.

There are in fact variants within weak sustainability. At one extreme in terms of "technology optimism" is a view that resource substitutability is more or less unlimited, and natural and ecological resources are fairly unimportant, especially in light of the very large contributions that technical innovation as well as simple resource substitution have made in moderating natural resource scarcity. It would be logically redundant to impose a sustainability constraint on an economy in practice if there is good reason to expect technical progress to enable economic growth to continue, with no long-run constraint from environmental resource depletion and degradation, leaving future generations better off than today. This view, a normative extension of the analysis presented by Stiglitz (1974), holds that market forces will automatically ensure sustainability, at least if innovation is sufficiently robust, even without policy intervention that internalizes environmental externalities and thus moves the economy to a PV-optimal path. The depletion of specific "natural capital" is not a cause for concern, as there will be enough offsetting increases in other forms of capital (including knowledge) to maintain or increase overall well-being over time. This is broadly Beckerman's (1994) view, and is supported for example by calculations in Weitzman (1997, 1999). More moderate versions of weak sustainability would put more importance on the need to efficiently allocate natural resources by overcoming market failures, including externalities in the use of nonmarket environmental and ecological resources, but would still accept the inherent feasibility of sustained growth as long as resources are allocated more efficiently.

There are also many variants of the strong sustainability view, which is often labeled "ecological" or "environmental" sustainability. These are typically derived from natural science arguments that capital–resource substitutability is either a self-evidently impossible concept—indeed, some argue that capital and natural resources are aggregate complements rather than substitutes—or subject to strict and fairly imminent limits. This view implies that economic growth over a few centuries inevitably will require higher material throughput, not just improved efficiency in the generation of disembodied economic value from the same menu of material inputs. Since these inputs (including available energy and waste-holding capacity) are both inherently limited in their availability and must result in material outputs, strong sustainability leads to both a positive and normative problem in balancing people's aspirations for more output as a way to improve living standards, with binding physical limits.[6]

A well known outline of a strong sustainability view is by Daly (1990), who regarded as self-evident that sustainability requires

- ecological services critical to life support to be maintained and pollution stocks to be prevented from increasing beyond certain critical loads;
- renewable resource stocks (or at least aggregates of these stocks) to be used no faster than they are renewed; and
- depletion of nonrenewable resources to be offset by investment in the production of comparable services from renewable resources (e.g., in the switch from fossil to renewable energy sources).

From this perspective flow different variants of the *precautionary principle* (Howarth 1997)—a priori constraints on resource degradation or depletion, distinct from what might be justified on cost–benefit grounds.

To be sure, no advocate of strong sustainability argues for preserving *all* stocks. This would lead to the absurd implication that no depletable resource should ever be touched. Rather, in a development of Daly's ideas, advocates of strong sustainability argue for maintaining the key *functions* of natural resource stocks (see also Common and Perrings 1992; Ekins 1996). Much heat and some light has been generated in discussing the extent to which the first (material conservation) and second (increasing entropy) laws of thermodynamics underpin various strong sustainability views (see Pezzey and Toman 2002, section 4.1, for a survey), but here we only report one view.

Ayres (1999) shares the concern with the adverse consequences of excessive material throughout, but he rejects the notion that entropic dissipation of materials is an inherent limit to growth. He argues that although a large flow of renewable energy is available directly or indirectly from the sun, this energy (along with natural biological processes that already act to concentrate material like nutrients) can be used to recycle dissipated materials. From this perspective, Ayres argues that the key challenge is not materials conservation, but a more rapid development and improvement (effectively by diverting some of the fossil fuel endowment from satisfying current demands) of *renewable energy*, so that the required energy is available to work with more dilute materials streams. But he would also support materials conservation as a way to take stress off natural systems (including the human species) that are adversely affected by harmful residuals.

Measuring Sustainability

We have discussed the various ways that economic approaches to sustainability address issues of intergenerational equity and the potential paradox that can arise. We also note that economic analyses of sustainability generally start from a premise that natural and other forms of capital are adequately substitutable for each other to make nondecreasing well-being over time achievable. In other words, broadly speaking, sustainability is held to be feasible, though it is by no means guaranteed by the operation of unfettered markets or policies that focus only on internalization of current environmental spillovers.

In this section we first briefly summarize and interpret some key findings from more than 25 years of economic research on sustainable and unsustainable economic trajectories, as well as the practical limitations on what insights these findings can provide in the context of scarcity and growth. Much of the relevant literature is highly technical and even arcane; all we do here is provide a flavor of the material. We then discuss equally briefly some noneconomic approaches to sustainability measurement.

Measuring Sustainability in Economic Analysis

In order to summarize recent and general writing in economic theory that explicitly focuses on sustainability, we need to introduce some of the jargon of income accounting. Standard *national income* is the sum of flows of *consumption* and *investment* goods, the latter being *flows* that add to *capital stocks*. In practice, because real economies are composed of many different goods, income, consumption, and investment are measured in *value* units: they are the sums of physical flows weighted by unit values, or price. *Prices* indicate the *relative* values per unit of the various flows. *Green* (national) income and product accounting extends the standard framework to include marketed but normally omitted natural resource flows, such as fossil fuel depletion, and nonmarket stocks and flows, such as biodiversity and pollution damages. The nonmarket elements can affect both utility (improved biodiversity increases recreational values) and production (improved air quality improves human health, and hence labor output). The implicit or *shadow* prices of these nonmarket flows represent their relative values, like "regular" prices. The practical difficulties of measuring nonmarket prices and quantities are profound, but the theoretical principles are well understood in economics. Indeed, the shadow prices could in principle be derived in the process of solving a model of the economy for a PV-optimal growth path.

Less obvious, but also important and relevant to sustainability accounting, is that *time* itself can be a source of rising economic value. In particular, if the economy is spontaneously ("exogenously") increasing in productive capacity over time without added investment in stocks of extra knowledge and tools, then, as shown by Stiglitz (1974), opportunities for sustaining consumption and well-being also will grow over time, other things being equal. We will refer in what follows to *augmented national income* as the measure of national income that includes the value of exogenous improvements over time to productive capacity, as well as the non-market "green" elements outlined above. In other words, "augmented" national income is the most comprehensive possible definition of the flow of generalized economic product in the economy. Finally, we define *augmented net (national) investment* as the value of all net stock changes plus the "value of time" just mentioned. Unaugmented net national investment is also called *genuine saving* (see for example Hamilton and Clemens 1999).

The results we are about to present assume that an economy is following a PV-optimal path, subject to a sustainability constraint, notwithstanding the apparent paradox already noted in reconciling PV-maximization and sustainability as public policy concerns.

With this set of definitions and concepts, we can now paraphrase a key result from neoclassical sustainability economics (we refer again to Pezzey and Toman 2002 for technical details). *If augmented national income is **not** rising at a particular time, or equivalently if augmented net national investment is **not** positive at that time, then the economy is **un**sustainable at that time.* As defined above, to say the economy is unsustainable means that the current level of well-being is above what can be sustained over the longer term, so that some decline in well-being is inevitable. The economy is then "living beyond its means." Moreover, constant augmented income or wealth can imply *unsustainability*, even though some heuristic discussions of sustainability target the goal of maintaining national wealth.

It is theoretically possible in this result regarding income and sustainability that the exogenous time component in augmented net investment, reflecting spontaneous exogenous increase in productive capacity, could be large enough to offset "dissaving" in natural capital as well as other capital. This restates the point, made earlier in defining weak sustainability, that a sustainability constraint would be logically redundant if spontaneous economic progress is robust enough to make the economy sustainable as a whole. Aside from the previously mentioned debate about substitutability this engenders, the point also draws attention to the need for a better empirical understanding of what this spontaneous, exogenous time component of economic progress in national income data really is. If it is in fact not really exogenous, but reflects a mismeasurement of the fruits of economic investment in acquiring and applying knowledge and tools, then the economy is less sustainable than it might appear, because the measured progress is only resulting from the diversion of goods and services to investment in innovation.

The converse of the above key result is generally *not* true: one *cannot* assume from an observation of positive augmented net investment that the economy is sustainable.[7] Information about an economy at a point in time is not enough to judge that the economy is sustainable; the entire time path of augmented national income and its components must be assessed. This obviously limits the practical ability to use sustainability tests with existing economic data: in particular, sustainability cannot be assessed just by valuing most items in a nation's balance sheet with prices defined on world markets. Another limitation on the use of market prices is that they may suggest that any one nation can largely deplete its natural resources now, become a "knowledge nation" reliant on human capital, import most of its resources in the far future, and perhaps remain sustainable. However, this is not an option for the global economy, since not every nation can be a resource importer (Brekke 1997, 72; Pezzey 1998), and if they all tried to be, resource prices would rise dramatically.

A second important result, using the concept of "sustainability prices" introduced in the preceding section, is this: *If augmented net investment evaluated using a set of sustainability prices is zero forever, then well-being stays at a constant level forever.* This result is a generalization of a famous theorem found by Hartwick (1977) and subsequently much elaborated by Hartwick and others. To understand the weaknesses as well as the accomplishments of economics in assessing sustainability, it is crucial to realize what the Hartwick result does *not* say. It does not say that zero or even growing augmented net investment at one point in time implies sustainability at that time; there are counterexamples that demonstrate the opposite (Asheim 1994). And the result holds only for augmented net investment evaluated at hypothetical sustainability prices (as defined above). It is *not* enough for sustainability just to observe nonnegative net investment along a market equilibrium price trajectory (after perhaps adjusting for current environmental spillovers), despite suggestions of this in the literature (Solow 1986, *147*; Hartwick 1997, *511*; Aronsson et al. 1997, *101*; Aronsson and Lofgren 1998, *213*). This further underscores the practical challenge of measuring sustainability.

Empirical Economic Measures of Sustainability

Most empirical economic measurements of sustainability estimate (after heroically wrestling with serious data deficiencies) the growth rate of green national income, or net national investment as a proportion of national income, but in both cases excluding the exogenous impact of time on income. Such calculations are incomplete, since time-augmented investment could be positive (and thus the economy perhaps sustainable) even if conventional or green net investment is negative. The problem arises because most authors ignore technical progress and shifts in world prices of traded goods, both of which cause exogenous (time-augmented) shifts in production possibilities.

Measuring sustainability has often been done using just a measure of green net investment, for example by Pearce and Atkinson (1993), Pearce et al. (1993), Hamilton (1994), Atkinson et al. (1997) and Hamilton and Clemens (1999), of which we report just on the first, widely cited paper. Pearce and Atkinson use data for 18 countries, ranging from the United States to Burkino Faso, of which they rely on savings instead of investment data to calculate the net increase in built capital. The value of changes in natural capital are calculated by using data on net changes in resource stocks valued at current market prices. Rough adjustments are also made for the flow of different environmental disamenities.

All the European countries and Japan were judged to be definitely sustainable, essentially because of high savings rates and low resource depletion rents, the latter perhaps being because these countries have relatively few resources left to deplete. By contrast, all the African countries were judged to be definitely unsustainable, because of low savings rates and high depletion rents. The United States was only marginally sustainable, because of a much lower savings rate than Europe or Japan.

Papers like Weitzman (1997), focusing on exogenous technical progress in the United States, and Vincent et al. (1997), focusing on exogenous shifts in oil export prices facing Indonesia,[8] are the exception rather than the rule in that they deal with exogenous influences on sustainability. Weitzman's analysis in turn can be seen as connected to a huge rebirth of interest within mainstream economics in sustainable growth, though it is usually called the economics of *endogenous growth* (for a survey see Aghion and Howitt 1998). The name of this new growth economics comes from its focus on how technical progress arrives not just exogenously from time passing. Endogenous technical progress results from people's and firms' economic (though often suboptimal) decisions to invest in accumulating human capital from education and knowledge or improved product quality from research and development.[9] Endogenous growth economics can add important new elements to net investment calculations. For example, Hamilton et al. (1998) reckoned that endogenous growth accounts for most of the "technical progress premium" otherwise attributed to exogenous forces. This reckoning would greatly reduce an economy's apparent sustainability relative to measures that treat technical progress as exogenous.

Weitzman's (1999) analysis of resource scarcity took a different tack than the approach followed by Pearce and Atkinson. Weitzman asked a seemingly complicated question: What is the total cost to the U.S. economy of minerals being nonrenewable, so that eventually the economy must forgo at least some of its currently provided mineral inputs? Weitzman (1976) shows that *if* one accepts the neoclassical premises of this article on "green income" accounting, then the answer is simple. The present value "scarcity premium" relative to total social wealth is just the current value of the total resource rent, divided by the current flow of conventionally measured national income (i.e., income without deduction for depletion rent). Using some admittedly rough calculations, Weitzman finds that according to this formula, the total cost of resource nonrenewability is relatively quite low: very roughly 1 percent of total wealth.

This result has an intuitive appeal, since a resource scarcity rent, as indicated by the unit price of in situ reserves, is in fact the present value of future benefits from an additional unit of resource stock today. What is striking about the model, and not necessarily so general, is the applicability of this reasoning to *total* changes in resource stocks.[10] So one can question Weitzman's interpretation of his results as "pricing the limits to growth." If resource substitution and innovation do not provide large cushions against scarcity, and some critical and nonsubstitutable resources are depleted, then such nonmarginal (i.e., total) scarcity ought to be hugely expensive; but this kind of scarcity will not be picked up in Weitzman's model.

Two other papers will suffice to show that many authors extend economic measures of sustainability beyond conventional neoclassical bounds. In valiantly estimating and comparing seven sustainability measures, some noneconomic or at least nonneoclassical, during the period 1980–1993 for Scotland, Hanley et al. (1999) helped to highlight both progress and problems in sustainability empiricism. The two economic sustainability indicators, green income and investment or "genuine savings," yielded fairly different results, though both indicated that Scotland was moving in the direction of being *more* sustainable. There was no accounting for Scottish trade in goods, resources, and financial capital, the last of which allows savings to diverge widely from investment. By contrast, two broader, "sociopolitical" measures, the Index of Sustainable Economic Welfare (ISEW) and the Genuine Progress Indicator (GPI), were falling over the period, mainly because of a worsening income distribution, a measure of intragenerational equity that purely economic measures ignore.[11]

The main focus of Proops et al. (1999) was to show that trade in both resources and resource-intensive goods is significant in analyses of sustainability for most countries. Their calculations for various countries centered on the difference between "closed economy sustainability," analogous to Pearce and Atkinson's measure, and the nonneoclassical notion of "open economy" sustainability, which replaces calculations of the capital and resources used "by" an economy, with the capital and resources used "for" or "attributable to" an economy, as calculated from an input–output analysis of world trade flows. Moving from the closed to the open measure dramatically increases the calculated sustainability of resource-rich regions like the Middle East and reduces it for OECD (industrial) regions. However, despite the contention of Proops et al. (1999, 77) that "industrialized countries appropriate the carrying capacity of other countries (e.g., by importing natural resources), therefore benefiting at the expense of their trading partners," Proops et al. drew no conclusions from their calculations for either national or international policymaking. Also absent was any explicit discussion of how and to what extent free trade can be unfair and exploitative.

Physical Measures of Sustainability

Critics of neoclassical, market-based sustainability measures have argued that they are overoptimistic, and that physical measures can give warnings about possible long-run limits to growth that better reflect the great uncertainties involved, and hence the need for a precautionary approach. This is at least plausible where non-owned, large-scale ecological resources are concerned. Nevertheless, physical indicators also have serious shortcomings, not least in their sometimes arbitrary methodologies. The definitions and boundaries of physical indicators (what is included or excluded, and how to meaningfully add together very different types of hectares of land or tons of material), are often unclear or arbitrary. Different divisors or none are used: is it physical impact per person, impact per dollar of real value, or total impact that is reported? The indicators are not robust under changes in technology, and often provide little insight into the degree of environmental risk incurred, the best remedial policies to use, and where these should be applied. They may not, for example, distinguish between a hectare of new paved road in an industrial zone or in pristine wilderness or between a kilogram of biologically inert material or of toxic heavy metal.

The simplest physical measures are those measured in area, mass, or energy units.[12] Various *ecological footprints* (Wackernagel and Rees 1996; *Ecological Economics* 2000) measure the land area of ecosystems required to produce a population's consumed resources and assimilate its wastes. *Materials intensity per unit of service* (Hinterberger et al. 1997) aggregates tons of material moved during economic production. Two very different types of energy measures are *energy return on investment* (EROI)—the amount of direct and embodied energy invested to produce a unit of useful output or energy (see Kaufmann and Cleveland 2001 for a recent exposition)—and the *human appropriation of net primary (photosynthetic) production* (HANPP) (see Vitousek et al. 1986). An EROI of less than one for produced energy, or a HANPP of more than 100 percent for a whole economy, would clearly be physically unsustainable, but appropriate aggregation techniques for these measures are quite contentious. A subtler physical measure is that of "sustainability gaps," or distances of an economy in years from prespecified but ultimately arbitrary environmental targets (Ekins and Simon 2001). Without a firmer grounding in some broad notion of relative risk or opportunity cost that can be used to weigh the significance of all these physical measures, it is difficult to use them to assess sustainability or redirecting policy (Ayres 1995, 2000). But by their very existence and popularity with noneconomists, they present a challenge to economic approaches to sustainability.

Conclusions

What can we say about the accomplishments of economics research directed at sustainability over the past two to three decades and its ability to illuminate debates on scarcity and growth? Theoretical work in the field has in one sense greatly outstripped empirical work, but this is ultimately because the theoretical work has not yet produced readily usable empirical measures. This is unfortunate, as judgments about the importance of sustainability ultimately are empirical.

The debate between weak and strong sustainability proponents, while little more settled than it was a decade ago, has again drawn valuable attention to what physical limits to growth may exist because of limited substitutability, in concept and in practice. It has helped stimulate a new examination of the powers and limits of technical change as a means of relaxing these limits, as well as to focus attention on how physical laws interact with economic principles.

Empirically, the "green accounting" efforts initiated by Repetto et al. (1989) and Pearce and Atkinson (1993) have begun to flesh out economic measurement of sustainability based on "weak" assumptions of capital–resource substitutability. Despite their theoretical and practical shortcomings, such studies have undoubtedly stimulated collection and codification of data, without which progress would be impossible. There are much more raw data, if of often unknown quality, available on estimated resource quantities and values for many countries (see for example World Bank 2001). The studies have also pointed up some gross disparities between different countries' sustainability measures, disparities that are likely to remain even after many of the known shortcomings in the empirical work are identified and corrected. It should even be possible in time to crudely test the theoretical prediction that the augmented national income of an economy that is currently measured as severely unsustainable will eventually fall.

We have pointed out a number of lingering issues in defining sustainability conceptually in economics, in particular the coexistence of present value maximization and the implications of discounting and concern for the further future. From an empirical perspective, perhaps the greatest challenge in using current economic theory is that even if PV-optimality is achieved, it tells us nothing precise about sustainability, because of the chicken-and-egg problem that "sustainability prices" are observed only once sustainability has already been achieved. We need to be able to estimate the gap that theory shows to exist between green national income based on current market observations or environmental valuations and sustainable income. How accurately can we frame sustainability policies under market conditions that by their nature do not reveal sustainability prices? Thus, the extent to which weak sustainability approaches can be empirically substantiated and operationalized remains an open question.

One area of particular interest that we have omitted in this paper is the question of sustainability in poor countries. This was a prime inspiration for the popularizing by WCED (1987), and more recently under the Millenium Development Goals (UNDP 2003), of the very notion of sustainability and the connections between sustainability and the development process.

A multigenerational time scale requires both realistic psychology to be built into the economic representation of choice through a utility function and realistic physics, biology (including neurology), and engineering to be built into the production function. The whole question of what biophysical limits to production growth might exist still needs more work. One approach would be to use forms of production functions that have built-in (but not immediate) limits to substitution. This may mean much more reliance on numerical methods and less stress on mathematically elegant but empirically questionable paths of balanced growth. Efforts to this end would also invigorate and make more rigorous empirical work in the strong sustainability tradition, by more clearly identifying where resource availability and substitutability constraints stretch thin the neoclassical paradigm, rather than by just making blanket applications of the precautionary principle.

Another important area for advance is in designing policies for sustainability. As already noted, if long-term economic growth conditions do not automatically make sustainability policies logically redundant, then there is a need to consider policies not just to correct externalities but also to foster greater savings rates, either in general (weak sustainability) or with respect to natural capital (strong sustainability). One option for doing so is a variant of the *safe minimum standard* idea, which builds a greater presumption of care into social decisionmaking as temporal and spatial scales grow (see, among many, Barbier et al. 1990; Toman 1994; Chichilnisky et al. 1997; Norton and Toman 1997; Woodward and Bishop 1997; Farmer and Randall 1998). Making this kind of approach operational (as opposed to a metaphor), however, relies either on relatively objective measures of risk whose absence is the heart of the problem or on some kind of social intuition. The approach sketched in Norton and Toman (1997) and Toman (1999) attempts to draw together some of these threads by envisaging combinations of technical analysis with processes that more directly engage public participation in shaping the technical analysis. These include not just cost–benefit analysis and distributional impacts analysis, but also noneconomic analysis of physical impacts and risks and other disciplines that help inform and shape public values.

In all these areas, only a suitable balance of theory and data, and a reduced role for ideological preconvictions, will move us toward what we take to be a common goal: a better economic science of sustainability, with a greater ability to identify future limits to growth.

Acknowledgments

This chapter was prepared while Toman was a senior fellow at Resources for the Future. The authors are grateful to Anil Markandya, Stephen Polasky, anonymous reviewers, and other participants in the workshop that gave rise to this volume for a number of suggestions that helped to improve the chapter.

Notes

1. This phrase is deliberately chosen. The sustainability theories we outline use the mainly "neoclassical" techniques of conventional, mainstream economics. However, the relevant publications are still mainly located in more specialized environmental economics outlets; outside of this literature, sustainability issues largely are ignored altogether.

2. Unfortunately, this jargon is not entirely consistent among economists, as many writers use "welfare" to mean instantaneous well-being.

3. Despite popular confusion, "unlimited" and "perfect" substitutability are not the same. If production is the multiple of powers of the stock of (human-made) capital and the flow of resource depletion, then capital is an unlimited substitute for resource flow: however small (but still positive) the latter, combining it with a large enough capital stock will produce a given output. But if production is a linear combination of capital and resources, then production can be sustained even with zero resources, so capital is then a perfect substitute for resources.

4. This was actually proved in a static (timeless) context, but it can be generalized to the dynamic (intertemporal) context of this chapter. Note that this economic definition of efficiency is different from an engineering or technological definition, for example, in relation to energy efficiency. It may be economically efficient to select technologies that use more energy per unit of output if the opportunity cost to society of achieving the lower energy intensity is higher than the value of the energy conserved.

5. Debate on this issue in economics actually goes back much further than this. In his early contribution to economic growth theory, for example, Ramsey (1928) argued that discounting the welfare of future generations just on the basis of the impatience of the current generation was ethically indefensible. We would not go as far as Marglin and say that "The Economic Man and the Citizen are for all intents and purposes two different individuals," because Economic Man can still maximize self-interest (seek optimality) within the bounds (sustainability) that the Citizen lays down. However, protests in 2000 over fossil fuel prices in many Western European

countries, despite those countries having recently signed the Kyoto Protocol, which aims to limit greenhouse gas emissions, may perhaps be a sign that the schizophrenia is real and can lead to quite disconnected behavior.

6. For further discussions of these considerations, see Stern (1997), O'Connor (1998), and Beard et al. (1999).

7. Only under certain conditions involving a small open economy does this converse hold (Asheim 2000). The arguments surrounding these results unfortunately are not always that intuitive, so as noted previously we can only paraphrase the economics literature in this nontechnical overview.

8. The need for such a calculation is underscored by the fact that resource prices have typically been flat or falling, not rising as in most theoretical models.

9. There has been only limited cross-fertilization between the endogenous growth and sustainable development literatures, mainly from neoclassical (new) growth economists adding pollution or resource depletion to their models' production or utility functions. For examples, see Stokey (1998), chapter 5 of Aghion and Howitt (1998), and Smulders (2000); see also the chapter by Smulders in this volume.

10. A similarly critical remark applies to a polar opposite calculation by Costanza et al. (1997) of the *total* "value of the world's ecosystem services and natural capital."

11. See Neumayer (2000) for a critique of both ISEW and GPI methodologies.

12. See Pezzey and Toman (2002, *205–209*) for a fuller survey.

References

Aghion, P., and P. Howitt. 1998. *Endogenous Growth Theory.* Cambridge, MA: MIT Press.

Aronsson, T., P.-O. Johansson, and K.-G. Lofgren. 1997. *Welfare Measurement, Sustainability and Green National Accounting: A Growth Theoretical Approach.* Cheltenham, UK: Edward Elgar.

Aronsson, T., and K.-G. Lofgren. 1998. Green Accounting: What Do We Know and What Do We Need to Know? In *The International Yearbook of Environmental and Resource Economics 1998/99,* edited by T. Tietenberg and H. Folmer. Cheltenham, UK: Edward Elgar.

Arrow, K.J., and G. Debreu. 1954. Existence of an Equilibrium for a Competitive Economy. *Econometrica* 22(3): 265–290.

Asheim, G.B. 1994. Net National Product as an Indicator of Sustainability. *Scandinavian Journal of Economics* 96(2): 257–265.

———. 2000. Green National Accounting: Why and How? *Environment and Development Economics* 5(1): 25–48.

Asheim, G.B., W. Buchholz, and B. Tungodden. 2001. Justifying Sustainability. *Journal of Environmental Economics and Management* 41(3): 252–268.

Atkinson, G., et al. 1997. *Measuring Sustainable Development: Macroeconomics and the Environment.* Cheltenham, UK: Edward Elgar.

Ayres, R. 1995. Life Cycle Analysis: A Critique. *Resources, Conservation and Recycling* 14: 199–223.

———. 1999. The Second Law, the Fourth Law, Recycling, and Limits to Growth. *Ecological Economics.* 29: 473–484.

———. 2000. Commentary on the Utility of the Ecological Footprint Concept. *Ecological Economics* 32(3): 347–350.

Barbier, E.B., A. Markandya, and D.W. Pearce. 1990. Environmental Sustainability and Cost–Benefit Analysis. *Environment and Planning* 22(9): 101–110.

Barnett, H., and C. Morse. 1963. *Scarcity and Growth: The Economics of Natural Resource Availability.* Baltimore: Johns Hopkins University Press for Resources for the Future.

Beard, T., R. Lozada, and G. Lozada. 1999. *Economics, Entropy, and the Environment: The Extraordinary Economics of Nicholas Georgescu-Roegen.* Cheltenham, UK: Edward Elgar.

Beckerman, W. 1994. "Sustainable Development": Is It a Useful Concept? *Environmental Values* 3(3): 191–209.

Brekke, K.A. 1997. *Economic Growth and the Environment: On the Measurement of Income and Welfare.* Cheltenham, UK: Edward Elgar.

Camerer, C.F., and H. Kunreuther. 1989. Decision Processes for Low Probability Events: Policy Implications. *Journal of Policy Analysis and Management* 8(4): 565–592.

Chichilnisky, G. 1996. An Axiomatic Approach to Sustainable Development. *Social Choice and Welfare* 13(2): 231–257.

Chichilnisky, G., G.M. Heal, and A. Vercelli (eds.). 1997. *Sustainability: Dynamics and Uncertainty.* Dordrecht: Kluwer Academic Publishers.

Common, M., and C. Perrings. 1992. Towards an Ecological Economics of Sustainability. *Ecological Economics* 6(1): 7–34.

Costanza, R., et al. 1997. The Value of the World's Ecosystem Services and Natural Capital. *Nature* 387: 253–260.

Daly, H.E. 1990. Toward Some Operational Principles of Sustainable Development. *Ecological Economics* 2(1): 1–6.

Daly, H.E., and J.B. Cobb. 1989. *For the Common Good: Redirecting the Economy toward Community, the Environment and a Sustainable Future.* Boston: Beacon.

Dasgupta, P.S., and G.M. Heal. 1974. The Optimal Depletion of Exhaustible Resources. *Review of Economic Studies, Symposium on the Economics of Exhaustible Resources,* 3–28.

Dasgupta, P.S., and K.-G. Maler. 2000. Net National Product, Wealth, and Social Well-Being. *Environment and Development Economics* 5(1): 69–94.

Ecological Economics. 2000. Forum: The Ecological Footprint. 32(3, March).

Ekins, P. 1996. Towards an Economics for Environmental Sustainability. In *Getting Down to Earth: Practical Applications of Ecological Economics,* edited by R. Costanza et al. Washington, DC: Island Press.

Ekins, P., and S. Simon. 2001. Estimating Sustainability Gaps: Methods and Preliminary Applications for the UK and Netherlands. *Ecological Economics* 37(1): 5–22.

Farmer, M.C., and A. Randall. 1998. The Rationality of a Safe Minimum Standard. *Land Economics* 74(3): 287–302.

Faucheux, S., and G. Froger. 1995. Decision-Making under Environmental Uncertainty. *Ecological Economics* 15(1): 29–42.

Frederick, S., G. Loewenstein, and T. O'Donoghue. 2002. Time Discounting and Time Preference: A Critical Review. *Journal of Economic Literature* 40(2): 351–401.

Hamilton, K. 1994. Green Adjustments to GDP. *Resources Policy* 20(3): 155–168.

Hamilton, K., G. Atkinson, and D. Pearce. 1998. Savings Rules and Sustainability: Selected Extensions. Paper presented at the 1st World Congress of Environmental and Resource Economists, Venice, Italy, June.

Hamilton, K., and M. Clemens. 1999. Genuine Savings Rates in Developing Countries. *World Bank Economic Review* 13(2): 333–356.

Hanley, N., et al. 1999. Measuring Sustainability: A Time Series of Alternative Indicators for Scotland. *Ecological Economics* 28(1): 55–74.

Hartwick, J.M. 1977. Intergenerational Equity and the Investing of Rents from Exhaustible Resources. *American Economic Review* 67(5): 972–974.

———. 1997. Paying Down the Environmental Debt. *Land Economics* 73(4): 508–515.

Hinterberger, F., F. Luks, and F. Schmidt-Bleek. 1997. Material Flows vs. "Natural Capital": What Makes an Economy Sustainable? *Ecological Economics* 23(1): 1–14.

Howarth, R.B. 1996. Status Effects and Environmental Externalities. *Ecological Economics* 16(1): 25–34.

———. 1997. Sustainability as Opportunity. *Land Economics* 73(4): 569–579.

———. 1998. An Overlapping Generation Model of Climate–Economy Interactions. *Scandinavian Journal of Economics* 100(3): 575–591.

Howarth, R.B., and R.B. Norgaard. 1992. Environmental Valuation under Sustainable Development. *American Economic Review* 82(2): 473–477.

Kaufmann, R.K., and C.J. Cleveland. 2001. Oil Production in the Lower 48 States: Economic, Geological, and Institutional Determinants. *Energy Journal* 22(1): 27–49.

Koopmans, T.C. 1960. Stationary Ordinal Utility and Impatience. *Econometrica* 28: 287–309.

Marglin, S.A. 1963. The Social Rate of Discount and the Optimal Rate of Investment. *Quarterly Journal of Economics* 77(1): 95–111.

Mourmouras, A. 1993. Conservationist Government Policies and Intergenerational Equity in an Overlapping Generations Model with Renewable Resources. *Journal of Public Economics* 51(2): 249–268.

Neumayer, E. 2000. On the Methodology of ISEW, GPI and Related Measures: Some Constructive Suggestions and Some Doubt on the "Threshold" Hypothesis. *Ecological Economics* 34(3): 347–361.

Norton, B.G., and M.A. Toman. 1997. Sustainability: Ecological and Economic Perspectives. *Land Economics* 73(4): 553–568.

O'Connor, M. 1998. Ecological-Economic Sustainability. In *Valuation for Sustainable Development*, edited by S. Faucheux and M. O'Connor. Cheltenham, UK: Edward Elgar.

Pearce, D.W., et al. 1993. *Blueprint 3: Measuring Sustainable Development*. London: Earthscan.

Pearce, D.W., and G.D. Atkinson. 1993. Capital Theory and the Measurement of Sustainable Development: An Indicator of "Weak" Sustainability. *Ecological Economics* 8(2): 103–108.

Pezzey, J.C.V. 1989. Economic Analysis of Sustainable Growth and Sustainable Development. Environment Department Working Paper No. 15. Published as *Sustainable Development Concepts: An Economic Analysis*, World Bank Environment Paper No. 2, 1992. Washington, DC: World Bank.

———. 1995. Concern for Sustainable Development in a Sexual World. Discussion Paper 95-02. Department of Economics, University College, London.

———. 1997. Sustainability Constraints versus "Optimality" versus Intertemporal Concern, and Axioms versus Data. *Land Economics* 73(4): 448–466.

———. 1998. Stripping Resources and Investing Abroad: A Path to Sustainable Development? In *Environmental Valuation, Economic Policy and Sustainability*, edited by M. Acutt and P. Mason. Cheltenham, UK: Edward Elgar.

———. 2004. Sustainability Policy and Environmental Policy. *Scandinavian Journal of Economics* 106(2): 339–359.

Pezzey, J.C.V., and M.A. Toman. 2002. Progress and Problems in the Economics of Sustainability. In *International Yearbook of Environmental and Resource Economics 2002/3*, edited by T. Tietenberg and H. Folmer. Cheltenham, UK: Edward Elgar, 165–232.

Proops, J.L.R., et al. 1999. International Trade and the Sustainability Footprint: a Practical Criterion for Its Assessment. *Ecological Economics* 28(1): 75–98.

Rabin, M. 1998. Psychology and Economics. *Journal of Economic Literature* 36(1): 11–46.

Ramsey, F.P. 1928. A Mathematical Theory of Saving. *Economic Journal* 38(152): 543–559.

Rawls, J. 1971. *A Theory of Justice*. Cambridge, MA: Harvard University Press.

Repetto, R., M. Wells, C. Beer, and F. Rossini. 1989. *Wasting Assets: Natural Resources in the National Income Accounts*. Washington, DC: World Resources Institute.

Riley, J.G. 1980. The Just Rate of Depletion of a Natural Resource. *Journal of Environmental Economics and Management* 7(4): 291–307.

Roughgarden, T., and S.H. Schneider. 1999. Climate Change Policy: Quantifying Uncertainties for Damages and Optimal Carbon Taxes. *Energy Policy* 27(7): 415–429.

Smith, A. 1776. *The Wealth of Nations*. Reprinted 1937. New York: The Modern Library.

Smith, V.K., and J.V. Krutilla. 1979. The Economics of Natural Resource Scarcity: An Interpretive Introduction. In *Scarcity and Growth Reconsidered*, edited by V.K. Smith. Baltimore: Johns Hopkins University Press for Resources for the Future.

Smulders, S. 2000. Economic Growth and Environmental Quality. In *Principles of Environmental Economics*, 2nd edition, edited by H. Folmer and L. Gabel. Cheltenham, UK: Edward Elgar.

Solow, R.M. 1974. Intergenerational Equity and Exhaustible Resources. *Review of Economic Studies, Symposium on the Economics of Exhaustible Resources*, 29–46.

———. 1986. On the Intergenerational Allocation of Natural Resources. *Scandinavian Journal of Economics* 88(1): 141–149.

Spofford, W.O., Jr. 1979. Foreword. In *Scarcity and Growth Reconsidered*, edited by V.K. Smith. Baltimore: Johns Hopkins University Press for Resources for the Future.

Stern, D.I. 1997. Limits to Substitution and Irreversibility in Production and Consumption: A Neoclassical Interpretation of Ecological Economics. *Ecological Economics* 21(3): 197–216.

Stiglitz, J.E. 1974. Growth with Exhaustible Natural Resources: Efficient and Optimal Growth Paths. *Review of Economic Studies, Symposium on the Economics of Exhaustible Resources*, 123–137.

Stokey, N.L. 1998. Are There Limits to Growth? *International Economic Review* 39(1): 1–31.

Toman, M.A. 1994. Economics and "Sustainability": Balancing Trade-offs and Imperatives. *Land Economics* 70(4): 399–413.

———. 1999. Sustainable Decisionmaking: The State of the Art from an Economics Perspective. In *Valuation and the Environment: Theory, Methods, and Practice*, edited by M. O'Connor and C.L. Spash. Cheltenham, UK: Edward Elgar.

UNDP (United Nations Development Programme). 2003. *Human Development Report 2003: Millenium Development Goals. A Compact among Nations to End Human Poverty.* New York: Oxford University Press for UNDP.

Vincent, J.R., T. Panayotou, and J.M. Hartwick. 1997. Resource Depletion and Sustainability in Small Open Economies. *Journal of Environmental Economics and Management* 33(3): 274–286.

Vitousek, Peter M., et al. 1986. Human Appropriation of the Products of Photosynthesis. *Bio-Science* 36(6): 368–373.

Wackernagel, M., and W. Rees. 1996. *Our Ecological Footprint: Reducing Human Impact on the Earth.* Gabriola Island, BC, Canada: New Society Publishers.

Weitzman, M.L. 1976. On the Welfare Significance of National Product in a Dynamic Economy. *Quarterly Journal of Economics* 90(1): 156–162.

———. 1997. Sustainability and Technical Progress. *Scandinavian Journal of Economics* 99(1): 1–13.

———. 1999. Pricing the Limits to Growth from Mineral Depletion. *Quarterly Journal of Economics* 114(2): 691–706.

Woodward, R.T., and R.C. Bishop. 1997. How to Decide When Experts Disagree: Uncertainty-Based Choice Rules in Environmental Policy. *Land Economics* 73(4): 492–507.

World Bank. 2001. *World Development Indicators.* Washington, DC: World Bank. Also available on CD-ROM, http://publications.worldbank.org/ecommerce.

WCED (World Commission on Environment and Development). 1987. *Our Common Future.* Oxford: Oxford University Press.

CHAPTER 7

Resources, Scarcity, Technology, and Growth

Robert U. Ayres

SCARCITY AND GROWTH ARE the core concerns of this book. At RFF they have been linked, in the past, by concern whether scarcity of natural resources would act as a brake on economic growth. In the study by Barnett and Morse (1963) that initiated this line of research 40 years ago, scarcity was identified with rising prices (in real terms). The key empirical result of their work was that, by the test of rising prices, resource scarcity in the United States (except for lumber) had not increased since the end of the nineteenth century. They attributed this favorable trend to technological progress in the exploration for and processing of raw materials. When Smith and Krutilla (1979) revisited the topic, they reached much the same conclusions, despite the sharp increases in petroleum prices in the early 1970s.

If the same methodology were applied today, based on data for the past two decades, similar conclusions would undoubtedly be reached once more. Indeed, before the dramatic price rises in the summer and fall of 2004, at least as many experts feared the economic effects of a glut of oil in the near future as they did a shortage.

However, other issues have arisen in the last two decades, and a reformulation of the underlying problem is now appropriate. In this century, the scarcest resources are more likely to be common property resources, such as benign and stable climate, clean air, environmental waste assimilative capacity, biodiversity, tropical forests, fisheries, and fresh water. Moreover, the pressure causing environmental degradation comes from industrial activity, land use, and especially emissions of waste residuals to air, water, and soil. The emissions, in turn, are from the end of the "life cycle" of materials previously harvested from the biosphere or extracted from the earth's crust, processed, and discarded either during the production process or after final consumption.

To understand whether or how global society will respond to this "New Scarcity," we need to investigate in more detail the process by which technological change occurs. The neoclassical growth model as articulated by Robert Solow (1956, 1957) might be credited more with naming a phenomenon than with explaining it. The unexplained "residual" that is often joined with his name suggests that something else (other than accumulation of capital) has accounted for almost 90 percent of the per capita growth in output over the past half-century. Many economists since Solow have simply postulated an exogenous driver called "technical progress"—or "multi-factor productivity" (MFP)—to explain why the growth of output (GDP) has been much faster than growth of inputs. Recently the notion that growth is attributable to the enhancement of "human capital" has gained in popularity, although no satisfactorily measurable proxy for the latter has been suggested. These views are significant. If technical progress—hence growth—is independent of resource use, then "decoupling" growth from resource consumption is conceptually easy because they were never "coupled" in the first place.

In this chapter I start from a very different thesis, namely that resource consumption and economic performance are strongly coupled and that this coupling is essential to understanding economic growth. I conceptualize the economy as a processor and converter of material resources, first into an intermediate good called "useful work." Useful work then converts nonfuel raw materials into finished materials, material products, and eventually nonmaterial services.

Useful work is precisely defined in thermodynamics, but intuitively it can be thought of in terms of exerting a force to overcome gravity or friction. More precisely, there are three major categories of useful work:

- muscle work by humans or animals, which is still the dominant source of work in many developing countries;
- mechanical work done by so-called prime movers, such as windmills, hydraulic turbines, or heat engines, mostly utilizing fossil fuels. The obvious examples of heat engines are steam turbines (stationary) or internal combustion engines (mostly mobile); and
- heat per se, delivered to a point of use. High-temperature heat (mostly from combustion) drives many chemical and metallurgical processes, while low-temperature heat provides hot water, cooking, and space heating.

Because of electricity's extraordinary importance in the modern economic system, it is important to note that it is a form of useful work—increasingly the dominant form. Electricity is also the only form of useful work that is bought and sold as a commodity, with an explicit price. Other forms of work are associated with processes or services, as in transportation, where the price (value) of the mechanical work done by the engine of a car is not explicitly separable from the price of the transport service provided.

Technical progress since the invention of the steam engine can be conceptualized as the substitution of useful work done by machines or by combustion processes for work done by muscles. The ratio between useful work output and raw energy input from fuels or other natural sources is the *conversion efficiency*.[1] Among the work categories mentioned above, stationary electric power generation is the most efficient: 33 percent on average in the United States, but up to 60 percent if natural gas is available as a fuel for so-called combined-cycle units. High-temperature heat is also generated quite efficiently in specialized chemical or metallurgical facilities, although numerical calculations are scarce. By comparison, mobile power systems range from 10 percent to 30 percent, while low-temperature heat is very inefficient (4 percent to 5 percent in modern insulated buildings), and animal muscle power (based on horses and mules) is of the same order of magnitude, about 4 percent (Ayres and Warr 2003).

The substitution process (machines for muscles) since the beginning of the industrial revolution, has been driven by a combination of (a) rising costs for human labor and (b) declining costs for the primary materials and fuels themselves and for the useful work and finished materials that are produced from raw materials, including primary fuels. (RFF has played a major role in documenting this long-term trend, especially in the famous 1963 study by Barnett and Morse). Evidently the main reason for declining resource costs is technical progress in the discovery and extraction process. In the same spirit, I argue hereafter that declining costs throughout the process chain, driving growth, are attributable largely to declining costs of useful work. Technical progress, in this conceptualization, is essentially equivalent to increasing efficiency of converting raw resources, such as coal, into useful work, such as the output of a steam engine used for pumping the water out of a coal mine back in the time of Newcomen and Watt or a steam turbine that generates electric power today.

Elsewhere we have shown that historical improvements in energy-to-work conversion efficiency, in a production function formulation, can quantitatively account for the major component of economic growth traditionally explained by "technological progress" (Ayres and Warr 2002, 2005).[2] However, it is also important to realize that the gradual, continuous, and homogeneous technical progress normally assumed in economic models cannot explain key aspects of economic growth.

A Closer Look at Technological Change

Technological change in the real world is not gradual, continuous, or homogeneous. Nor is it fungible in the sense that progress in one field is applicable elsewhere. It is intermittent and, for the most part, narrowly specialized. There are two fundamentally different modes of technical progress. There is a "normal" mode, in which technological improvements occur incrementally and more or less automatically as a result of accumulated experience and learning. Spillovers can contribute to this mode of progress, but rarely in a major way. In this mode there is a simple positive feedback between increasing consumption, increasing investment, increasing scale, and learning by doing (or experience). These result in gradually declining costs and prices, stimulating further increases in consumption, hence economic growth.

However, these forces operate economy-wide and do not distinguish between sectors, hence they cannot explain structural change. In reality, there is not one single aggregate technology of production for a single composite universal product. (Nor is there a single unique and unchanging technology for each product, as assumed by activity analysis.) Two cases are distinguishable. In the first case, some generic technologies (like rolling, mold casting, machining, or spray painting) are applicable to a variety of different products. Product improvements do not require a new production technology, although the production technology itself may also gradually become more efficient. But in the other case there are multiple competing and evolving production technologies. Examples of radical changes range from basic substitution of light metals for iron and steel to plastics and synthetic fibers for natural materials to transportation equipment (e.g., automobiles replacing horses and carriages, air transport displacing railways). The substitution of one broadly applicable general purpose technology for another can be both very productive and very traumatic.

Both gradual improvement within a generic technological trajectory and radical innovations among competing technologies have been likened to biological evolution. But the biological analogy is inappropriate. Evolution is driven by random mutation and recombination, whereas technological progress is the result of conscious effort directed to a specific goal.

The small changes that characterize gradual improvement can be adequately characterized as consequences of a constrained search process. However, few major discoveries, and essentially no radical innovations, can be regarded as random or accidental. External factors often play a role in stimulating research in a particular direction. Geopolitical events and wars have had important technological consequences, as has the prospect of some future scarcity. But the process can be entirely endogenous. If improvement of the existing dominant technology for a product or product group slows down while demand for the accompanying function continues to grow, the economic incentives to find an alternative likewise increase. If the demand for continued improvement is sufficiently powerful, there may be enough R&D investment to achieve a "breakthrough," that is, some radically new innovation capable of displacing the older technique (Ayres 1988). Schumpeter's evocative term for this process was "creative destruction" (Schumpeter 1934). Spillovers from radical general purpose technological innovations since the industrial revolution have been the most potent drivers of economic growth.

So much for generalities. Examples below demonstrate that radical improvements in energy conversion possibilities, as represented by the introduction of steam piston engines, steam turbines, gasoline and diesel engines, gas turbines, and electric power, have had huge and far-reaching impacts on economic growth. In particular, each of these technological innovations led to sharp decreases in the cost of primary products (such as iron and steel, aluminum, and plastics) and in "useful work" and useful power itself.[3] As noted already, declining costs result in declining prices, increased demand, and increasing scale of production, both of which push costs down further. This cyclic scheme can be regarded as the primary "engine" of economic growth.

But the historical engine of growth, predicated as it was on increasing demand for resources, is problematic in our era. There are physical limits to conversion efficiency gains and limits to the primary resources upon which we can call. Among those limited resources are the capacity of air, water, and earth to absorb and mitigate the dangerous byproducts of energy exploitation. Radical innovations are sometimes developed in response to a perceived need to *economize* or *conserve* scarce resources.

Evidently radical innovations are necessary for continued long-term growth. Radical innovation has played and still plays a salient role in material resource consumption and especially energy conversion technology. Responding to the new shortage of environmental waste assimilation capacity—especially in regard to combustion products—without stopping or reversing the engine of economic growth will almost certainly require radical innovations in the technologies for producing useful work in the near future. Protecting existing technologies and the industries depending on them from change does not encourage radical innovation. The policy implication of these observations is that regulatory pressures intelligently applied can have the effect of lowering the barriers to innovation that established industries erect to protect themselves.[4]

Barriers, Breakthroughs, and Radical Technological Innovations

Radical innovations are different in kind from the gradual incremental improvements that arise from scale economies, learning by doing, or learning by using. Breakthroughs presuppose barriers. Barriers often create needs. Barriers may be implicit in the limited sensory or manipulative abilities of humans, the properties of materials, or the capabilities of tools, measuring instruments, or manufacturing processes. Examples of each type are numerous.

Needs may be created by the very success of a technology. For instance, the success of airmail and air taxi services in the 1920s and 1930s created a need for instrumentation to enable aircraft to fly safely at night and in bad weather. The buildup of German bomber forces prior to World War II created an urgent need in Britain to detect approaching aircraft at a distance to allow interception by fighters; the eventual answer was radar (Jewkes et al. 1958, 345).

Needs may also arise from threats to health and safety, such as water pollution by sewage in the late nineteenth century that prompted the development of public health measures. One innovation in response was the use of chlorine for water treatment systems. Infectious diseases and wartime injuries prompted numerous medical innovations, from vaccinations to antiseptics and anesthetics, from DDT to antibiotics. Finally, needs may arise from geopolitical circumstances, such as trade embargos or conflict. Natural rubber suddenly became unavailable in the United States at the outset of World War II after the Japanese conquest of Indochina and Malaysia, the sites of major rubber plantations; but synthetic rubbers from petrochemicals soon replaced most of the natural rubber. Petroleum became scarce in German-controlled Europe during both the First and Second World Wars. In response, the Germans learned how to produce synthetic liquid fuels by coal hydrogenation.

Resource scarcities, actual or anticipated, have kicked off major efforts to find substitutes or alternatives. There have been many cases of actual resource scarcity—or even exhaustion—usually limited to a particular resource or country. Charcoal became scarce in western Europe, especially England, in the seventeenth century, as trees were cut to build warships and to clear land for agriculture and grazing.[5] Coal came into general use in Britain as a substitute for charcoal in the eighteenth century. Sperm whales, the preferred source of lamp oil and tallow for candles, were becoming scarce by the mid-nineteenth century; whaling ships in those days were often away for as long as three years. Kerosene, derived from "rock oil" (petroleum) was the eventual choice from among several possibilities, including lard oil, turpentine, and camphene (Williamson and Daum 1959). It is interesting to note that gasoline was a low-value byproduct of kerosene (illuminating oil) until about 1900. Kerosene output exceeded gasoline output until 1908 or 1909.

In the early 1940s, executives at Bell Telephone Laboratories noticed that the demand for telephone service was growing so fast that electric power requirements for switching would, within two or three decades, consume a large share of the projected electric power production of the United States. Their response was to initiate a deliberate search for alternative switching devices that would consume much less power. Semiconductors seemed like a promising avenue for research. The outcome of that search, in 1947, was the transistor (Jewkes et al. 1958, *399*).

In ENIAC and other first-generation computers, vacuum tube diodes were used as switches (Ayres 1984, *145–151*). Engineers quickly realized that the complexity and hence the computational power of computers was limited by the failure rate of the short-lived vacuum tubes. Transistors and ferrite cores soon replaced tubes and raised the limit on computing power. Exponential increases in design complexity to cope with increasing computational demands, requiring a similar increase in the number of circuit elements, foreshadowed another crisis. It was avoided by the invention of integrated circuits. Integration was soon followed by Large Scale Integration (LSI), Very Large Scale Integration (VLSI), and Ultra Large Scale Integration (ULSI) (Noyce 1977). Gordon Moore propounded his famous "law" of exponential increase in 1965, and it still holds (Schaller 1997).

A radically new industry based on solid-state electronics was born in the 1950s and 1960s. It began with a response to a perceived *scarcity* but it also had general-purpose capabilities leading to eventual spillovers into other sectors. Nobody alive then could have predicted its growth or the ways it has metamorphosed. But electronic computers, together with applications to digital communications and the Internet, are among the industries (along with biotechnology) most likely to drive economic growth in this century.

Spillovers and Economic Growth

At the macroscale, the growth mechanism can be understood as a positive feedback (auto-catalytic) cycle. Declining resource costs lead to declining prices, which enable new uses and drive increased consumption. That, in turn triggers investments in new capacity, resulting in increased economies of scale, or R&D aimed at cutting production costs or increasing product performance. The entire cyclic process also results in learning by doing, which also increases efficiency and cuts costs. New applications and new product markets emerge as prices fall. Increased demand leads to still further economies of scale and learning.

To be sure, this analysis applies perfectly well to incremental improvements driven by learning (knowledge accumulation) and economies of scale. What it does not illustrate is that most, if not all, technologies (in the narrow sense of the term) eventually exhaust the possibilities for further improvement at affordable costs. To put it another way, the returns to R&D decline as the technology approaches a barrier. This creates a need for breakthroughs in the form of radical innovations.

The qualitative history of energy technology since 1700 confirms the importance of radical innovations, together with consequent spillovers and feedbacks, as drivers of economic growth. Newcomen's invention of the first crude steam engine (really a pump) in the early years of the eighteenth century was the first substitution of work performed by a machine powered by heat for work done by animals. As such, it can be regarded as the precursor, if not the beginning, of the industrial revolution. As the early coal (and copper-tin) mines penetrated below the water table, they were constantly being flooded, and water had to be removed. Before Newcomen the only way to do this was by means of an endless chain of buckets, lifted (literally) by horsepower. Newcomen replaced the horses by a heat engine driven by combustion of coal from the mine itself. This cut the cost of coal and, hence, all the downstream uses of coal, such as iron smelting and forging (Landes 1969, *101*).

Newcomen's reciprocating pump engines were thermodynamically very inefficient, however. James Watt improved the simple Newcomen pump so much, in several ways, that he is regarded as the true inventor of the steam engine. His engines were much smaller, they delivered rotary rather than reciprocating motion—which enabled many new uses—and they used far less fuel per unit of work done.

The first Watt & Boulton engine was sold to John Wilkinson, an iron smelter, who put it to work driving a boring machine to make cannons. The same boring machine later found a use drilling cylinder holes for Watt & Boulton's steam engines, thus completing a classic feedback loop. The second Watt & Boulton engine sold was put to work pumping air into a blast furnace to increase the blast temperature and reduce the quantity of fuel needed to make iron. Some of that cheaper iron went into more steam engines, as well as iron rails for the carts carrying coal from the mines and—in due course—the steam-powered railways. So the steam engine reduced the cost of coal, which cut the cost of making steam. Later it cut the cost of iron and the cost of metalworking. It made possible railways, which cut the cost of transportation sharply and further cut the cost of coal and iron delivered to consumers in the cities, encouraging still more uses of metal and greater demand.

By the end of the eighteenth century, England had replaced Sweden as the low-cost producer of iron and steel. Rapid technological progress in the British iron industry continued: the coal requirement per ton of iron decreased by a factor of three between 1830 and 1860, cutting costs dramatically. Meanwhile, demand for wrought iron—from the growing British railway industry—increased tenfold (Jevons 1865, 1871)

The growth of the wrought iron industry was followed by breakthroughs by William Kelly in the United States (1847) and Henry Bessemer in the United Kingdom (1856), who independently discovered a revolutionary way to convert pig iron directly to carbon steel in liquid form suitable for casting or rolling. This was quickly commercialized and triggered a remarkable further decline in the price of steel rails, which was matched by spectacular growth in demand.

Demand for steel rails slacked off in the early twentieth century, but demand for other steel products, such as structural shapes, barely begun by 1888, outstripped demand for rails by 1914. By 1957 demand for rails was roughly one-sixth of that for structural shapes for the building industry. Still later, rolled steel products for the automobile industry and for a host of other products drove further growth of the steel industry.

This history of the interplay between steam power, coal, iron, steel, railways, and other inventions illustrates a couple of important phenomena. First, such developments are all interrelated, with one playing off another, and all cycling back to enhance the impact of the original innovation. Steam engines made it possible to recover coal from mines that would otherwise have been waterlogged, which made smelting iron cheaper, which made railroads—which, of course, used steam locomotives—feasible, which made it cheaper to carry coal back to the plant that machined the steam engine parts.

In another context—current concern with energy conservation—this feedback has recently been termed a "rebound effect." Coal was expensive, at first, because in the absence of machinery, it was difficult to mine. This expense presented a barrier to further growth until a technological breakthrough—the steam engine—emerged. This breakthrough spawned so many other complementary innovations, however, that total consumption of coal soon vastly exceeded what it would have absent the steam engine.

The same feedback phenomenon can be observed in other breakthroughs. Consider electric power. The efficiency of generation and delivery of electric power increased roughly tenfold from 1900 to 1960 (Figure 7-1), with the most rapid increase occurring after 1920. As a result, retail prices to all users declined in real terms from about 42 cents per kWh (in 1992$) to about 3 cents today. A considerable part of the demand for electric power in the late nineteenth and early twentieth centuries resulted from the substitution of electric light for gas light and electric motors for stationary prime movers like steam engines or gas engines in factories (Devine 1982) or the replacement of horses by trams in urban transport. The price reduction for retail users has been especially dramatic.[6] Indeed, the subsequent applications of electric power to a wide variety of new products and services probably constitute the most important example of the growth-inducing impact of spillovers from radical innovations.

The above capsule histories are reassuring demonstrations of the ability of technology in a market economy to find or create substitutes for scarce materials or, more generally, to find suitable answers to emerging problems. When a looming crisis stimulates the search for a new technology, the resulting innovation often has spillover effects that create wholly new products and markets. The need for a substitute for charcoal in the seventeenth century triggered the innovation of steam power and all that followed. The need for a substitute for sperm whale oil in the 1850s triggered the development of a petroleum industry and incidentally created the conditions for encouraging the use of internal combustion engines, which led to automobiles and aircraft (Yergin 1991). The popularity of automobiles, especially after Ford introduced the Model T, led to a rapidly growing demand for gasoline—a volatile liquid mixture of light hydrocarbons. This resulted from the need to "crack" and reconstitute the heavy molecules that predominate in petroleum. The resulting availability of byproduct ethylene, propylene, and butylene led to the plastics industry of today. The need for a substitute for natural rubber in World War II was also a huge boost to the petrochemical industry.

The search for a better source of light, which began at the dawn of history, led eventually to the combination of innovations (both in the light source and the generation and distribution system, many by Thomas Edison) that created

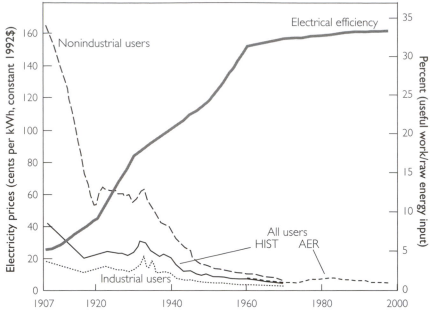

Figure 7-1. *Electrical Efficiency and the Price of Electricity in the United States, 1907–1998*

Sources: (left scale) Nonindustrial users, 1907–1970, *Historical Statistics of the United States* (HIST), series S-116; industrial users, 1907–1970, HIST, series S-118; all users, 1907–1970, HIST, series S-119; all users, 1960–1998, *Annual Energy Review* (AER), Table 8.13. *(right scale)* Electrical efficiency, 1907–1998, HIST, series S-107 and AER Table 8.02.

an electrical engineering industry. Apart from important metallurgical applications (aluminum) and motor drive, electric power enabled all the applications of electromagnetism—including telephones, radio, TV, radar (microwaves), and even lasers—that have emerged during the succeeding century. Burgeoning demand for electric power for conventional electromechanical telephone switching networks, in turn, set off the semiconductor revolution and made electronic computers and mobile telephony possible. Computers, in turn, have enabled digitization of analog communications, including television and image processing. The merger of digital telecommunications with computer technology resulted in the Internet, which many see as the principal driver of economic growth in the coming decades.

Environmental Implications of Increasing Technical Efficiency: The Rebound Effect

In case after case, energy efficiency and service delivery improvements have been significant. However, although I have not attempted a thorough material-by-material analysis, it appears that efficiency improvements have rarely, if ever, resulted in reduced aggregate energy (including materials) consumption. The usual response is increased demand, as amply illustrated in the preceding sections. This response is part of the positive feedback cycle that drives economic growth. It is described in some circles as the rebound effect. Steps taken initially to economize on the use of one resource often result in far more profligate use of that resource or another. Some economists, notably Daniel Khazzoom and Leonard Brookes, have argued that the rebound effect is so general that energy conservation policies can have little benefit (Brookes 1979, 1990, 1992, 1993; Saunders 1992; Khazzoom 1980, 1987).

The most recent example, of course, is that of computers and semiconductor products. Efficiency gains have indeed been very great, but the growth of demand for materials (and electric power) consumed in the manufacture of computers and related goods has been far greater than the efficiency savings (Williams et al. 2003). According to Energy Information Administration statistics (EIA 1999), information and computer technology (ICT) now accounts for around 3 percent of gross U.S. electric power consumption, and conservative projections suggest that this could increase to 10 percent or more within a dozen years unless some radical new technology (such as superconductivity) comes along soon to reduce the energy requirements of ICT. Moore's Law, now nearly 40 years old, is also an expression of the rebound phenomenon.

Nevertheless, there are important exceptions to the Khazzoom-Brookes postulate. The most obvious is the U.S. passenger car fleet, which, thanks to the application of CAFE (corporate average fuel economy) regulatory standards, consumes significantly less fuel than it would have in the absence of the regulations (Schipper and Grubb 1998). In fact, in the late 1980s, the fleet was using less fuel than a decade earlier. The gradual introduction of compact fluorescent lights in place of incandescent bulbs is likely to be another exception.

When an efficiency breakthrough occurs during the early stages of a general "infant" technology development—as it did for iron and steel, electric power, aluminum, and computers—the rebound effect is virtually automatic. Indeed, it is nothing less than the main historical driver of economic growth. On the other hand, when the gain in energy efficiency is achieved late in the life of a mature industry, and where improvement affects energy costs but has little impact on overall operating costs, the rebound effect will be minor or negligible. That is the case with automobiles today: lower fuel costs may have a small impact on vehicle miles driven per year, but it will sell few if any additional (or bigger) cars. Similarly, substituting compact fluorescent lights for existing incandescent lights will cut operating costs for existing installations significantly, but will not induce extensive rewiring to add new lighting units.

All of this has serious implications for future economic growth. The prospects for decoupling economic growth from fossil fuel consumption must be regarded as dependent on continued progress in converting "raw" energy inputs to work, including all the downstream applications of work (i.e., metallurgical, chemical, and information-related processes). But, since the use of conventional sources of energy—notably fossil fuels—cannot be continued at present levels and costs, still less multiplied, new sources of energy inputs, or new ways of doing useful work, are top priority for the coming decades. Luckily, the long-term prospects seem moderately favorable (see Azar in this volume).

The threat of global climate change (or, more accurately, "climate chaos") has created a need for alternatives to fossil fuels. The spectrum of possibilities on the supply side ranges from wind turbines to photovoltaic (PV) to fuel cells. All of these (and others) do useful primary work. But primary conversion efficiency, especially of electric power, is already approaching its limit. The greatest opportunities for technological innovation lie in the domain of secondary conversion (use of electric power for lighting, refrigeration, manufacturing, and so on), or better still, in the domain of final services, such as transportation and construction.

In a truly competitive market environment, this need should kick off a new round of radical inventions and innovations. In turn, past experience suggests, such innovations will have significant spillover effects that will trigger new growth areas. This will happen more quickly if the government intervenes to support appropriate environmental policies to discourage carbon-based technologies on the one hand, and to encourage noncarbon alternatives on the other.

The conventional wisdom in conservative circles— that government intervention can only inhibit innovation—is flatly wrong. A number of important innovations, including radar, jet engines, nuclear power, and computers, have resulted from wartime exigencies. Military research has prompted subsequent innovations, including ARPANET, the precursor of the Internet. Many peacetime developments, such as the Japanese and French high-speed rail systems and the German-Japanese mag-lev systems, could never have been developed by the private sector, and the most ambitious program of all, space travel (of which the economic benefits are still probably far in the future), would have been far beyond the resources of profit-seeking firms.

Conversely, the current policy of the George W. Bush administration has been designed by the established coal and hydrocarbon industries to protect them from competition and thereby inhibit the development of competitive alternatives. This policy of protecting the old established energy sector will also delay the radical innovations on which long-run economic growth depends.

Notes

1. The term "efficiency" is used (and misused) in several ways in the historical literature. For purposes of what follows, we define efficiency in the so-called second-law sense, as distinguished from the more familiar "first law" usage, as spelled out in the report of an important study sponsored by the American Physical Society (Carnahan et al. 1975). Very briefly, the first law efficiency (normally used in engineering work) refers to the ratio of "useful" outputs to inputs. This term is often used to describe the efficiency of a combustion process in terms of the fraction of the chemical energy in the fuel that is converted into heat available for further use, that is, captured by a heat exchanger or—in the case of a stove or fireplace—radiated into the room. Thus a reported efficiency of 80 percent for a gas-fired furnace means that only 20 percent of the heat generated is lost up the chimney. Second law efficiency is the ratio of the minimum theoretical requirement for energy to the energy actually used in practice to achieve the function. The second law efficiency for space heating is particularly low because the desired temperature of the room is roughly the same as the waste heat from an electric power generating plant, which means that the same fuel could have been used to generate electricity before it was finally used to heat the room. High efficiencies can be achieved by cascading uses, starting with uses that require the highest temperatures, followed by lower and lower temperature uses.

2. The term "we" is meant literally. Benjamin Warr, my research assistant for the past four years, is co-author of several publications in which the role of energy conversion in growth theory is more completely elaborated.

3. Power is defined as work per unit time. Efficiency, in this context, is the ratio of actual work done to the maximum theoretical possible work output. Electricity can be regarded as a pure form of work.

4. This is a version of the so-called Porter hypothesis that received some attention a few years ago (Porter and van der Linde 1995). A possible future example would be the targeted deregulation of the centralized electric power sector (which was originally regulated on the presumption that electric power generation is a natural monopoly) to encourage decentralized application of a technology called combined heat and power (CHP).

5. A reviewer has added an interesting sidelight. In Britain (and probably elsewhere), "royal" oaks were reserved for the crown (i.e., for ships), and farmers were not allowed to cut them. As a probable consequence, farmers fed the acorns to pigs and otherwise discouraged the growth of seedlings, thus contributing to the ultimate shortage.

6. Most big industrial users of electric power in the early twentieth century located near sources of inexpensive hydroelectric power, such as Niagara Falls. Retail consumers had to wait for high-voltage long-distance power transmission systems to share the benefits of large hydroelectric facilities.

References

Ayres, R.U. 1984. *The Next Industrial Revolution: Reviving Industry through Innovation*. Cambridge, MA: Ballinger Publishing Company.

———. 1988. Barriers and Breakthroughs: An Expanding Frontiers Model of the Technology Industry Life Cycle. *Technovation* 7: 87–115.

———. 1990. Technological Transformations and Long Waves. *Journal of Technological Forecasting and Social Change* 37. Part I, 1–37; Part II, 114–137.

Ayres, R.U., and B. Warr. 2002. Economic Growth Models and the Role of Physical Resources. In *Unveiling Wealth: On Money, Quality of Life, and Sustainability*, edited by P. Bartelmus. Dordrecht: Kluwer Academic Publishers.

———. 2003. Exergy, Power and Work in the US Economy 1900–1998. *Energy—The International Journal* 28: 219–273.

———. 2005. Accounting for Growth: The Role of Physical Work. *Structural Change and Economic Dynamics* 16(2): 181–209.

Barnett, H.J., and C. Morse. 1963. *Scarcity and Growth: The Economics of Resource Scarcity*. Baltimore: Johns Hopkins University Press for Resources for the Future.

Brookes, L. 1979. *A Low-Energy Strategy for the UK* by G. Leach et al.: A Review and Reply. *Atom* 269: 3–8.

———. 1990. Energy Efficiency and Economic Fallacies. *Energy Policy* 18(3): 199–201.

———. 1992. Energy Efficiency and Economic Fallacies: A Reply. *Energy Policy* 20(5): 390–392.

———. 1993. Energy Efficiency and Economic Fallacies: The Debate Concluded. *Energy Policy* 21(4): 346–347.

Carnahan, W., K.W. Ford, A. Prosperetti, G.I. Rochlin, A.H. Rosenfeld, M.H. Ross, J.E. Rothberg, G.M. Seidel, and R.H. Socolow. 1975. *Efficient Use of Energy: A Physics Perspective*. New York: American Physical Society.

Devine, W. D., Jr. 1982. *An Historical Perspective on the Value of Electricity in American Manufacturing*. Oak Ridge, TN: Oak Ridge Associated Universities Institute for Energy Analysis.

EIA (Energy Information Administration), Office of Energy Markets and End Use. 1999. *Annual Energy Review 1998*. Washington, DC: U.S. Department of Energy, Energy Information Administration.

Jevons, W.S. 1865. *The Coal Question: An Inquiry Concerning the Progress of the Nation, and the Probable Exhaustion of Our Coal-Mines.* London: Macmillan.

———. 1871. *The Theory of Political Economy.* 5th edition. New York: Kelley.

Jewkes, J., D. Sawers, and R. Stillerman. 1958. *The Sources of Invention.* London: Macmillan.

Khazzoom, J.D. 1980. Economic Implications of Mandated Efficiency Standards for Household Appliances. *Energy Journal* 1(4): 21–39.

———. 1987. Energy Savings Resulting from the Adoption of More Efficient Appliances. *Energy Journal* 8(4): 85–89.

Landes, D.S. 1969. *The Unbound Prometheus: Technological Change and Industrial Development in Western Europe from 1750 to the Present.* Cambridge: Cambridge University Press.

Noyce, R.N. 1977. Microelectronics. *Scientific American* 237(3): 63–69.

Porter, M.E., and C. van der Linde. 1995. Green and Competitive: Ending the Stalemate. *Harvard Business Review* (Sept.-Oct.): 120–134.

Saunders, H. 1992. The Khazzoom-Brookes Postulate and Neoclassical Growth. *Energy Journal* 13(4): 131–148.

Schaller, R.R. 1997. Moore's Law: Past, Present and Future. *IEEE Spectrum* 34(6): 53–59.

Schipper, L., and M. Grubb. 1998. *On the Rebound? Using Energy Indicators to Measure the Feedback between Energy Intensities and Energy Uses.* Paper presented at the IAEE 1998 Conference in Quebec.

Schumpeter, J.A. 1934. *Theory of Economic Development.* Cambridge, MA: Harvard University Press.

Smith, V.K., and J. Krutilla (eds.). 1979. *Scarcity and Growth Revisited.* Baltimore: Johns Hopkins University Press for Resources for the Future.

Solow, R.M. 1956. A Contribution to the Theory of Economic Growth. *Quarterly Journal of Economics* 70: 65–94.

———. 1957. Technical Change and the Aggregate Production Function. *Review of Economics and Statistics* 39: 312–320.

Williams, E., M. Heller, and R.U. Ayres. 2003. The 1.7 Kg Microchip: Energy and Material Use in the Production of Semiconductor Devices. *Environmental Science and Technology* 36(24): 5504–5510.

Williamson, H.F., and A.R. Daum. 1959. *The American Petroleum Industry.* Evanston, IL: Northwestern University Press.

Yergin, D. 1991. *The Prize: The Epic Quest for Oil, Money and Power.* New York: Simon and Schuster.

CHAPTER 8

Endogenous Technological Change, Natural Resources, and Growth

Sjak Smulders

IN THE EARLY 1800s, England's industrial revolution seemed to be in danger of losing momentum as coal supplies dwindled, but rapid growth continued over the subsequent century. In developing countries, many poor cities that have grown at a fast pace have simultaneously experienced rapid growth in air pollution, yet air quality in the large cities of the Western world has improved in recent decades. Does growth lead to faster depletion of resources, or does it create the resources to clean up the environment? History, recent or otherwise, shows many possible interactions between growth and scarcity of resources.

The economic forces that shape the interaction between growth and scarcity are substitution and technological change. In their absence, each additional unit of output requires a given amount of resource use, for example, energy input, and creates a given amount of pollution; output cannot expand without reducing resource stocks and environmental quality. Almost all economists, neoclassical economists in particular, have stressed that the amount of resources needed to produce a given amount of output is not constant on an economy-wide level. Consumers can shift demand to goods that can be produced with less energy or pollution. Producers can switch to techniques that use less resources. As a result, the interaction between growth and scarcity is shaped by consumer preferences and willingness to adjust consumption patterns, as well as by technological possibilities and opportunities for producers.

Limits to growth are then determined by limits to substitution. However, even with ample substitution possibilities, resource substitution is inevitably constrained when resource availability falls: the productivity of human–made inputs falls by the law of diminishing returns. Only technological change can offset the diminishing returns: new opportunities for substitution are opened up by shifts to new, more productive technologies, less resource-dependent technologies, or technologies that rely on completely new resources.

The great waves of economic growth can be attributed to major breakthroughs followed by incremental improvements in technology. The connection to natural resources is obvious when we consider the role of waterpower, steam power (fueled by coal), and the internal combustion engine (powered by fossil fuels). Some technological developments have been due to luck or the genius of an individual. For the most part, however, the commercialization and diffusion of new technologies, as well as their subsequent improvements and alternative applications, have required deliberate investment and a rational business strategy. One of the major innovations in the twentieth century is in fact the introduction of the R&D department. All in all, it is fair to say that a substantial part of technological change is the result of economic investment decisions. Technological change therefore occurs as a reaction to economic incentives and opportunities to develop new technologies over time; that is, technology is endogenous.

Limits to growth must then not only be determined by substitution at a given time, but also by innovation incentives or opportunities for development. Recently, economists have begun to explore the implications of endogenous technological change in formal models of scarcity and growth. Increased scarcity may, through rising prices, stimulate firms to develop new technologies. Thus endogenous technological change may alleviate scarcity limits. But if technological change is costly, it can also be crowded out by resource scarcity. If lower availability of resources reduces the productivity of human-made inputs, it becomes less rewarding to develop complementary new technologies. Scarcity may provide the stimulus to develop new technologies that save on resource input, but at the cost of innovation projects in other directions (for example labor-saving technological change). Changes in the direction of innovation efforts may reduce the effect of aggregate innovation effort on economic growth. Allowing technological change to respond endogenously to scarcity does not necessarily lead to a more optimistic outlook with fewer episodes of scarcity.

This chapter discusses how scarcity can be alleviated by substitution with, or development of, less energy- or material-intensive and cleaner technologies. We treat both substitution and technological change as *endogenous* and sort out the determinants behind them. Why does substitution with clean or less resource-extensive technologies not occur when it is technically possible? When does faster growth speed up depletion? When does growth coincide with improved environmental quality? Must policies aimed at improving environmental quality or energy conservation restrict growth? Do more stringent environmental policies induce innovation?

The interaction between growth and scarcity is complex because both growth and scarcity are the result of economic decisions. The direction in which the economy grows determines the effect of growth on resource stocks. Conversely, resource availability shapes opportunities for growth and the rate of return on investment. Economists have analyzed the elementary forces behind depletion and economic growth with simplified models that abstract out many complexities of the real world. To understand the crucial economic forces, we focus on one natural resource at a time, first nonrenewable resources, such as energy and materials, then environmental resources, such as fish, forests, clean air, and water. We also aggregate economic activity into a single production activity that requires several inputs, among them natural resources. We focus on theoretical considerations, but we review empirical evidence related to scarcity and growth on the aggregate level.

We first study the effects of changes in production technology on resource use. It turns out that new technologies do not necessarily lead to less depletion. New technologies produce a certain amount of output using less resource—they facilitate substitution—but this implies that they also improve the productivity of resource use. Technological change may thus stimulate the demand for the resource. Second, we turn to the determinants behind technological change itself; that is, we treat technological change as an endogenous variable. We discover when the growth process comes to a halt because of scarcity, and how growth rates are affected by resource policies or environmental policy. Scarcity is likely to crowd out innovation, but knowledge spillovers may offset this tendency.

By moving from nonrenewable to renewable resources, and contrasting exogenous and endogenous technological change, we follow more or less the chronological developments in the literature on growth and resource scarcity. This chapter first reviews models of aggregate growth with a single resource, in the tradition of the economic growth literature (Stiglitz 1974; Dasgupta and Heal 1979). We then extend the scope of analysis to environmental problems and endogenous technological change, and see how the findings of the older literature have to be revised.

Two themes in the recent literature give endogenous technological change a prominent place. One springs from growth theory, in which the assumption of exogenous technological change is considered more and more unsatisfactory. It is unclear whether growth can be sustained if endogenous technological change requires investment, and natural resources are essential for production. If resource substitution makes the returns to human-made inputs fall, what happens to the incentive to invest in new technology, also a human-made input? On the other hand, environmental economists have also become interested in endogenous technological change. They realize that the environmental cost of growth might be substantially lower if substitution within given technologies were supplemented by development of new technologies.

We also study market failures in environmental resources, and how they change the role of technological change in mitigating scarcity limits. We discuss the forces behind endogenous technological change, and its implications. Here the central topic is how environmental resource scarcity affects the rate and direction of technological change, and the policy implications. Then we turn to growth models; we are mainly interested in the conditions under which growth can be unlimited despite dependence on natural resources. Finally we discuss how environmental policy affects economic growth.

The Neoclassical Perspective: Nonrenewable Resources

The debate on scarcity and growth has traditionally focused on the scarcity of nonrenewable resources, such as fossil fuels (oil, coal). "For there to be a meaningful natural resource problem, a resource must be in limited supply, must be non-renewable and nonrecyclable, essential, and without perfect substitutes," as Stiglitz (1979, 40) noted when reviewing the neoclassical view of the problem of the scarcity of natural resources. The seminal work of Stiglitz, Solow, and Dasgupta and Heal in 1974 established the benchmark neoclassical framework to study scarcity of nonrenewable resources.[1]

The Neoclassical Trinity: Substitution, Diminishing Returns, and Technological Change

In the neoclassical view, the economy can produce only if it extracts resource inputs. Each unit of resources used for production reduces the stock of available resources one-for-one and irreversibly. The stock of resources is privately owned and traded in markets, which also holds for other inputs that substitute for resource inputs (physical capital and labor). Inevitably, production depletes the resource stock. The question is whether such an increase in physical scarcity also implies an increase in economic scarcity: Must economic production ultimately fall?

The main message from the neoclassical literature is that substitution of human-made capital inputs for the resource alleviates the economic consequences of physical resource scarcity. The market provides incentives for this substitution: thanks to the existence of markets in which resources are traded, rising prices signal increased scarcity and trigger substitution of less resource-intensive techniques. Capital replaces resources, and limits to growth can be avoided if there are enough substitution alternatives.

The substitution mechanism itself tends to become less and less powerful, however, because the productivity of a piece of equipment tends to fall if a larger amount of capital is combined with fewer resources or other inputs. This law of diminishing returns makes capital accumulation less productive as the amount of available resource inputs falls. So while *substitution* mitigates the drag on growth from resource scarcity, *diminishing returns* constitute another drag on growth. The neoclassical model relies on a third assumption, the presence of ongoing exogenous *technological improvements*, by which growth can be sustained over time. Technological change exogenously improves the productivity of the factors of production—capital as well as resources. It offsets the diminishing returns so that growth can be sustained.

The empirical validity of the neoclassical model is subject to an ongoing debate, discussed elsewhere in this book. Many empirical studies of technological change and substitution on the macro level directly or indirectly support the neoclassical view. Energy use per unit of production has declined steadily in most industrialized countries over a very long period. Rates of technological change are almost universally impressive.[2] In two studies, Weitzman (1997, 1999) provides an indirect way of testing and calibrating the model. He finds that roughly 40 percent of annualized welfare is the result of technological change, while at most 1.5 percent of income could be gained if the limited nonrenewable resources we rely on were to remain available without limit at today's flow rates and extraction costs.

Depletion of Nonrenewable Resources and Technological Change: Basic Results

What determines the rate of depletion, and therefore scarcity, in a world of substitution? *The resource stock is, in fact, never fully depleted*, neither in the ideal (first-best) situation nor in a market with resource property rights—at least as long as the resource is necessary for production. Society as a whole ideally wants to avoid depletion, because when the resource stock is exhausted, production, hence consumption, is impossible. In extreme scarcity, a small amount of resources is extremely valuable. Full depletion can never be optimal, since society anticipates that it is necessary to preserve some resources to maintain production. Where full depletion looms, society can consume and deplete less today in exchange for a small increase in future consumption.

The avoidance of full depletion is not only optimal, but is also the likely outcome if resource markets function correctly. Individuals own the resource, and they can trade property rights. To secure future consumption, younger generations are willing to buy resource stock from the older generations. When the stock runs very low, the young are willing to pay a very high price, thus giving older generations an incentive not to fully deplete the stock. Forward markets guarantee the dynamic efficiency of the resource market. Market participants may hold incorrect expectations about future scarcity, which may lead to inefficient extraction of the resource. However, market participants are likely to systematically arbitrage away large discrepancies between expectations and realizations. Thus, with the existence of property rights and (forward) markets, there is not much reason for active resource policies.

Some readers will object, arguing that individuals may not care enough about the future to conserve resources. When agents discount the future at a higher rate, they tend to speed up depletion by increasing current consumption. Yet, even with discounting, the necessity of the resource for production prevents full depletion.

Exogenous technological change has an ambiguous effect on the rate of depletion. Consider the prospect of a new technology that allows a larger output level for a given resource and other inputs. Individuals who anticipate this technological change attach a higher value to the resource stock, because it will be more productive in the future (younger generations are willing to pay a higher price for property rights). This makes it attractive to conserve more resources for future periods in which the productivity will be higher (this is a substitution effect). However, at the same time, a given resource stock can produce more goods, which increases income and makes consumption more abundant in the future. With higher lifetime income due to future technological progress, current demand for resources goes up, because richer households want to consume more not only in the future but also today, that is, they want to smooth consumption (this is an income effect). Thus, substitution and income effects work in opposite directions; society's preferences will determine which one dominates. Most empirical research on changes in patterns of consumption indicates that income effects dominate substitution effects.[3] Hence productivity gains may raise depletion.

From Old to New Scarcity: Environmental Resources

Since the 1970s, attention has shifted from the scarcity of materials and energy resources to the scarcity of environmental resources, such as clean air, clean water, clean soil, forests, fish stocks, and rare species.[4] Economists have attempted to analyze the scarcity of environmental resources within the framework they developed for nonrenewable energy and mineral resources. Certainly, the two types of natural resources have some features in common—they are inputs in production processes, and they can in principle be depleted. But environmental resources are not priced or traded in markets, and their use has much in common with public goods. Consequently, environmental resource scarcity may pose bigger economic problems than scarcity in nonrenewable resource markets.

The Characteristics of Environmental Resources

When we claim that environmental resources are an input to production, the connection is less direct than that between energy or materials and production. Production creates pollution, a byproduct that diminishes environmental quality. Pollution can therefore be regarded as the inevitable depletion of a resource stock due to production. In this sense, pollution is similar to resource use in the standard neoclassical approach. Pollution is often linked to a particular input, for example chemicals, which cause toxic waste, or energy use, which pollutes the air. Substitution between polluting inputs and other inputs takes place because firms can choose to undertake abatement activities, which reduce pollution or mitigate its effects. Technological progress may result in cleaner production processes, products that generate less waste when consumed, or more efficient (cheaper) abatement technologies, such as filter, scrubbers, and other add-on technologies.

We commonly distinguish environmental resources from mineral resources because the former are renewable and the latter are nonrenewable. Pollution creates environmental resource scarcity, but the damage need not be long lasting, because nature has a capacity to neutralize pollution. For example, soil and river water pollutants can be diluted and flushed out by rainfall. The pollution absorption capacity of ecosystems, though not unlimited, makes environmental quality a renewable resource.

The stock of environmental resources may bear directly on society's welfare. We care about the environment as an amenity: unique landscapes and species, clean air, and attractive sites. We care about mineral resources only indirectly as inputs to production of goods that we value, but we care about environmental resources directly. Some industries might benefit from cleaner water supply: with lower water treatment costs, they might produce useful goods more cheaply. But clean water resources directly benefit consumers in the form of health benefits and amenity values.

Environmental resources are public goods. Although nonrenewable resources like oil and minerals are traded as private goods in markets and protected by property rights, it is impossible to define the owner or enforce property rights of environmental resources like clean air, ocean fish stocks, or the ozone layer, and it is not straightforward to charge users of these resources. The essential properties of public goods apply to environmental resources: access is hard to exclude, and consumption is nonrival. Here markets fail, and the price mechanism on its own cannot ensure that environmental resources are allocated and used in the best way for society: as long as users do not need to pay a price, they do not internalize the social cost of resource depletion and environmental degradation. This calls for public intervention, like pollution taxes or fish quotas, to correct the resource market externalities.

Environmental Degradation and Technological Change: Basic Results

We can modify the standard neoclassical model to reflect the connection between production and environmental resource depletion. We have to account for not only the renewable character of environmental resources, but also the public goods character: markets for resources are missing, but regulation may repair resource externalities.

To analyze environmental degradation, it is useful to deal separately with two cases. In the "first-best case," institutions and regulation ensure best use of environmental resources. Society trades off degrading the environment for the sake of production against maintaining environmental quality for the sake of its amenity value and its future resources. In the "unregulated market case," market prices fail to reflect scarcity. Firms are typically not charged the full cost of depletion of resources. If firms are allowed to pollute more, their unit production cost is lower and productivity is higher. They maximize profits by expanding production until the marginal product of additional natural resource use (read, pollution) is zero. Clearly, if the environmental asset is provided for free, undesirably high levels of production, and pollution, result. Between these extreme cases, regulation addresses some of the externalities, but is inadequate to attain the socially efficient situation.

Without adequate resource policy, environmental resources are easily overexploited. If the resource becomes scarcer, the price does not reflect this, and firms or individuals have no private incentive to reduce their consumption. Profitable low-cost options may be available to reverse environmental degradation, but individual agents would ride free on the investment of others in the public good. Without price signals, substitution possibilities are not exploited. Technological change may also fail: if agents do not pay to pollute, there is no incentive to develop cleaner technologies.

Without adequate resource policy, technological change is likely to speed up environmental degradation. With adequate resource policy, the effect of technological change on the environment is ambiguous, as we have seen in the conflict between the substitution effect and the income effect. However, in the absence of resource policy, firms extract resources without concern for future resource scarcity. They have no reason to conserve resources to exploit future productivity improvements. The substitution effect no longer applies, but the income effect does, which speeds up resource degradation.

Technological change can be *neutral* or *biased*. Neutral technological change raises the productivity of all conventional (nonenvironmental) inputs in the same magnitude. For example, firms may improve the organization of their production process so that more outputs are produced from the same inputs, or they may redesign products without changing input requirements, making them more attractive to customers. The marginal products of all inputs, including polluting inputs, will rise, and firms will want to consume more of all inputs, unless prices change. Without a price on pollution, however, pollution must increase. Neutral technological change makes it possible, in principle, to produce the same amount with *less* pollution, but this gives firms the incentive to pollute *more* because the productivity of pollution rises.

Biased technological change affects the productivity of one input more than another. An improvement in the design of gas-guzzling cars that draws customers away from other, less-polluting cars implies a higher productivity of energy, and polluting inputs in particular; such a resource-biased technological makes consumers tend to spend more on gasoline. In contrast, a cost reduction for production of low-emission vehicles might reduce the expenditure share of gasoline—a technological change classified as resource saving.

Without regulation, new technology improves environmental quality only if

- the new technology decreases the unit production cost (which excludes the environmental cost if the environment is not priced). If not, the firm is better off with the old technology and will not adopt the new one. Producers pass on the decrease in unit costs as lower prices, which result in larger production.
- the new technology substantially reduces the marginal productivity of polluting inputs. If not, firms may either increase or decrease pollution per unit of output, but because of the expansion of the scale of output, total pollution will still increase (the rebound effect).

Market-based instruments can trigger resource-saving innovations. In particular, incentives for adoption of cleaner technology change if regulation imposes a cost per unit of pollution, such as a pollution tax or a system of tradable pollution permits. Whether or not the regulation is stringent enough to produce the social optimum, firms have an incentive to avoid pollution up to the point where the marginal returns to pollution equal the pollution tax or the price of a tradable permit. Suppose regulation imposes a pollution tax, fixed at an arbitrary level. Now technological change occurs; if it takes the form of improvements in total factor productivity, pollution will increase because the marginal productivity of pollution is increased, but the cost remains the same.[5] On the other hand, if technological change results in lower costs of reducing pollution (improvements in abatement technology), firms reduce pollution.

It may well be optimal for a growing economy to aim at improving environmental quality. That is, society should expand environmental resource availability, rather than deplete it. In a growing economy, consumption goods become more abundant, so they are valued at a lower marginal utility relative to environmental amenities. This increases the demand (willingness to pay) for environmental quality. In other words, the demand for both consumption and environmental quality goes up with income (economists have called this the "normal goods" property). Economic growth calls for higher environmental quality as long as there is some "satiation" in preferences with respect to produced consumption goods (Lieb 2002). When the availability of produced goods is lower than environmental quality, as may be the case in poorer countries, society gives priority to increased production at the expense of environmental quality. When income grows large enough, demand should shift to environmental quality.

In the long run, environmental resources are conserved in the optimum. Whereas in the standard neoclassical model the nonrenewable resource stock is never fully depleted, because extractions from the stock are essential to production, in the case of renewable resources it is the stock itself that is essential. Environmental quality serves as an amenity in utility and as a productive asset in production: clean air is essential for health and for workers' productivity; soil quality is essential for agriculture. Environmental resource stock is not asymptotically depleted, but under normal conditions a nonzero optimal stock of environmental resources is approached in the long-run social optimum.[6]

Some Empirics: The Environmental Kuznets Curve

The main evidence on environmental resource scarcity and economic growth comes from the literature on the Environmental Kuznets Curve (EKC) hypothesis.[7] The relation between pollution and income is characterized as an EKC if pollution first rises with income, then declines when income exceeds a certain threshold.

There is no theoretical reason to expect pollution and growth to be unambiguously related, because both income and pollution are endogenous variables (Copeland and Taylor 2003). The pattern of growth, choice of technology, and nature of technological change determine how income and pollution evolve over time, and a host of underlying factors can affect both variables. However, our review of basic theory will help us to sort out the basic forces that affect pollution and environmental degradation in the process of growth.

Note that the theory predicts an EKC pattern under two specific alternative sets of circumstances:

- if environmental policies reflect social preferences for growth and environment, and thus boost the demand for environmental quality as "normal goods."
- if for low income the productivity of polluting inputs increases, but for higher income it falls.

Indeed for water pollution and several types of air pollution, for example, sulfur dioxide (SO_2), suspended particulate matter (SPM), and nitrogen oxides (NO_x), most studies agree that the relationship between per capita income and pollution per capita is an inverted-U. There is mixed evidence of an EKC pattern for deforestation, but here variation in income level seems to be relatively unimportant in explaining deforestation differences among countries. Municipal waste, carbon dioxide (CO_2) emissions, and aggregate energy use are all monotonically related with income. It should be noted that the results are far from conclusive. Most estimates stem from multicountry analyses, so we cannot immediately discern the relationship between growth and pollution. Moreover, the results are biased because of selective data availability: typically, data are collected only for pollutants that are considered a problem for sufficiently long a period in many countries.

The evidence that emissions fall with income growth is limited to a small number of pollutants—local pollutants with immediate health effects—and to higher income countries. The preceding theoretical considerations partly explain this. In an economy in which pollution is unregulated, whether pollution falls over time depends on the nature of growth and technological change. Indeed, if economies grow in early stages by accumulating polluting capital and in later stages rely on clean human capital, the EKC might emerge as a byproduct of the pattern of growth. Similarly, the EKC can be explained as a byproduct of the structural change that accompanies growth. Following a transition from agriculture to manufacturing, industrial pollution grows with income; a subsequent shift from manufacturing to services may explain the cleaning-up phase of the EKC. Empirically, this latter effect turns out to be weak, however. Structural change has lost most of its momentum in OECD countries in the last decades. Most reduction in pollution intensity takes place *within* the manufacturing sector. Moreover, the computerization of the service industry points out that services may become more energy- and material-intensive than suggested by the simple theory.

When we connect our insights about technological change and environmental resource scarcity to the empirical findings, two clear conclusions emerge. First, despite the discovery of an EKC for several pollutants, pollution will not decline automatically as an economy grows richer; we may find an EKC pattern only because richer economies implement more stringent environmental policies. Second, reduced pollution is more likely the result of a deliberate change in technology, rather than a byproduct of technological change or growth.

From Manna from Heaven to Innovation as an Economic Decision

Technological change is an essential driving force behind economic growth and a powerful mechanism to mitigate the cost of resource scarcity. So far, we have discussed only the effects of technological change. Now we turn to the determinants of, and the driving forces behind, technological change.

The economics view of technological change has itself changed markedly over the past few decades. Economists long treated technological change as too complex to explain on an economy-wide level starting from the economist's standard assumption of competitive markets. However, commercial research and development are increasingly important strategies in multinational corporations and small firms in new product markets. Industry leaders and national policymakers stress the role of innovation for national wealth and competitiveness. The way Japan in the 1960s and other Asian industrializing countries later on achieved rapid growth suggests that policy and economic incentives can influence the pace of technological change within a nation.

All this has led economists, and growth theorists since the late 1980s in particular, to view the pace and direction of technological developments as the outcome of economic decisions rather than an unexplained fact of life. The process of growth as well as the reactions to changes in economic environments (such as increasing scarcity) can be much better understood if technology is seen as an endogenous rather than exogenous variable. When the path of technological change was fixed, firms and individuals could react to changes in resource availability only by changing the allocation of economic activity; resource scarcity would trigger only substitution. However, with endogenous technology, innovation may be intensified or redirected in response to economic changes.

Economists have tried to incorporate endogenous technology into the neoclassical growth framework. This involves a major change—abandoning the idea of perfect competition. Economists have identified new market failures and public goods and property rights problems when shifting from exogenous technology to endogenous technology. These externalities interact with resources externalities introduced with the shift from modeling nonrenewable resources to environmental resources.

Innovation Incentives and Opportunities

Little technological change would take place if no effort were spent on innovation (in the form of, for example, inventive activity, R&D expenditures, building prototype factories). Deciding how much effort to spend on innovation requires the calculus of costs and benefits. The cost of innovation consists of the costs of inputs in the innovation process: laboratory equipment and tests, but mainly time and engineering labor. The returns are the discounted expected profits reaped once the innovation is put on the market. The development of new knowledge (a new idea, a blueprint for a new product or technology) typically has a fixed cost character. Incurring a one-time investment cost is sufficient to develop a new idea, which can subsequently be applied and put into practice many times at no additional cost. Hence, the size of the market determines the rate of return, which implies increasing returns.

Innovators balance costs and expected benefits in determining how much to spend on what kind of innovation projects. Hence, with endogenous innovation, the direction (bias) of technological change is endogenous. Innovators choose among different investment projects; some improve the productivity of resources, while others improve the productivity of capital or reduce the costs of extraction, and so on.[8] The higher the expected returns and the lower the innovation costs in a particular project (direction), the more innovation will take place on this project (in this direction). By this mechanism, high prices for certain inputs shift innovation efforts to projects that develop technologies that save on these inputs. Relative price may affect the direction of technological change, which is known as the *induced innovation hypothesis*.[9]

Market failures show up at various stages of the innovation decision, because of *monopolistic product markets*, *knowledge spillovers*, and *creative destruction*.

- Imperfections in the product market arise because innovation requires monopoly power: no firm or individual will invest in developing a new technology unless it can appropriate the returns by making users pay for it and exclude those who will not pay. The monopoly profits are the carrot for the innovator, and thus enhance dynamic efficiency. However, they burden society with prices above marginal production costs at the cost of static efficiency.
- Knowledge spillovers occur when agents can benefit from new knowledge developed by other firms or research institutions without (fully) paying for it. Knowledge is hard to exclude by means of tight property rights. Patent laws may keep producers from using blueprints to produce a specific product or use a specific technique, but it is hard to prevent use of the more general knowledge that can be inferred from the blueprints. Current innovators build on knowledge developed by earlier innovators, without compensating them. Imitation

and patent infringement is another reason for knowledge spillovers. Intertemporal knowledge spillovers mean that the innovator can appropriate only part of the social returns, and the incentive to research is suboptimally low.

- There are also inducements to overinvestment in R&D: several firms may race for the same patent, and duplication of research effort takes place. Innovating firms may replace other firms before these have recouped their investment costs. Such a process of creative destruction may impose a social cost, since the innovator does not internalize the cost it imposes on the other firms.

In theory, we cannot say which type of externality dominates. However, the consensus from the empirical literature is that the positive externalities dominate and that the social rate of return to innovation exceeds the private return.[10]

Resource Scarcity and Endogenous Technology: Basic Insights

A large endowment of resources has an ambiguous effect on the rate *of innovation.* On the one hand, abundant supply of production factors makes it more attractive to develop new knowledge that increases the productivity of these resources; the returns of R&D rise with the scale at which it is applied, but the cost of developing new knowledge is independent of scale: knowledge is nonrival, and development is a fixed-cost activity. On the other hand, however, the opportunity cost of R&D also rises: with more (nonlabor) resources available, the marginal product of labor in production grows, which makes it attractive to allocate labor to production rather than to research.

Induced technological change compensates for low substitution possibilities: the poorer substitution (between natural resource inputs and other inputs) is, the more likely the *direction* of technological change shifts to the scarce factor. Lower resource availability drives up the prices of marketed resources. This results in higher prices of resource-intensive goods as well, which makes it attractive to invest more in innovation in resource-intensive sectors. However, the price effect may be counteracted by a market size effect: lower resource availability reduces output in resource-intensive sectors and makes innovation in these sectors less attractive. With less production in the sector, the scale at which a new technology can be applied is smaller. Which effect dominates depends on the combined price and quantity effect of lower resource availability on revenue and profits in the sector, because innovation shifts to the sector in which profits of innovation increase. If goods from other sectors easily substitute for resource-intensive goods, the price increase due to lower resource availability will be small, revenues in resource-abundant sectors will fall, and innovation will shift away from these sectors. In contrast, if goods from other sectors poorly substitute for resource-intensive goods, revenues in the resource-intensive sector will rise and innovation will shift to these sectors. In both cases, however, innovation results in lower demand for energy: in the case of poor substitution, because innovations are directly energy saving, and in the case of good substitution, because innovation makes substitute goods in energy-extensive sectors cheaper and demand shifts further to these sectors.

Resource market failures may cause induced technological change to inefficiently speed up environmental degradation and depletion. Because market responses determine technology choices, before resource scarcity can stimulate innovation, resources markets must exist and function efficiently. Markets for fish exist, but excessive catch is likely to result because world fish stocks are not governed by property rights. If inadequate fishery management causes fish populations to decline, fish prices will go up. This could actually stimulate investment in new fishing technology and more powerful vessels. Hence, in a situation of excessive harvesting, induced technological change might even increase harvesting and depletion. The technological change moves in the "wrong" direction.

Technology responds to scarcity in an efficient way only if resource and environmental policies provide adequate price signals. But policy itself has to be in place. It may be that environmental problems induce policy changes, which in turn induce technological changes. The *induced policy response* seems to be empirically important, as is discussed above in the context of the Environmental Kuznets Curve.

Some Empirics on Energy, Environment, and Innovation

Studies of the correlation between resource endowments and economic growth have produced mixed results, in line with theory. Resource booms have often deteriorated rather than improved economic performance in, for example, Latin America (Sachs and Warner 2001). However, resource-rich countries like the United States and Norway provide counterexamples. Wright and Czelusta (2002) attribute the U.S. growth success to a combination of large resource availability and targeted investment in skills and new technologies. Institutional quality and the correct innovation incentives prove to be essential in transforming resource availability into wealth and coping with problems of resource scarcity. Easterly and Levine (2003) find that the apparently adverse effects identified by Sachs and Warner only arise when the concentration of resource wealth is associated with corruption and weak institutions.

Empirical studies into the link between environmental regulation and innovation typically find ambiguous results, again as predicted by theory. Research and development expenditures tend to rise with environmental compliance expenditures, but there is no correlation with innovative output as measured by patent applications (Jaffe and Palmer 1997).

There is some support for the induced innovation hypothesis for environmental innovation. Lanjouw and Mody (1996) find that increases in environmental compliance cost lead to increases in the patenting of new environmental technologies with a one- to two-year lag. This finding supports the poor substitution case: the price effect dominates the market size effect, which spurs innovation. There is also evidence that energy-saving technological change has been especially important in periods of high energy prices and oil shortages (Kuper and van Soest 2003).

Empirical studies point out that price changes and regulation explain only a relatively small part of the bias of innovation. Newell et al. (1999), for example, find evidence for the role of energy prices, regulation, and market size in directing innovation. However, up to 62 percent of the total change in energy efficiency must be attributed to other factors. They also find no effects of these three factors on the overall rate of technological change. Similarly, Popp (2001) finds that two-thirds of the change in energy consumption with respect to a price change is due to simple price-induced factor substitution, while the remaining third results from induced innovation. Popp (2002) finds evidence for knowledge spillovers: using patent citation data, he finds that innovations directed at energy improvements build on the total stock of knowledge embodied in the (quality-adjusted) stock of patents for energy efficiency improvements. He also finds, however, that there are diminishing returns with this stock of knowledge. One of the very few economy-wide studies on the bias of technological change, by Jorgenson and Fraumendi (1981), finds that the majority of sectors in the U.S. economy have experienced technological change that not only saved material but used energy.

In a famous article, Michael Porter (1991) argued—on the basis of case stud-ies—that environmental regulation often increases profits of firms, because of first-mover advantages or because of the elimination of waste of input use. The eco-nomics profession has reacted skeptically. Environmental regulation restricts firms in their behavior and reduces their menu of choices. The Porter hypothesis seems to claim that firms choose an action from this smaller menu that gives higher prof-its than the action they would have chosen from the larger choice set available in the absence of regulation. But then it becomes unclear why firms did not choose this action without the regulation. In a world with endogenous technological change and knowledge spillovers, the Porter hypothesis may be valid, however, because technology, productivity, and profits of an individual firm now depend on aggregate innovation activities and knowledge stocks, which may change in reaction to environmental regulation. Unregulated firms' R&D strategies are sub-optimal because of knowledge spillovers and other market failure in markets for technology. Environmental regulation may improve the incentives for innovation, thus improving not only social welfare, but perhaps also firms' profits.

Limits to Growth?

Despite substitution and technological change, dependence on limited resources may ultimately result in declining economic output. Limits to growth can be avoided if incentives to accumulate capital substitutes and to innovate new technologies con-tinue, even when resources become scarcer. To understand limits to growth, we have to examine long-run incentives to investment and innovation, and how they change if the economy grows and resource stocks change. Regulation affects incentives, so we must distinguish between situations without inadequate intervention and those with optimal policies that address market failures.

Long-run Growth, Capital Accumulation, and Exogenous Technological Change

Capital accumulation allows society to invest in substitutes for natural resources. Individuals choose to invest up to the point where the marginal product of capital equals their required rate of return, which reflects their impatience (utility dis-count rate). They accumulate more capital the more patient they are and the less quickly the returns to capital fall with accumulation. The following basic results emerge in the neoclassical model.

First, substitution of capital for the depleted resource can prevent falling output. But without technological change, output is still likely to fall unless very stringent conditions apply. If substitution of human-made capital for resources is poor, the accumulation of capital cannot prevent production falling in the long run. If substitution and the production elasticity of capital are sufficient, a constant level of production can in principle be sustained in the absence of exogenous technological progress (Solow 1974; Hartwick 1977). Even then, without government intervention individuals who maximize the discounted lifetime utility will not find it optimal to accumulate enough capital to sustain this constant level of income.[11] In other words, while nondecreasing production is feasible, it is not optimal. As we have seen before, the complementarity between resources and capital, together with the diminishing returns with respect to capital, imply that the returns to capital fall when more capital is used per unit of resource use. With the decline in the rate of return, investment falls and output ultimately declines.

Second, with a constant rate of exogenous technological change, growth and capital accumulation can be sustained. Technological change increases the productivity of capital and offsets the fall in returns due to capital–resource substitution. The presence of technological change is not sufficient. If resource–capital substitution is poor, the nature of technological change has to be resource saving, increasing the productivity of capital more than the productivity of the resource.[12] The technological change has to be rapid enough to counteract the fall in returns to capital. Accordingly, the faster capital accumulates and the poorer substitution is, the more rapid the change must be.

Third, ongoing technological change can sustain growth only if environmental policy is rigorous. In particular, in a growing economy, the tax on pollution must increase over time to prevent environmental degradation. Environmental resources are bounded, so pollution must be bounded to ensure that environmental resources are not completely depleted. In contrast, the stock of human-made assets expands continuously in a growing economy, driven by improvements in total factor productivity. With more human-made capital per unit of polluting input, the productivity of polluting inputs rises. To keep firms from increasing pollution, they have to face higher costs for pollution.

Fourth, with capital accumulation and depletion driven by exogenous technological change, preferences affect long-run growth and depletion rates. Lower discount rates reduce the pace at which the resource stock is depleted and speed up long-run growth. Faster technological change also boosts growth.

How to Interpret Capital and Technological Change in the Neoclassical Model

Capital, the key variable in the neoclassical model, is sometimes narrowly understood as machines and hardware. However, we would do more justice to the spirit of the neoclassical approach by interpreting capital in a broader sense, namely, as forgone consumption. Today's consumption is given up for the sake of new assets. Investment not only gives rise to a physically larger mass of machines, but also to better, more efficient machines and organizations, perhaps even to new social attitudes toward waste of energy. The assets allow production of at least the same services with less use of scarce nonrenewable resources. Assets thus comprise not only capital in a narrow sense, but also intangible assets and knowledge; similarly, not consuming implies investment and innovation.

The broad interpretation of capital clarifies the neoclassical assumption of substitution between capital and resources like oil and materials. An expanding capital stock is not necessarily a collection of more and more of the same type of machines. Instead, capital is knowledge frozen in material, not just material; capital embodies the knowledge stock, and this knowledge stock expands in the process of accumulation. With a larger capital stock, production might require less material or lower total energy inputs, because the replacement of old machines by new ones puts into practice new ways to harvest energy, to use materials, and so on. A larger capital stock produces new products and satisfies new wants, which may require less energy and materials. Thus conceptualized, capital does not so much provide the capacity to produce a given physical object as provide the capacity to create valuable things, where the exact nature and physical properties of these things may change.[13] Society can give up consumption in order to create assets that generate more future value and replace resources. The degree of substitutability determines the degree to which this dematerialization is possible.

This broad interpretation of capital brings up some problems, however. The standard neoclassical framework treats investment as a single homogenous activity. It does not explicitly model investment as a joint process of the creation of new knowledge and the embodiment of new knowledge in capital goods.

All investment is still treated as an activity that requires the production and trade of private goods for which there are well-defined markets—perfectly competitive markets even, according to the model. Although this may hold true for equipment and mass-produced machinery, this is less likely to apply to knowledge. As argued above, knowledge has a public good character, is subject to increasing returns, and gives rise to monopolistic markets. Thus the neoclassical model sweeps some important sources of market failures under the carpet.

In the broad interpretation of capital in the neoclassical model, forgone consumption results in not only physical capital accumulation, but also endogenous technological change. Paradoxically, according to the model, endogenous technological change cannot prevent growth from falling due to diminishing returns, while exogenous technological change does exactly the opposite and is introduced in the model to keep growth going. If no exogenous changes in technology took place, and all technological progress resulted from forgone consumption converted into intangible assets, diminishing returns to capital (now including knowledge capital) would cause incentives to accumulate (now including the incentives to innovate) to fall over time. Thus, endogenizing innovation seems to change the role technological change plays in alleviating scarcity limits.

We can imagine three solutions to this paradox. First, the pessimistic view is to simply conclude that scarcity puts limits to growth because innovation is not automatically arriving as manna from heaven; it requires effort, the returns of which inevitably fall with the depletion of resources and the need to reduce pollution. Thus the endogeneity of technological progress undermines one of the central results from the standard neoclassical approach—the power of the price mechanism using its own cornerstone, diminishing returns. As a second solution, recent developments in growth theory have suggested that returns to investment are no longer diminishing if we take into account the accumulation of intangible goods with the character of public goods, like knowledge. This indeed generates ongoing growth if we can abstract from natural resources. However, even with constant returns to the broad concept of capital in production, if production requires nonrenewable resource inputs, growth cannot be sustained unless there is another exogenously growing factor, for example, population growth or technological change (Groth and Schou 2002). Only if production requires only renewable resources, and if society keeps constant the overall stock of them, can growth be sustained with constant returns. In a third and more appealing solution, we treat innovation as a separate activity. That is, the production function of new ideas (the R&D technology) is completely different from the production function of equipment and physical capital goods (Bovenberg and Smulders 1995, 1996; Aghion and Howitt 1998). Then, if resource inputs are not important as an input in R&D, growth can be sustained. It is this approach that we consider next.

Endogenous Technological Change and Endogenous Growth

By explicitly introducing endogenous technological change into the neoclassical model, we can study the incentives for technological change. Technological progress requires considerable investment effort in the form of learning or research and development. Whether innovation is sufficiently fast to make growth sustainable thus depends on innovation opportunities and incentives.

The standard approach to endogenous technological change is the *endogenous growth framework*.[14] This assumes that a third asset is relevant for production: not only the stocks of resource and capital, but also the stock of knowledge. Expansions of the physical capital stock and the knowledge stock require different types of investment. New productive knowledge is created when firms undertake research and development. Knowledge is a nonrival factor of production: it raises the productivity of the capital and resource inputs. The production of new knowledge—the innovation process—requires that some consumption is forgone: workers have to devote their labor effort to research instead of final goods production, or some of the output of the economy serves as an intermediate input (research lab equipment) in R&D. In addition, knowledge is an input in R&D. Current research builds on the achievements of past research.

The key assumption in the endogenous growth framework is that forgoing consumption in order to accumulate productive human-made assets no longer runs into diminishing returns, because current research builds on past research. A higher stock of knowledge makes research so much easier that no matter how large the stock of knowledge grows, the returns to investment in new knowledge remain constant.[15] A society willing to spend enough on R&D can realize a steady rate of technological change sufficient to offset the diminishing returns from capital–resource substitution and sustain long-run growth. If the private returns to innovation are large enough, the economy can grow without bounds as in the standard neoclassical model, but without relying on exogenous technological change.

In most models of endogenous growth and natural resources, the market generates too little innovation, too little growth, and typically suboptimal depletion. Whether depletion is too slow or too fast depends on the income and substitution effects. Inadequate innovation affects the incentives to deplete, as we discussed above in the context of an exogenous change in innovation. In the empirically relevant case, in which income effects dominate intertemporal substitution effects, this results in depletion at a slower pace than in the social optimum. The best policy is to subsidize research and development, which will increase growth. The optimal technology policy can be expected to speed up both depletion and growth as well as innovation: individuals anticipate higher consumption levels and respond with faster depletion to smooth consumption.

In the endogenous growth framework, society can grow without bound and at the same time maintain a stable level of environmental quality. However, there is a trade-off between the rate of growth and environmental quality. Society may prefer low growth, or even constant output, in order to maintain high environmental quality. On the other hand, societies that care little about the future (discount at high rates) may optimally choose to produce at high current levels, deplete resources, and let environmental quality decline to low levels. So, while technological opportunities are less constraining because of the absence of diminishing returns, society's preferences and willingness to take action, its capability to implement resource and technology policies, become much more crucial for scarcity.

Conclusions

This chapter has reviewed several aspects of endogenous technological change and how it affects the tension between growth and scarcity. Substitution and technological changes are the main ways to alleviate scarcity in the neoclassical approach. Since resource substitution is likely to run into diminishing returns, the offsetting force of technological change is necessary to sustain growth.

Technological change does not guarantee a win-win outcome, though. When technological change comes free, it is not necessarily conducive to resource conservation or environmental improvements. Technological change that improves the productivity of resource inputs alleviates scarcity and boosts growth, but it also increases the demand for resources and may raise total depletion or pollution by a rebound effect. The form of technological change is crucial: technological change in abatement technologies, for example, does reduce pollution.

Technological change is to a large extent the result of economic decisions, so the rate and direction of technological change is affected by resource scarcity. Recent developments in endogenous growth theory have changed our understanding of scarcity and growth, because it treats technological change as a costly process, in which innovators trade off costs and expected benefits, and which is subject to market failure and spillovers. Technological improvements necessary for sustainable growth cannot be assumed to continue to arrive without cost. Given market failures, market responses cannot be expected to ensure technological change at a sufficient rate and in the right direction; regulation must be invoked. First, policies should create efficient resource markets and impose marked-based instruments, such that prices correctly reflect scarcity of natural resources. Then profit-maximizing entrepreneurs and innovators will have an incentive to develop cleaner, resource-saving technology. Second, technology policy is needed since the returns of innovation are hard for private investors to secure; compensating innovators for spillovers by means of research subsidies or innovation rewards will more accurately reflect the social value of innovation.

If technological change responds to market signals, policies can induce technological change in response to increased resource scarcity. But increased scarcity of resources may slow the overall rate of endogenous technological change, which could aggravate scarcity. The productivity of human-made capital and other inputs falls if complementary resource inputs decline. As a result, the returns to investment fall, not only with respect to capital investment, but also investment in new technology.

Endogenous technology makes policy important, much more than is suggested by the standard neoclassical approach to scarcity. Finding the exactly right policy is that much harder, and policy mistakes have potentially large adverse effects. Policy not only affects depletion and substitution directly, but may also crowd out innovation or shift technological change in the wrong direction, which in turn affects depletion and substitution. Differences in policy among countries also have more persistent effects: once an economy has adopted and adjusted to certain technologies, it is costly to change to fundamentally different technologies. Countries may become locked in to resource-intensive production structures, a problem if resources are depleted rapidly or demand for environmental amenities rises. Countries that have chosen a growth strategy based on human capital rather than natural capital may be better off in the long run. Such scenarios differ markedly from the standard neoclassical view on growth, in which technology is a public good, freely available for all and producing convergence among countries.

Technological change is essential to sustain growth, especially in the presence of resource scarcity. Technological change has been pervasive and effective in the past. There is no reason to expect this will change as long as human creativity, flexibility, and adaptability, combined with knowledge spillovers, provide us with new ways of production and organization. However, economies cannot flourish until policymakers acknowledge scarcity and market failures, apply innovation incentives to solve natural resource problems, and translate vision into policies for a sustainable economy.

Acknowledgments

The author thanks Cees Withagen for useful comments on an earlier version. The author's research is supported by the Royal Netherlands Academy of Arts and Sciences.

Notes

1. See Dasgupta and Heal (1979) and Withagen (1991) for surveys of the standard neoclassical model and its ramifications.

2. Berndt and Wood (1975) pioneered the estimation of substitution and technological change in energy use. Kemfert (1998) and Kuper and van Soest (2003) provide recent contributions. See Neumayer (2003, 64–65) for a comparison of estimates of substitution elasticities across studies. Jones (2002) summarizes stylized facts on energy use in the U.S. postwar economy.

3. For a classical overview, see Hall (1988). More recent estimates that take into account limited asset market participation are provided by Vissing-Jorgensen (2002).

4. In the sequel we will mainly deal with environmental resources, but most arguments also apply to biological resources.

5. The same will happen when technology does not change, but production factors are accumulated that are complementary to polluting inputs. We discuss factor accumulation in more detail below.

6. Krautkraemer (1985) has introduced amenity values as a motive for resource conservation in the neoclassical model. The level of the optimum long-run resource stock has become known under the "green-golden-rule level." See Beltratti et al. (1995). Smulders (2000) extends the analysis to an endogenously growing economy.

7. The literature started with Grossman and Krueger (1995). Excellent surveys are in Lieb (2003), Ansuategi et al. (1998), and De Bruyn (2000).

8. If technological change increases the productivity of, for example, resource inputs relative to other inputs, economists define technological change to be *biased* toward resources. In terms of costs, technological change is biased to a particular factor if it reduces that factor's share in production costs. Resource-biased technological change implies resource-augmenting technological change if substitution is poor. Neutral (unbiased) technological change increases the productivity of all factors to the same degree. If the elasticity of substitution between factors is unity, no bias in technological change can arise.

9. The hypothesis goes back to Hicks (1932). It was introduced in a growth context by Kennedy (1964) and Samuelson (1965). The approach has been criticized heavily for its lack of clear microeconomic foundations (see Ruttan 2001 for a survey). A new model of induced innovation builds on the microeconomics of technological change and knowledge spillovers as developed in endogenous growth theory (Acemoglu 2002, 2003). The empirics of induced technological change acquired a new impetus by moving to the micro level. Applying a product characteristics approach, Newell et al. (1999) study innovation and substitution at the level of different vintages of energy-using household durable goods and are able to identify the effects of prices and regulation on substitution, the rate of innovation, and the direction of innovation.

10. Jones and Williams (1998).

11. Intervention may change accumulation incentives in such an economy to guarantee "sustainability" of income, defined as a constant income level. These policies are basically policies to stimulate savings and investments, rather than interventions in resource markets. Thus, sustainability policy is different from resource policy. See Pezzey (2004).

12. That is, technological change must be "resource augmenting": it makes the resource effectively more abundant.

13. Notice the difference in aggregation levels. At the level of an individual production process, thermodynamic principles impose limits in terms of output per unit of energy input (Cleveland and Ruth 1997). These limits become less important the higher the level of aggregation: at a macroeconomic level, substitution between processes, goods, or even lifestyles becomes possible.

14. Seminal contributions in this tradition are Romer (1990), Grossman and Helpman (1991), and Aghion and Howitt (1992, 1998).

15. Schou (1999, 2000); Aghion and Howitt (1998, chapter 5); Scholz and Ziemes (1999); Barbier (1999).

References

Acemoglu, D. 2002. Directed Technical Change. *Review of Economic Studies* 69: 781–810.
———. 2003. Labor- and Capital-Augmenting Technical Change. *Journal of European Economic Association* 1: 1–40.

Aghion, P., and P. Howitt. 1992. A Model of Growth through Creative Destruction. *Econometrica* 60: 323–351.

———. 1998. *Endogenous Growth Theory*. Cambridge, MA: MIT Press.

Ansuategi, A., E. Barbier, and C. Perrings. 1998. The Environmental Kuznets Curve. In *Economic Modeling of Sustainable Development: Between Theory and Practice*, edited by J.C.J.M. van den Bergh and M.W. Hofkes. Dordrecht: Kluwer Academic Publishers, 139–164.

Barbier, E.B. 1999. Endogenous Growth and Natural Resource Scarcity. *Environmental and Resource Economics* 14: 51–74.

Beltratti, A., G. Chichilnisky, and G. Heal. 1995. Sustainable Growth and the Green Golden Rule. In *The Economics of Sustainable Development*, edited by I. Goldin and L.A. Winters. Cambridge: Cambridge University Press.

Berndt, E.R., and D.O. Wood. 1975. Technology, Prices and the Derived Demand for Energy. *Review of Economics and Statistics* 57(3): 259–268.

Bovenberg, A.L., and S. Smulders. 1995. Environmental Quality and Pollution-Augmenting Technological Change in a Two-Sector Endogenous Growth Model. *Journal of Public Economics* 57: 369–391.

———. 1996. Transitional Impacts of Environmental Policy in an Endogenous Growth Model. *International Economic Review* 37(4): 861–893.

Cleveland, C.J., and M. Ruth. 1997. When, Where, and by How Much Do Biophysical Limits Constrain the Economic Process? A Survey of Nicholas Georgescu-Roegen's Contribution to Ecological Economics. *Ecological Economics* 22: 203–233.

Copeland, B., and S. Taylor. 2003. *International Trade and the Environment: Theory and Evidence*. Princeton: Princeton University Press.

Dasgupta, P., and G.M. Heal. 1974. The Optimal Depletion of Exhaustible Resources. *Review of Economic Studies* 42 (Symposium): 3–28.

———. 1979. *Economic Theory and Exhaustible Resources*. Cambridge: Cambridge University Press.

De Bruyn, S.M. 2000. *Economic Growth and the Environment: An Empirical Analysis*. Economy and Environment 18. Dordrecht: Kluwer Academic Publishers.

Easterly, W., and R. Levine. 2003. Tropics, Germs, and Crops: How Endowments Influence Economic Development. *Journal of Monetary Economics* 50: 3–39.

Grossman, G.M., and E. Helpman. 1991. *Innovation and Growth in the Global Economy*. Cambridge, MA: MIT Press.

Grossman, G.M., and A.B. Krueger. 1995 Economic Growth and the Environment. *Quarterly Journal of Economics* 112: 353–377.

Groth, C., and P. Schou. 2002. Can Non-Renewable Resources Alleviate the Knife-Edge Character of Endogenous Growth? *Oxford Economic Papers* 54: 386–411.

Hall, R.E. 1988. Intertemporal Substitution in Consumption. *Journal of Political Economy* 96: 339–357.

Hartwick, J.M. 1977. Intergenerational Equity and the Investing of Rents from Exhaustible Resources. *American Economic Review* 67: 972–974.

Hicks, J.R. 1932. *The Theory of Wages*. London: MacMillan.

Jaffe, A.B., and K. Palmer. 1997. Environmental Regulation and Innovation: A Panel Data Study. *Review of Economics and Statistics* 79: 610–619.

Jones, C.I. 2002. *Introduction to Economic Growth*. 2nd edition. New York: W.W. Norton.

Jones, C.I., and J.C. Williams. 1998. Measuring the Social Return to R&D. *Quarterly Journal of Economics* 113: 1119–1135.

Jorgenson, D.W., and B.M. Fraumendi. 1981. Relative Prices and Technical Change. In *Modeling and Measuring Natural Resource Substitution*, edited by E.R. Berndt and B. Field. Cambridge, MA: MIT Press, 17–47.

Kemfert, C. 1998. Estimated Substitution Elasticities of a Nested CES Production Function Approach for Germany. *Energy Economics* 20: 249–264.

Kennedy, C. 1964. Induced Innovation and the Theory of Distribution. *Economic Journal* 74: 541–547.

Krautkraemer, J. 1985. Optimal Growth, Resource Amenities, and the Preservation of Natural Environments. *Review of Economic Studies* 52: 153–170.

Kuper, G.H., and D.P. van Soest. 2003. Path-Dependency and Input Substitution: Implications for Energy Policy Modeling. *Energy Economics* 25(4): 397–407.

Lanjouw, J.O., and A. Mody. 1996. Innovation and the International Diffusion of Environmentally Responsive Technology. *Research Policy* 25: 549–571.

Lieb, C.M. 2002. The Environmental Kuznets Curve and Satiation: A Simple Static Model. *Environment and Development* 7: 429–448.

———. 2003. The Environmental Kuznets Curve: A Survey of the Empirical Literature and of Possible Causes. Discussion paper series, University of Heidelberg, Department of Economics, 391.

Neumayer, E. 2003. *Weak versus Strong Sustainability: Exploring the Limits of Two Opposing Paradigms.* 2nd edition. Cheltenham, UK: Edward Elgar.

Newell, R.G., A.B. Jaffe, and R.N. Stavins. 1999. The Induced Innovation Hypothesis and Energy-Saving Technological Change. *Quarterly Journal of Economics* 114: 941–975.

Pezzey, J.C.V. 2004. Sustainability Policy and Environmental Policy. *Scandinavian Journal of Economics* 106(2): 339–359.

Popp, D. 2001. The Effect of New Technology on Energy Consumption. *Resource and Energy Economics* 23(4): 215–239.

———. 2002. Induced Innovation and Energy Prices. *American Economic Review* 92: 160–180.

Porter, M.E. 1991. America's Green Strategy. *Scientific American* 264: 168.

Romer, P.M. 1990. Endogenous Technological Change. *Journal of Political Economy* 98: S71–S103.

Ruttan, V.W. 2001. *Technology, Growth, and Development: An Induced Innovation Perspective.* New York: Oxford University Press.

Sachs, J.D., and A.M. Warner. 2001. The Curse of Natural Resources. *European Economic Review* 45: 827–838.

Samuelson, P. 1965. A Theory of Induced Innovations along Kennedy-Weisacker Lines. *Review of Economics and Statistics* 47: 444–464.

Scholz, C.M., and G. Ziemes. 1999. Exhaustible Resources, Monopolistic Competition, and Endogenous Growth. *Environmental and Resource Economics* 13: 169–185.

Schou, P. 1999. Endogenous Growth, Nonrenewable Resources, and Environmental Problems. Ph.D. diss., Copenhagen University.

———. 2000. Polluting Nonrenewable Resources and Growth. *Environmental and Resource Economics* 16: 211–227.

Smulders, S. 1998. Technological Change, Economic Growth, and Sustainability. In *Theory and Implementation of Economic Models for Sustainable Development*, edited by J. van den Bergh and M. Hofkes. Dordrecht: Kluwer Academic Publishers, 39–65.

———. 2000. "Economic Growth and Environmental Quality." In *Principles of Environmental Economics*, edited by H. Folmer and L. Gabel. Cheltenham, UK: Edward Elgar, 602–664.

Solow, R.M. 1974. Intergeneration Equity and Exhaustible Resources. *Review of Economic Studies* 41 (Symposium): 29–45.

Stiglitz, J.E. 1974. Growth with Exhaustible Resources: Efficient and Optimal Paths. *Review of Economic Studies* 41 (Symposium): 123–137.

———. 1979. A Neoclassical Analysis of the Economics of Natural Resources. In *Scarcity and Growth Revisited*, edited by V.K. Smith. Baltimore: Johns Hopkins University Press for Resources for the Future.

Vissing-Jorgensen, A. 2002. Limited Asset Market Participation and the Elasticity of Intertemporal Substitution. *Journal of Political Economy* 110(4): 825–853.

Weitzman, M.L. 1997. Sustainability and Technical Progress. *Scandinavian Journal of Economics* 99(1): 1–13.

———. 1999. Pricing the Limits to Growth from Minerals Depletion. *Quarterly Journal of Economics* 114(2): 691–706.

Withagen, C. 1991. Topics in Resource Economics. In *Advanced Lectures in Quantitative Economics*, edited by F. Van der Ploeg. London: Academic Press.

Wright, G., and J. Czelusta. 2002. Exorcizing the Resource Curse: Minerals as a Knowledge Industry, Past and Present. Stanford University Economics Department, Working Paper 02.008.

CHAPTER 9

Evolutionary Analysis of the Relationship between Economic Growth, Environmental Quality, and Resource Scarcity

Jeroen C.J.M. van den Bergh

T HE ANALYSIS OF ECONOMIC growth is dominated by neoclassical aggregate models of exogenous and endogenous growth in equilibrium, as is
the application of growth theory to environmental problems and resource scarcity. Although formal models of economic growth have generated many clear
insights, they suffer from two problems. First, they fail to address some relevant
issues related to growth, because they omit certain elements in their descrip-
tion of reality: processes out of equilibrium; "choice" between multiple equilibria;
and structural changes in the economy. The latter is the more surprising because
economic growth hardly ever occurs without structural change. Growth theory
makes many convenient but erroneous assumptions, so its results are questionable
at best. Representative agents, rational behavior, perfect information, an aggregate
production function, growth in equilibrium, and reversible growth are all debat-
able to say the least. Moreover, under these assumptions certain aspects of policy
disappear from the analysis.

This chapter starts from an alternative set of assumptions offered by evolutionary growth theory, which is part of evolutionary economics. Evolutionary
economics can be characterized in various ways: interaction of innovation and
selection, changing populations of heterogenous agents, the impact of economic
distribution on economic dynamics, and agents characterized by adaptive routines and imitation. Attention is focused on the link between evolutionary growth
theory and scarcity. The analysis of resource scarcity and environmental pollution
in the framework of growth theory during the 1970s occurred entirely within the
domain of standard growth theory. Since the 1980s, growth has been studied from
the perspective of evolutionary theories of growth and technological change. This
has, however, scarcely influenced environmental and resource economics.

Evolutionary growth theory employs micro-level descriptions of populations
of firms characterized by nonequilibrium: differential growth with continuous

interaction between innovation and selection of diversity. This allows for a subtle, realistic, long-run relationship between resource scarcity, environmental conditions, and economic growth. It acknowledges that growth virtually always entails changes in the underlying distributions of technologies and firms. The notion of coevolution is relevant for both historical and future growth analysis in that the economy is seen as adaptive to the environment and vice versa. It will further be shown that (co)evolutionary growth should not be equated to progress. Finally, evolution means that the link between a social optimum and market equilibrium is lost, implying that optimal public policy—focusing on resource exploitation and environmental externalities—receives little attention in evolutionary economics. On the other hand, evolutionary theory is well able to address the dynamic implications of population diversity and distribution, as well as policies for innovation incentives and (altering) selection forces.

This chapter first discusses general characteristics of evolutionary systems and briefly reviews the relevant elements of evolutionary economics. It considers evolutionary analysis in environmental and resource economics, introduces evolutionary growth theories, and compares these with neoclassical growth theories. The core of the chapter is an analysis of the relationship between growth and environment from an evolutionary perspective. The next section studies the question of whether growth can be considered as progress and offers some policy suggestions. The chapter ends by offering some conclusions.

Evolutionary Economics

Evolution, whether genetic or nongenetic—as in economics—involves a number of complementary core elements and processes (similar terms encountered in the biological and economic literature are shown in parentheses):

1. *Diversity (variety, variation):* populations of dissimilar agents, strategies, products, or technologies.
2. *Selection:* processes that reduce diversity.
3. *Innovation (adoption):* processes that generate new diversity.
4. *Inheritance (transmission):* replication through reproduction or copying (imitation); the cause of durability and cumulative processes.
5. *Bounded rationality:* individual and organizational (group) behavior according to adapted or selected habits and routines; imitative, myopic behavior.

Any evolutionary theory has to start with a population. This immediately clarifies an essential difference from traditional microeconomics, where the assumption of a representative agent is central. Contrary to common belief, such a microeconomics is not really as "micro" as is possible. In fact, evolutionary theories are more micro, because they describe populations with behavioral or technological diversity among individuals or firms.

A population approach can be operationalized in three different ways (van den Bergh 2003). One uses aggregate variables, as is common in evolutionary game theory. This assumes that diversity is limited, or can be simplified, to subpopulations, each of which is assumed to be homogenous. Another describes population distributions and changes therein. The third is disaggregate and represents the most thoroughly micro approach. It takes the form of multiagent systems, in which each individual is explicitly described and can be assigned unique features. The agents can operate in a setting of entirely random interactions ("gaseous cloud") or systematic interactions through a network structure or a spatial grid ("lattice"). The traditional multiagent (general equilibrium) and multisector (macroeconomic) models in economics, which are based on complementary and representative agents, are essentially different from multiagent population models.

The fundamental mechanism of any evolutionary process, genetic or otherwise, can be modeled as an accordion. Evolution is powered by opposite forces or causal processes. One is the creation or generation of variation (or variety or diversity), which can be considered a disequilibrating force. The other is selection or reduction of variety, which can be considered an equilibrating and directive force. The result of these opposite forces is similar to the movement of an accordion. Its dynamics depend on the existing diversity and in turn change it. The consequence of a sustained accordion movement is that structure and complexity emerge along a nonequilibrium path of change. This is best illustrated by computer simulations in the field of evolutionary computation and modeling, which show that surprisingly complex structures can be generated from simple models of interactive innovation and selection (Bäck 1996).

Evolutionary change implies that a system has so much diversity that it is extremely unlikely that it will revisit a previous state. In economics this is known as path dependence. In effect, it means that history is introduced. Indeed, a unique and important feature of evolutionary thinking is that it can integrate theory and history. Evolution has become one of the most powerful ideas in science, with the capacity to synthesize a wide range of phenomena (Ayres 1994; Dennett 1995).

Evolutionary economics is very much the legacy of Joseph Schumpeter, the most influential of the early evolutionary economists, who also wrote much about growth-related issues. He questioned the static approach of standard economics and showed a great interest in the dynamics of economies, in particular the capitalist system, in all of his major works. He considered qualitative economic and technological change in a wider context of social change, focusing on the impact of the innovative "entrepreneur" (Schumpeter 1934). Schumpeter described economic (capitalistic) change as "creative destruction"—the result of revolutionary forces from within the economy, which destroy old processes and create new ones. This allowed for discrete or nongradual changes, through clusters of derived innovations following a major invention. These themes were elaborated in his studies of business cycles (long waves). Schumpeter shares with Marx, Mill, and Ricardo the general idea of a final steady state. In Schumpeter's work, this state is characterized by technological progress as the result of carefully planned research under a socialist society. Like Marx, Schumpeter gave thought to the process of change from a capitalist to a socialist economy. Although he realized that discontinuities play a role, he did not assign to them the critical role that they have in Marx's theory. Instead, he believed that political responses would lead to a gradual transition.[1]

Since the 1950s, there has been a slow increase in the number of publications on economic evolution. This can be partly explained by the success of evolutionary biology, the limits of neoclassical economics, and the search for evolutionary underpinnings of optimizing behavior as assumed by neoclassical economics. The most cited work since the 1950s has been that of Nelson and Winter (1982). The three building blocks of their theory of microevolution are organization routines, search behavior, and selection environment. A routine can be considered as the equivalent of the gene in biological evolution, having some durability and being subject to change due to selection. A routine consists of a complex set of skilled individuals. Interactions between them are crucial and depend on earlier contacts (learning, adaptation) and organization-specific "language." Routines create a constancy or continuity in the firm's behavior, manifested as organizational politics, avoiding conflicts, vested interests, financial costs of change, and management control. Change in routines follows two routes—organized search through R&D and nondirected, accidental change due to solving problems in the organization's performance, including replacement of employees. The theory supports bounded rationality as a general model.

Various other, authentically evolutionary approaches have been proposed—perhaps with less impact (so far), but not necessarily less relevant. The most important recent proposal concerning the direction evolutionary economics should follow is Potts (2000). Potts presents an axiomatic foundation of evolutionary economics. In his view, economic systems are complex "hyperstructures," that is, nested sets of connections among components. Economic change and growth of knowledge are in essence a process of changes in connections. New technologies, products, firms, sectors, and spatial structures arise that are more complex than the old ones. Growth of firms and economies is the creation of a more complex organization or new connections, as well as the grouping of those connections. In line with the idea of changing connections, Potts calls for a new microeconomics based on discrete, combinatorial mathematics, for instance graph theory.

Neo-Schumpeterian theories of technological change currently dominate the evolutionary approach in economics (Dosi et al. 1988; Metcalfe 1998). They state that innovation causes asymmetry in technology among firms, sectors, and countries, leading to exchange and trade. Comparative advantages change with innovation and diffusion. Trade itself stimulates diffusion of knowledge. In addition, technological change affects the division of labor, the organization of intra-firm and interfirm relationships, and thus the industrial structure and patterns of intermediate deliveries. Within the neo-Schumpeterian literature on technological evolution, the notion of path dependence has received much attention (Arthur 1989). Path dependence is a result of increasing returns, which may be due to learning by using, bandwagon demand side effects (imitation), network externalities (e.g., telecommunication), informational increasing returns (the more adopted, the better known), and technological interrelatedness or complementarity. One consequence of increasing returns, or path dependence to one of multiple potential equilibria, is that inefficient equilibria can arise and an inefficient technology can become locked-in.[2]

Evolutionary game theory is also becoming more influential. It is also known as "equilibrium selection theory," because it solves the problem of multiple Nash equilibria common in nonlinear economic equilibrium models (Friedman 1998). Evolutionary game theory is analytical. It adopts an aggregate approach to describing evolutionary economic phenomena. Usually, two groups are distinguished, reflecting minimal diversity. Groups are considered to consist of identical individuals; in this sense, evolutionary game theory really is a compromise between representative agent and full-fledged evolutionary models. Interactions among individuals and between individuals and their environment are usually described through an aggregate replicator equation. This formalizes the idea that individuals with above-average fitness will increase their proportion of the population. Evolutionary games give rise to asymptotic equilibria, because no process of regular generation of diversity is assumed. As a result, selection completely dominates system dynamics. In other words, there is none of the interaction between innovation and selection that characterizes evolution. A more suitable name for this approach would be "selection game theory."

The Intersection between Evolutionary and Environmental and Resource Economics

Economic evolutionary theories are incomplete because they neglect environmental dimensions. Important phases of economic history cannot be well understood without resorting to environmental or resource factors. Environmental and resource economics has been dominated by equilibrium theories in which individuals are assumed to maximize utility or profits, markets are clear, and any changes are aggregate and mechanistic. This holds for all three components: environment policy theory, monetary valuation, and resource analysis. The recent adoption of the notion of "sustainable development" has meant a more explicit long-term focus, which invites researchers to apply evolutionary perspectives, notably to the complex role of structural and technological change in the conflict between economic growth and environmental preservation (Mulder and van den Bergh 2001; Gowdy 1999). Norgaard (1984) proposed to use the biological notion of coevolution as a joint, interactive evolution of nature, economy, technology, norms, policies, and other institutional arrangements. Gowdy (1994) combines the notion of coevolution with macroevolutionary elements, noting that economic evolution is a process at multiple scales, a view consistent with hierarchical approaches to economic evolution.

Recently, Munro (1997) has added evolutionary elements to the standard problem of renewable resource harvesting. Harvesting affects not only the quantity of the resource but also its quality, or composition in genetic terms. Examples can be found in agriculture (monocultures, the use of pesticides and herbicides), fisheries (mesh size, season of fishing), ecosystem management (control of groundwater level, fire protection), and health care (use of antibiotics). The genetic-selective effects of resource use and habitat destruction provide a link to biodiversity loss. Munro formulates a dynamic optimization problem based on the theory that the use of insecticides raises the fitness of resistant insects relative to their susceptible competitors. The optimal use of insecticide is influenced by the evolutionary-selective dynamics of the system. Compared with this, the traditional optimal plan, which neglects evolution, myopically prescribes too high a level of pesticide use.

Environmental economics has paid much attention to the risk of overexploitation of common property or common-pool resources, such as fisheries. As in open access, with which common property is often confused, the risk of overexploitation is also serious, depending on the type of common property regime in force. A fundamental choice is whether to respond to resource conflicts and overuse with strict government policies or to rely on endogenous formation of use regimes. An evolutionary perspective has been used to analyze the latter, given the premise that such regimes are supported by enough individuals, or in other words, that a single norm evolves. Much of this literature suggests that externally imposed rules and monitoring can reduce and destabilize cooperation or even completely destroy it. Instead, it is preferable to have a norm supported by communication among the resource users. External regulation is desirable only if an effective system of monitoring and sanctions can be implemented. However, self-organization in its most fundamental and general form is probably still not entirely understood. For instance, the size of the respective group seems to be important, but it is not clear what determines a critical size for norms or institutions to arise. Instability in the evolutionary equilibrium can arise when certain parameters change (e.g., the resource price), or rules are implemented by an external regulator. In the latter case, norms may erode, ultimately leading to resource extinction. Equilibrium can also breakdown when sanctions decrease, harvesting technology becomes more productive (technological progress), or the price of the resource increases (Ostrom 1990). These issues have been examined using a wide range of methods, including analyses based on evolutionary game theory (Sethi and Somanathan 1996; Noailly et al. 2003), laboratory experiments, and empirical field studies.

The examples given above might suggest that environmental and resource economics have incorporated evolutionary elements. But these examples are exceptions; generally, environmental economics has neglected evolutionary issues, and evolutionary economics has neglected environmental issues (van den Bergh and Gowdy 2000).

Evolutionary Growth Theory

At the base of an evolutionary theory of economic growth is the notion of a population of heterogenous firms. This gives rise to differential growth, which can be seen as a change in the frequencies of all possible individual characteristics. Nelson and Winter (1982, part 4, chapter 9) developed the first formal evolutionary model of economic growth, which is compared with Solow's descriptive growth model from 1957. Its purpose is to explain patterns of aggregate outputs, inputs, and factor prices. Changes in the state of a sector follow probability rules, modeled as a Markov process with time-dependent probabilities, that depend on search behavior, imitation, investments, entry, and selection. If firms make sufficient profits, then they do not search or imitate others, otherwise they do. Search is local, implying small improvements or deviations from the present production technique. Imitation is based on either the average or the best practice. The output generated qualitatively resembles the real data used by Solow. Note that this evolutionary growth theory is based on the evolutionary theory of firms and industry structure also proposed by Nelson and Winter and consisting of routines, search, and selection.

A central concept in neoclassical (exogenous and endogenous) growth theory is the aggregate production function. Nelson and Winter argue that "movements along the production function into previously inexperienced regions—the conceptual core of the neoclassical explanation of growth—must be rejected as a theoretical concept." Of course, the Cambridge capital debate has already assessed that it is a theoretical construct that gives rise to the internal inconsistency of growth theory (the implications for environmental economics are discussed in van den Bergh 1999). Neither single firms nor the aggregate of all firms can move along an aggregate and continuous production function, because they possess only information or knowledge about a limited and discreet number of production techniques. This idea is also recognized by the neo-Austrian approach, formalized in an activity analysis model (Faber and Proops 1990). The conclusion is that an aggregate production function, a standard and necessary element of neoclassical growth theory, is an artifact with no clear link to reality. It is certainly not in line with microfoundations (van den Bergh and Gowdy 2003). Instead, evolutionary theories propose, as an alternative to an aggregate production function, a diversity of production relationships at the level of individual firms.

Nelson and Winter criticize growth accounting for explaining only about 20 percent of productivity growth, based on movements along an aggregate production function due to factor input changes, and leaving 80 percent as unexplained residual, often referred to as "technological change," although sometimes partly attributed to environmental and resource factors. Instead, Nelson and Winter's approach integrates the micro and macro aspects of technology and its change over time. It yields results consistent not only with firms' decisionmaking (routines, search), but also with empirical observations, such as aggregate data on factor level and efficiency features across sectors, and patterns of innovation and diffusion.

More recently, other formal evolutionary models of growth have been proposed (Conlisk 1989; Silverberg et al. 1988). Conlisk works with a probability distribution of productivity of firms. The growth rate can be analytically derived as being dependent on the rate of diffusion of innovations and the size of innovations as indicated by the standard error of the productivity probability distribution. Silverberg et al. have proposed an evolutionary growth model that starts from the Goodwin model, which revolves around a formalization of the Philips curve, the relation between wage change and unemployment level: the higher employment is, the higher the wage increase (inflation). Modeling a large population of firms and their behavior in terms of fixed rules generates the industry dynamics. An important behavior rule is that new capital follows from profit accumulation, where profit is redistributed so that more profitable types of capital accumulate more quickly. This can be likened to selection—in that a technique with a relatively high fitness spreads quickly—combined with a growing "population of technologies" through accumulation. In order to complete the evolutionary dimensions of the model, selection is complemented by a mechanism of innovation. The introduction of new firms and technologies in the economy follows from firms undertaking R&D to improve labor productivity, the outcome of which is stochastic (a Poisson process). The stochastic character is a way to reflect, in the form of surprises and ignorance, uncertainty associated with innovation. Spillovers, through which a firm can profit from other firms' R&D, are taken into account as economy-wide R&D. Firms can employ two strategies for innovation: mutation or imitation. The probability of imitation depends on the gap between the firms' own profit rate and the maximum profit rate in the population. This follows the general model of innovation and imitation as developed by Iwai (1984).

Empirical research on diversity has centered on statistical analysis of country differences (Fagerberg 1988), using input measures like R&D expenditures and output measures like the rate of patenting. By combining these indicators with levels of productivity (income per capita), information clusters of countries can be identified. R&D and patenting are weakly correlated with productivity, R&D does not guarantee successful patenting, and growth rates can be inversely related to levels of productivity in the same period. The latter suggests that technological gaps are closed through imitation. The general Schumpeterian nonequilibrium approach emphasizes the interaction between opposing forces, consistent with the accordion model: innovation that increases technological differences among countries; and imitation and diffusion that reduce such differences.

Evolutionary versus Endogenous Growth Theory

Both neoclassical and evolutionary theories endogenize technological change (R&D) by using outlays on R&D as a core variable. Neoclassical theory defines R&D at the aggregate level. Evolutionary theory derives production as well as technological change from the population of firms, thus explicitly modeling R&D as being undertaken inside productive firms. Neoclassical models also depend on an aggregate production function, thus reflecting macro-level theories, whereas evolutionary models start from firm populations, thus really reflecting micro-level theories.

With the population approach, evolutionary models can address behavioral and technological diversity or heterogeneity, whereas in neoclassical theories representative, identical agents are assumed. The latter is implicit in the argument that micro-level production functions can be replicated, resulting in constant returns to scale at an aggregate level (in the absence of positive R&D externalities).

Evolutionary models further assume bounded rationality, usually in the form of routines and learning through imitation. Neoclassical models assume, by definition, individual rationality (marginal decision rules) and social (intertemporal) optimality. This explains neoclassical growth theory's concentration on equilibrium growth paths, as opposed to the nonequilibrium features of evolutionary growth. Aghion and Howitt (1992) incorporate some elements of heterogeneity and destructive or vertical innovation—creative destruction à la Schumpeter—in a neoclassical type of model, but maintain the assumption of the rational agent. Mulder et al. (2001) refer to this as a "neo-classical Schumpeterian approach."

Both theories can address uncertainty and irreversibility, although this is more common in evolutionary models. Evolutionary theories incorporate a particular type of irreversibility: namely, path dependence (see above). Neoclassical growth models, at their macro-level of aggregation, cannot address path dependence, because it requires the use of a population model in which the distribution of characteristics follows a historical path along which the distribution of characteristics (firms, technologies) irreversibly changes. Stochastic elements are common elements of evolutionary models, notably to specify the timing of innovations.

Neoclassical endogenous growth theory focuses on public externalities in technological innovation through the public good nature of knowledge and technology. In contrast, evolutionary theory focuses on the barriers and delays in diffusion of innovations, as well as the imperfect nature of replication and diffusion. Imperfect imitation itself is a cause of innovation.

As illustrated by Nelson and Winter (1982) and Conlisk (1989), evolutionary models can generate patterns close to those of neoclassical models with particular assumptions of technological change. However, evolutionary growth models can generate other patterns as well as include more phenomena subject to public policy, such as diffusion (imitation) rates, firm-specific innovation factors, selection forces, and lock-in. Therefore, they can provide information about a wider set of policy instruments.

These insights are consistent with the general idea that evolution does not always imply growth, and vice versa. To understand the former, consider an evolutionary process in which the diversity of firms changes but the total output (in monetary or physical terms) is constant or even falling. The latter is illustrated by a (hypothetical) perfect replication of all productive activities in an economy, as is a common assumption underlying endogenous growth models.

With no provision for structural change, neoclassical growth theory cannot address time horizons beyond several decades. Stiglitz (1997) has suggested a time horizon on the order of 60 years, far below that of long waves (60 years is about the periodicity of the Kondratieff cycle). Evolutionary growth theory, in contrast, can address longer time horizons.

In summary, the most important feature of evolutionary growth models is their ability to track structural changes underlying growth by analyzing "differential growth." It is fair to say, however, that several of their "promises" are still to be delivered, and evolutionary growth analyses still suffer from ad hoc specifications or a lack of standards.

Evolutionary Growth, Environmental Quality, and Resource Scarcity

Surprisingly, the debate on growth versus environment and the recent related literature on sustainable development have neglected evolutionary considerations (van den Bergh and Hofkes 1998). The dominant literature in economics on sustainable development focuses on deterministic equilibrium growth theory, in which development is reduced to a nonhistorical and reversible process characterized by the accumulation of a one-dimensional capital stock (Toman et al. 1995).

In neoclassical economics, steady states and equilibria dominate. John Stuart Mill introduced the concept of a "stationary state economy," which was later adapted to an environmental and resource context by Daly (1977). Evolutionary theory, however, suggests that sustainable development as a stationary state is unrealistic. Selection and innovation will irreversibly change the structure of the economy at all levels: processes, products, firms, individuals, groups, and regions. "Weak sustainability" and "sustainable growth," allowing for substitution between economic capital and "natural capital," may be flawed guides to policy when uncertainty, irreversibility, and coevolution are taken into consideration (Ayres et al. 2001).

As argued in the previous section, evolutionary theory can extend the time horizon of analysis beyond decades and even centuries, as seems to be required by sustainable development. A distant time horizon is especially valuable for research on climate change and biodiversity loss, as these are bound to significantly affect both natural and cultural-economic evolution.

Not surprisingly, climate change research is one of the few areas where (optimal) growth models have been applied (Nordhaus 1994), drawing considerable criticism (e.g., Demeritt and Rothman 1999; Azar 1998). The issues of uncertainty and irreversibility have been addressed in the traditional economic growth theory context by Kolstadt (1994), who added stochastic elements to Nordhaus's DICE (Dynamic Integrated Climate Economy) model. Economic irreversibility due to overinvestment in greenhouse gas abatement techniques is more worrisome than irreversibility of natural processes like greenhouse gas accumulation, climate change, and ecological impacts. This is understandable given the focus on economic efficiency of economic growth in a very simple, aggregate economy and the neglect of uncertainty caused by future economic development.

A few studies have pursued evolutionary modeling in this area. Janssen (1998) and Janssen and de Vries (1998) have incorporated evolutionary elements in climate modeling by allowing adaptive agents to change their behavioral strategies in response to persistent changes in global climate, represented by the global mean temperature of the atmosphere. These strategies include "hierarchist"—complete control; "individualist"—adaptive management; and "egalitarian"—preventive management. The distribution of these strategies among the population of agents (voters in a democracy) changes according to a selection process modeled as a replicator equation based on an agent's fitness. This is a function of the difference between expected temperature change and actual temperature change. In other words, when persistent changes cannot be made consistent with expectations, the agent's strategy adapts. This approach therefore tries to address the lack of complete and correct understanding of climate issues.

Faber and Proops (1990) propose a neo-Austrian approach with evolutionary elements, to emphasize the role of time. They allow for irreversibility of changes in the sector structure of the economy, for uncertainty and novelty, and for a teleological detailed sequence of production activities, known as "roundaboutness." The long-term relation between environment, technology, and development is characterized by three principles:

- The use of nonrenewable natural resources is irreversible in time, so a technology based on this must ultimately cease to be viable.
- Inventions and subsequent innovations lead to both more efficient use of currently used resources and substitution with resources previously not used.
- Innovation requires that a certain stock of capital goods with certain characteristics be built up.

Faber and Proops construct a multisector model with the production side formulated in terms of activity analysis, which allows study of the effect of invention and innovation on the transition from simple to complex production activities. Complex activities use multiple technologies. For instance, food production has evolved from agriculture with labor, through agriculture with labor and capital, to a large food-processing industry with many intermediate deliveries. This process can be extended with the technology effects of resource scarcity as indicated above. It can then simulate economic and environmental history from a pre-industrial agricultural society to an industrial society using fossil fuels and capital. It can combine continuous change in technological efficiency with discrete jumps in the number of sectors and interdependencies among sectors.

An important question in the context of growth is whether technological innovation is subject to diminishing returns to scale. If technological change retrieved innovations from a limited set of potential innovations, then it would be subject to diminishing returns. But this seems doubtful in the view of Potts (2000), who observes that evolution means additional connections and higher levels in systems. Hence, there does not seem to be a scarcity of innovations. Moreover, various strategies, such as learning, enlarging the scale of activity, opening new markets, or seeking new applications, can counter diminishing returns. In addition, market mechanisms and profit-seeking help solve problems of diminishing returns. When marginal returns from additional innovation start to fall rapidly, firms will shift to new lines of R&D (Nelson and Winter 1982, *258*), be selected against (exit), or be taken over.

Some authors have argued that we already possess the technical knowledge to increase the efficiency of material and energy use by a factor of 4 to 10. Nevertheless, larger system changes seem necessary, and these cannot be framed merely as design issues. Ehrlich et al. (1999) scrutinize the growth–optimist view that knowledge and technology increases will resolve environmental problems almost automatically. Horgan (1996) and Maddox (1998) present opposite views about the "knowledge explosion," with Horgan claiming that the rate of important scientific discoveries is decreasing and Maddox representing an optimistic perspective. More people than ever are doing basic and applied research and diffusing innovations through journals and the Internet, creating new combinations or connections. But Ehrlich et al. argue that much of it is disinformation—inaccurate or superfluous—and that information is lost in the form of biological and cultural diversity.

Economic and Environmental History from a Coevolutionary Angle

An earlier section argued that evolutionary economics allows us to link theory to history. This section considers such a link in the context of growth and environment.

An initial model of long-run historical change and environmental degradation may represent important socioeconomic transitions in human history, such as from hunting and gathering to agriculture to industrial societies. Some have argued that these are consistent with the evolutionary theory of punctuated equilibrium, although so far this is no more than a loose analogy (Somit and Peterson 1989; Gowdy 1994).

Another notion, coevolution, reflects an integration of elements from ecology and evolutionary biology. Although initially used at the level of species interactions, coevolution has since been invoked to model a variety of interactions:

biological–cultural, ecological–economic, production–consumption, technology–preferences, and human genetic–cultural (Norgaard 1984; Durham 1991; Gowdy 1994; van den Bergh and Stagl 2003). An interesting typology of evolution comes from Durham (1991) in the context of genetic–cultural interactions:

- *Genetic mediation:* Genetic changes affect cultural evolution.
- *Cultural mediation:* Cultural changes affect genetic evolution.
- *Enhancement:* Cultural change reinforces natural evolution.
- *Opposition:* Cultural change goes against natural evolution.
- *Neutrality:* Cultural change is independent of biological evolution or selection.

According to Wilson (1998, *128*), "The quicker the pace of cultural evolution, the looser the connection between genes and culture, although the connection is never completely broken." Nevertheless, it is difficult to prove that cultural change is independent from genes, since the indirect effects of certain aspects of culture on a population level cannot be easily traced empirically.

The above classification might be extended to other types of coevolution, including the interaction of evolutionary economic and ecological systems. But coevolution is often used in a loose manner, without including aspects of populations and diversity.

Georgescu-Roegen (1971) identifies three "Promethean" innovations that significantly altered the relationship between humans and their natural environment: fire, agriculture, and the steam engine. It has been suggested that the invention of fire served to lengthen the day and stimulated late evening communication among humans, thus contributing to sociocultural evolution. This process accelerated after the last Ice Age (about 13,000 years ago), when the development of sedentary agriculture (the "Neolithic Revolution") led to division of labor and specialization. Other major inventions, or "macromutations" (Mokyr 1990), include the windmill, the mechanical clock, the printing press, the casting of iron, the combustion engine, the airplane, and the Green Revolution in agriculture.

Environmental factors may have influenced crucial changes during the sociocultural history of humankind: local and global climate, diversity of soil conditions, scarcity of fuels (notably fuelwood), and available local plants and animals with sufficient concentrations of proteins, carbohydrates, fats, and vitamins. Diamond (1997) summarizes the literature that supports the theory that climatic change and the availability of animal and plant species stimulated early domestication and thus agriculture and settlements. He emphasizes that sufficient diversity of agricultural experimentation was possible only on continents whose major axis was oriented east-west, as this would allow for the spread of agricultural technologies among regions with similar climates. This is an important reason for the early "economic success" of Eurasia. Diamond's theory explicitly relates early economic development to a combination of environmental resource and geographical factors.

Wilkinson (1973) has developed an ecological theory of economic development in which he relates the industrial revolution to natural resource factors. He recognizes a number of human strategies to respond to resource scarcity, such as using new techniques, exploration of new resources, product innovation, and migration. Wilkinson superimposes an environmental perspective on the origins of the industrial revolution at the end of the eighteenth century. Agriculture and the use of fuelwood in iron smelting led to a loss of forest cover in England. A shortage of wood, reflected in a higher price, stimulated the early use of coal. Coal mining first occurred on outcrops at the surface, but soon shifted to deep mining. For this purpose groundwater needed to be pumped out, which meant the first serious application of the steam engine. Large-scale use led to the refinement of the steam engine, which in turn stimulated various spin-offs, notably in the textile industry and transport (ships, locomotives).

Galor and Moav (2002) argue that "the struggle for survival that had characterized most of human existence generated an evolutionary advantage to human traits that were complementary to the growth process, triggering the take-off from an epoch of stagnation to sustained economic growth." This view fits in the "enhancement mode" of Durham's coevolution discussed above. At first one might think that unlike genetic evolution of certain physical features that depend on variations of a single or a few genes (lactose and gluten tolerance, sickle cell trait), the interaction between human genetic evolution and economic growth finds little support in evolutionary biology and theories of cultural evolution. The evolution of human behavior involves so many genes that its timescale does not match that of economic growth. In particular, Galor and Moav's view seems to overlook the fact that economic growth is a phenomenon that arose long after *Homo sapiens* had evolved (at least several hundred thousand years ago), and even much later than the rise of agriculture (about 13,000 years ago). Significant economic growth did not begin until the end of the Middle Ages, and sustained growth not until the industrial revolution, some 300 years ago.

Nevertheless, selection (and possibly recombination) effects may have changed the distribution of certain parental care characteristics, notably the trade-off between quantity of offspring and quality of parental care. In modern economic growth nomenclature, such quality improvements were a type of investment in human capital. In particular, the emergence of the smaller family after the rise of agriculture may have played an important role. Hitherto, larger groups built around one or more extended families dominated human evolution. Galor and Moav argue that organization around smaller families fostered a parental investment in quality of offspring, which might, for instance, involve education. Together with a large communicating population, this led to technological innovation, which fed the industrial revolution. In other words, the authors propose an "endogenous evolutionary theory" of the industrial revolution. Selection pressure was effective during the preceding "Malthusian era" because the majority of people were living on a subsistence consumption level.

The authors cannot rule out that the change in parental care evolved culturally rather than genetically. This theory needs to be tested empirically, probably a difficult if not impossible task. But perhaps this is not a real problem, because the theory works in quite a similar way for both cultural and genetic selection and may even be formulated to include both. Finally, since the industrial revolution, the evolutionary incentives have changed through institutionalized educational systems and requirements, as well as through incomes and consumption levels far exceeding subsistence levels. As a result, a new evolutionary regime applies nowadays, at least in the developed part of the world.

Table 9-1. *Environmental and Resource Aspects of Long Waves*

Phase	Key Resources	Main Environmental Impacts
Hunters and gatherers	Wild animals and plants	Forest fires
Early agriculture	Solar energy	Soil erosion
Late Middle Ages	Wind, water	Local desiccation and water pollution
Early industrial revolution	Coal	Urban pollution
Steam power and railways	Coal	Factory pollution of water and air, large-scale infrastructure
Mass production	Oil, synthetics, heavy metals, fertilizers	Factory and car-related (noise, exhaust pollution, road infrastructure), toxic substances, acid rain
Second half of twentieth century	Oil, gas, heavy metals, tropical wood	Biodiversity loss, global warming
Future	Genetic resources, water ?	Genetic pollution, climate change, large-scale extinctions?

Besides growth trends, a complete view of macrohistory involves cycles or long waves. Long waves are caused by major shifts in methodology due to fundamental advances in science. A rough classification of waves since the industrial revolution is shown in Table 9-1. Long waves have been accompanied by a number of changes: the average size of firms has increased; the research (R&D) and innovation process has advanced from firm to international levels; firm interactions and industry structure have altered; and new key resources and related production sectors have appeared. In addition, each period has its peculiar environmental impacts.

Progress and Policy in an Evolutionary Growth Context

Whether evolution is identical to progress has no simple answer, in part because evolutionary progress has been defined in many different ways (Gowdy 1994, chapter 8; Gould 1988; Maynard Smith and Szathmáry 1995):

- *Increasing diversity:* Diversity is often considered to entail evolutionary potential or adaptive capacity in the face of environmental change.
- *Increasing complexity:* This can apply to the number of components, the number of connections among components, and the levels of nesting of such connections (Potts 2000).
- *New ways of transmitting information:* Human communication has passed through numerous phases, from speech to writing to the Internet. This has gone along with a larger population of communicating individuals.
- *More extended division of labor:* One aspect of the increase in complexity is the extended division of labor, in natural as well as socioeconomic evolution.
- *Population growth:* From an evolutionary perspective, a species is successful if it dominates competitors, meaning dominance in ecosystems and control of its direct environment. This often goes along with growth in the size of population(s).
- *Increasing efficiency of energy capture or transformation:* Both in economic and biological systems, evolution can be related to energy processes (Buenstorf 2000).

From an ecological-evolutionary perspective, a rise in energy efficiency means less scarcity and less selection pressure, thus creating opportunities for (population) growth. Schneider and Kay (1994) state that open natural and economic systems tend to evolve into more complex arrangements, so as to reduce energy degradation and dissipation. This involves more energy capture, more cycling of energy and material, more complex structure, more energy stocking (biomass), and more diversity.

Maynard Smith and Szathmáry (1995) suggest that evolutionary history is better depicted as a branching tree rather than progress on a linear scale. There are indeed many reasons why evolution does not lead to progress (extending Campbell 1996, *433*):

- Selection is a local search process, which leads to a local, not global, optimum.
- Organisms are locked into historical constraints. In economics, this is treated under the headings of path dependence and lock-in (Arthur 1989).
- Adaptations are often compromises stimulated by a multitude of selection forces.
- Not all evolution is adaptive: randomness, (molecular) drift, coincidental founder effects, and so on, all play an important role. In addition, macroevolution creates boundary conditions for adaptation and may destroy outcomes of evolution and, in a way, set time back ("initialize").
- Coevolution means adaptation to an adaptive environment. All straightforward notions of static or dynamic optimization are lost; in the adaptive landscape metaphor, the landscape changes underneath adaptive agents.

Sen (1993) notes that evolution as improving species does not imply improving the welfare or quality of life of each individual organism. Fitness is not a useful criterion for human progress, as it does not imply a happier or more pleasant life. Moreover, evolution as continuous change in diversity implies that inequality will arise again and again. Distributional change and inequality are inherent to evolution. Repeated selection for fitness implies that populations and species are continually stimulated to improve their fitness, since otherwise they are taken over by others (known as the "Red Queen hypothesis"; Strickberger 1996, *511*). Welfare beyond basic needs is to a large extent relative, dependent upon the income and other qualities of individuals in a reference group. Without significantly changing the distribution of these qualities among individuals, economic growth is not necessarily equivalent to progress.

Policy Issues

A fundamental consequence of evolutionary features like bounded rationality, nonequilibrium, and path dependence is that the normative part of neoclassical economics no longer holds. In particular, the correspondence of market equilibrium and the social welfare optimum (Pareto efficiency), formalized in the two fundamental theorems of welfare economics, is lost. It is impossible to formulate an ideal blueprint of economic reality, to be implemented through planning or market approaches, since equilibrium theory on its own does not lead to a preference for either. Bounded rationality or alternative models of individual behavior lead to various policies that deviate from the standard economic theory of environmental policy (van den Bergh et al. 2000).

Important questions in the context of evolutionary analysis of growth are: How do regime shifts occur, and how can they be stimulated? How can we avoid lock-in of inefficient or undesirable technologies, or once this has occurred, how do we unlock them? Undesirable policies cannot be reversed by "correcting prices." To do so requires a combination of strategies, among them,

- Reduce policy uncertainty.
- Set clear overall goals.
- Correct selective pressures.
- Create semiprotected niches.
- Stimulate pathway technologies.
- Stimulate diversity of R&D.
- Stimulate complementary technologies.
- Encourage technologies with flexible design and multiple options.
- Communicate with stakeholders to create a broad basis for learning and selection.

Setting general goals and policies, such as the Zero Emission Mandate of California, provides a much stronger incentive for innovation than traditional environmental policy. Given the high degree of uncertainty faced by innovators, policymaking should create clear long-term goals and contexts, including at the global level.

Evolutionary reasoning indicates that, to assure adaptive potential in the face of a changing environment, diversity at all levels—firms, technology, knowledge, R&D efforts, and "schools" in science—should be fostered. Fisher's theorem is worth mentioning here: "The greater the genetic variability upon which selection for fitness may act, the greater the expected improvement in fitness" (Strickberger 1996, *510*). This also implies that variability will increase, that is, variability itself is selected. Focusing on a single best-available-technology is risky—knowledge of potential impacts is always incomplete, and other, preferable technologies may be overlooked.

Evolutionary policy implications often overlap and complement, rather than conflict with, traditional policy implications. For instance, price-based instruments, intent on "dynamic efficiency," are insufficient. Of course, if prices do not reflect positive or negative externalities, too little R&D will be undertaken. However, much more than price policy is needed to guide R&D. The tension between appropriability of the rewards of inventions and diffusion of inventions leads to their patentability. Avoiding early lock-in of technologies with uncertain social and environmental effects requires, in addition to the above measures, policies to stimulate fair competition. Market structure has received much attention, both in neoclassical and evolutionary theory. Both theories recognize that large firms with market power are essential to support R&D at the scale necessary for some types of innovation. The combined need for sufficient appropriability (market power) and diversity (competition) of R&D leads to a preference for an oligopolistic (supply side) market.[3] The liberalization of energy markets, currently pursued by many countries, may be inconsistent with this view, and may in fact slow down innovation in renewable energy technologies.

Beyond a certain scale, governments have to take control of R&D, through universities. This makes the link between R&D and profit making too indirect or uncertain. Basic university research provides the basis of major technological changes, such as pathway technologies (macromutations, long waves), and can help avoid diminishing returns. Relevant innovations from an environmental perspective include decentralized energy production based on renewables (solar and wind energy); precision-biological agriculture and genetic technology; low pressure/temperature catalytic chemistry; nanotechnology (dematerialization, waste and emission reduction); and battery electric vehicles. Social or organizational innovations, such as carpools or mixed car–public transport systems, may need governmental support.[4] Pathway technologies deserve more attention; for instance, energy storage supports renewable energy use, solutions to electricity peak demand, and zero emissions car technology.

Conclusions

Evolutionary growth theory cannot be developed by simply incorporating new elements into existing growth theory. It requires a completely different set of assumptions. Exogenous and endogenous neoclassical growth theories are really macrotheories that lack explicit microrelationships. The aggregation problem means that there is no unique mapping from micro- to macro-level relationships. This is illustrated by the specification of an aggregate production function and an aggregate cumulative innovation indicator. The evolutionary perspective, on the other hand, starts from micro-level descriptions of populations of firms that work according to routines, search, and selection. This results in nonequilibrium—differential growth with continuous interaction between innovation and selection of diversity. This in turn leads to a more intricate, more long-term relationship among resource scarcity, environmental conditions, and economic growth than that described by neoclassical growth theory. This has several implications:

- Growth is virtually always based on underlying structural change, at the level of changing distributions of technologies and firms. The long-term relationship between the economy and the environment needs to explicitly account for such structural changes, as these can lead to new patterns and interactions that offer solutions to pressing environmental and resource problems that impede sustainable growth.
- Scarcity of energy or material resources and lack of environmental regulation directly affect the distribution of firm and technology characteristics and indirectly aggregate economic activity. Since economic agents in evolutionary theory are characterized by bounded rationality, they will inefficiently use resources and environmental opportunities, as well as opportunities to substitute for scarce resources. Environmental regulation will not lead to optimal social welfare. Transitions from inefficient or outdated to new technologies are hampered by technological and organizational lock-in. All in all, the evolutionary perspective on environmental and resource limits to growth is much less optimistic than one based on neoclassical growth-cum-environment analysis.

- If technological change is retrieving innovations from a limited set of potential innovations, then it is subject to diminishing returns. If, however, evolution is seen as "additional connections" or more complex technologies and organizations, then there is no obvious limit to innovations. However, more complex systems may become less stable and more difficult to manage. Moreover, they may require more and more education of inventors and technicians, who are limited by an individual lifespan. Fruitful innovations may become increasingly scarce, and innovation may slow down.

- Major historical transitions, such as the rise of agriculture and the industrial revolution, have been influenced by environmental and resource conditions. Coevolution is relevant for both historical and future growth analysis. Both the economy and the environment consist of diverse elements that change through innovation and selection. They are, moreover, interactive: the economy is adaptive to the environment and vice versa.

- Evolutionary growth models can explain phenomena subject to public policy, but outside the realm of neoclassical growth models. Examples are diffusion (imitation) rates, firm-specific innovation factors, selection forces, and lock-in. As a result, such evolutionary models can support a broader range of policies than are traditionally examined within neoclassical (exogenous and endogenous) growth theory.

- Evolutionary growth is not identical with progress. Coevolution, local adaptation, and path dependence, among other things, suggest that evolution, at best, is caught in a local optimum. Furthermore, progress depends not only on the absolute or average size of the economy (per capita), but also on changes in distribution, which are inevitable. Mere income growth is a too crude indicator to capture the wide variety of structural changes in an evolving economy.

- Changes in population distributions of technology give rise to path dependence, a historical process, which in turn can create the problem of lock-in of an undesirable technology. Path dependence and lock-in require a different response than that suggested by the economic theory of environmental policy. Decentralized policy based on externality taxation (price regulation) is generally insufficient to unlock the system, except in the rare case where the marginal external cost is so high that the regulated price becomes prohibitive. Comparative statics results, which are common in equilibrium economics, provide insufficient information to formulate policy, as desired equilibrium states may not be reachable from the present state.

- Because the link between a social optimum and a market equilibrium is lost, optimal social welfare policies play a less prominent role in evolutionary economics than they do in neoclassical economics. On the other hand, evolutionary theory, because of its focus on population diversity, can address distribution issues better than neoclassical economics.

This chapter has introduced a range of evolutionary economic ideas that, although not elaborated in great detail, together suggest a fresh perspective on the relationship between growth, environment, and resource scarcity. Further theorizing, modeling, and empirical research are needed.

Acknowledgment

The author thanks Joelle Noailly for helpful comments.

Notes

1. Given the number of social welfare states in the West with a mixture of private and public activities, markets and social institutions to redistribute income, and labor markets, unions, and legislation, it seems that Schumpeter was closer to the truth than Marx. Of course, this state of affairs does not hold in the broader international setting or for certain undeveloped countries. Here a more pure form of capitalism is found, characterized by extremely skewed distributions of income and power.

2. Evolutionary reasoning itself can be invoked to explain the slow spread of evolutionary thinking in economics. What follows is that neoclassical economics is a case of lock-in at the level of scientific ideas. In fact, the Kuhnian notion of a paradigm is consistent with the notion of lock-in.

3. Nelson and Winter (1982, *390*), however, note that an oligopolistic market may also combine the worst features of monopoly and competition, since much R&D tends to be defensive, that is, focused on imitating competitors.

4. In addition, governments can change the selection environment for car producers, for example, through technical limits on motor size, speed, and power. This would stimulate technological innovations leading to slower and lighter cars. This would have several advantages. Besides reducing the amount of materials and energy in production, less energy would be involved in collisions. This in turn might shift the attention in car construction from passenger or single-car safety to system safety by taking into account interactions among all cars.

References

Aghion, P., and P. Howitt. 1992. A Model of Growth through Creative Destruction. *Econometrica* 60: 323–351.

Arthur, B. 1989. Competing Technologies, Increasing Returns, and Lock-in by Historical Events. *Economic Journal* 99: 116–131.

Ayres, R.U. 1994. *Information, Entropy and Progress: Economics and Evolutionary Change.* New York: American Institute of Physics.

Ayres, R.U., J.C.J.M. van den Bergh, and J.M. Gowdy. 2001. Strong versus Weak Sustainability: Economics, Natural Sciences and "Consilience." *Environmental Ethics* 23(1): 155–168.

Azar, C. 1998. Are Optimal CO_2 Emissions Really Optimal? *Environmental and Resource Economics* 11: 301–315.

Bäck, Th. 1996. *Evolutionary Algorithms in Theory and Practice: Evolution Strategies, Evolutionary Programming, Genetic Algorithms.* Oxford: Oxford University Press.

Buenstorf, G. 2000. Self-Organization and Sustainability: Energetics of Evolution and Implications for Ecological Economics. *Ecological Economics* 33: 119–134.

Campbell, N.A. 1996. *Biology.* 4th edition. Menlo Park, CA: The Benjamin/Cummings Publishing Company.

Conlisk, J. 1989. An Aggregate Model of Technical Change. *Quarterly Journal of Economics* 104: 787–821.

Daly, H.E. 1977. *Steady-State Economics.* San Francisco: Freeman.

Demeritt, D., and D. Rothman. 1999. Figuring the Costs of Climate Change: An Assessment and Critique. *Environment and Planning A* 31: 389–408.

Dennett, D. 1995. *Darwin's Dangerous Idea: Evolution and the Meanings of Life*. New York: Simon and Schuster.

Diamond, J. 1997. *Guns, Germs and Steel: The Fates of Human Societies*. New York: W.W. Norton.

Dosi, G., C. Freeman, R. Nelson, G. Silverberg, and L. Soete (eds.). 1988. *Technical Change and Economic Theory*. London: Pinter Publishers.

Durham, W.H. 1991. *Coevolution: Genes, Culture and Human Diversity*. Stanford: Stanford University Press.

Ehrlich, P.R., G. Wolff, G.C. Daily, J.B. Hughes, S. Daily, M. Dalton, and L. Goulder. 1999. Knowledge and the Environment. *Ecological Economics* 30: 267–284.

Faber, M., and J.L.R. Proops. 1990. *Evolution, Time, Production and the Environment*. Heidelberg: Springer Verlag.

Fagerberg, J. 1988. Why Growth Rates Differ. In *Technical Change and Economic Theory*, edited by G. Dosi, C. Freeman, R. Nelson, G. Silverberg, and L. Soete. London: Pinter Publishers.

Friedman, D. 1998. Evolutionary Economics Goes Mainstream: A Review of the Theory of Learning in Games. *Journal of Evolutionary Economics* 8: 423–432.

Galor, O., and O. Moav. 2002. Natural Selection and the Origin of Economic Growth. *Quarterly Journal of Economics* 117 (November).

Georgescu-Roegen, N. 1971. *The Entropy Law and the Economic Process*. Cambridge, MA: Harvard University Press.

Gould, S.J. 1988. On Replacing the Idea of Progress with an Operational Notion of Directionality. In *Evolutionary Progress*, edited by M. Nitecki. Chicago: University of Chicago Press.

Gowdy, J. 1994. *Coevolutionary Economics: The Economy, Society and the Environment*. Dordrecht: Kluwer Academic Publishers.

———. 1999. Evolution, Environment and Economics. In *Handbook of Environmental and Resource Economics*, edited by J.C.J.M. van den Bergh. Cheltenham, UK: Edward Elgar.

Horgan, J. 1996. *The End of Science*. Reading, MA: Addison-Wesley.

Iwai, K. 1984. Schumpeterian Dynamics, part 1: An Evolutionary Model of Innovation and Imitation. *Journal of Economic Behavior and Organization* 5(2): 159–190.

Janssen, M. 1998. *Modelling Global Change: The Art of Integrated Assessment Modeling*. Cheltenham, UK: Edward Elgar.

Janssen, M., and B. de Vries 1998. The Battle of Perspectives: A Multi-Agent Model with Adaptive Responses to Climate Change. *Ecological Economics* 26: 43–66.

Kolstadt, C.D. 1994. The Timing of CO_2 Control in the Face of Uncertainty and Learning. In *International Environmental Economics*, edited by E.C. van Ierland. Amsterdam: Elsevier Science Publishers.

Maddox, J. 1998. *What Remains To Be Discovered*. New York: Free Press.

Maynard Smith, J., and E. Szathmáry. 1995. *The Major Transitions in Evolution*. Oxford: Oxford University Press.

Metcalfe, J.S. 1998. *Evolutionary Economics and Creative Destruction*. Graz Schumpeter Lectures, 1. London: Routledge.

Mokyr, J. 1990. *The Lever of Riches: Technological Creativity and Economic Progress*. Oxford: Oxford University Press.

Mulder, P., and J.C.J.M. van den Bergh. 2001. Evolutionary Economic Theories of Sustainable Development. *Growth and Change* 32(4): 110–134.

Mulder, P., H.L.F. de Groot, and M.W. Hofkes. 2001. Economic Growth and Technological Change: A Comparison of Insights from a Neoclassical and an Evolutionary Perspective. *Technological Forecasting and Social Change* 68: 151–171.

Munro, A. 1997. Economics and Biological Evolution. *Environmental and Resource Economics* 9: 429–449.

Nelson, R., and S. Winter. 1982. *An Evolutionary Theory of Economic Change*. Cambridge, MA: Harvard University Press.

Noailly, J., J.C.J.M. van den Bergh, and C.A. Withagen. 2003. Evolution of Harvesting Strategies: Replicator and Resource Dynamics. *Journal of Evolutionary Economics* 13(2): 183–200.

Nordhaus, W.D. 1994. *Managing the Global Commons: The Economics of Climate Change*. Cambridge, MA: MIT Press.

Norgaard, R.B. 1984. Coevolutionary Development Potential. *Land Economics* 60: 160–173.

Ostrom, E. 1990. *Governing the Commons: The Evolution of Institutions for Collective Action.* New York: Cambridge University Press.

Potts, J. 2000. *The New Evolutionary Microeconomics: Complexity, Competence, and Adaptive Behavior.* Cheltenham, UK: Edward Elgar.

Schneider, E.D., and J.J. Kay. 1994. Life as a Manifestation of the Second Law of Thermodynamics. *Mathematical and Computer Modelling* 19(6–8): 25–48.

Schumpeter, J.A. 1934. *The Theory of Economic Development.* Original German edition, 1911. Cambridge, MA: Harvard University Press.

Sen, A. 1993. On the Darwinian View of Progress. *Population and Development Review* 19(1): 123–137.

Sethi, R., and E. Somanathan. 1996. The Evolution of Social Norms in Common Property Resource Use. *American Economic Review* 86(4): 766–788.

Silverberg, G., G. Dosi, and L. Orsenigo. 1988. Innovation, Diversity and Diffusion: A Self-Organization Model. *Economic Journal* 98: 1032–1054.

Somit, A., and S. Peterson. 1989. *The Dynamics of Evolution: The Punctuated Equilibrium Debate in the Natural and Social Sciences.* Ithaca, NY: Cornell University Press.

Stiglitz, J.E. 1997. Reply. *Ecological Economics* 22: 269–270.

Strickberger, M.W. 1996. *Evolution.* 2nd edition. Sudbury, MA: Jones and Bartlett Publishers.

Toman, M.A., J. Pezzey, and J. Krautkraemer. 1995. Neoclassical Economic Growth Theory and "Sustainability." In *Handbook of Environmental Economics,* edited by D. Bromley. Oxford: Blackwell.

van den Bergh, J.C.J.M. 1999. Materials, Capital, Direct/Indirect Substitution, and Materials Balance Production Functions. *Land Economics* 75(4): 547–561.

———. 2003. Evolutionary Modeling. In *Integrated Modeling in Ecological Economics,* edited by J. Proops and P. Safonov. Cheltenham, UK: Edward Elgar.

van den Bergh, J.C.J.M., and J.M. Gowdy. 2000. Evolutionary Theories in Environmental and Resource Economics: Approaches and Applications. *Environmental and Resource Economics* 17: 37–57.

———. 2003. The Microfoundations of Macroeconomics: An Evolutionary Perspective. *Cambridge Journal of Economics* 27(1): 65–84.

van den Bergh, J.C.J.M., and M.W. Hofkes (eds.). 1998. *Theory and Implementation of Economic Models for Sustainable Development.* Dordrecht: Kluwer Academic Publishers.

van den Bergh, J.C.J.M., and S. Stagl. 2003. Coevolution of Institutions and Individual Behaviour: Towards a Theory of Institutional Change. *Journal of Evolutionary Economics* 13(3): 289–317.

van den Bergh, J.C.J.M., A. Ferrer-i-Carbonell, and G. Munda. 2000. Alternative Models of Individual Behaviour and Implications for Environmental Policy. *Ecological Economics* 32(1): 43–61.

Wilkinson, R. 1973. *Poverty and Progress: An Ecological Model of Economic Development.* London: Methuen and Co.

Wilson, E.O. 1998. *Consilience.* New York: Alfred Knopf.

CHAPTER 10

Environmental Policy as a Tool for Sustainability

David Pearce

PREVIOUS VOLUMES FROM Resources for the Future on scarcity and growth (Barnett and Morse 1963; Smith 1979) have struck a generally optimistic note about the capacity of economic systems to deal with natural resource depletion. In so far as scarcity exists at all, adaptive mechanisms exist to ensure that more plentiful resources are substituted for scarcer ones, and the efficiency with which resources are used improves over time. Efficiency here refers to the ratio of output to input: more gross national product (GNP) per unit of primary energy, for example. Substitution and efficiency are, in turn, induced by the workings of the market system. Resource prices rise with the onset of scarcity, inducing substitution and technological change, which improves efficiency. These forces can be augmented by deliberate policy, for example, by raising the price of energy with an energy or carbon tax. The debate then reduces to (a) alternative views of the evidence for scarcity as shown by resource price or cost (or royalty) trends; (b) differing viewpoints about the right balance of free market forces and policy intervention; and (c) the right mix of intervention measures.

Setting the Scene: The "New Scarcity"

The substantial literature on the resource scarcity theme tend to focus on natural resources in a fairly traditional sense: fossil fuels, minerals, agricultural products. This can be termed the "Old Scarcity," which was the subject of the "optimal depletion" literature of the twentieth century (Gray 1914; Hotelling 1931), and even of one of the earliest essays on resource depletion and its implications—Jevons (1865) on the "coal question" in the mid-nineteenth century. Concerns about the Old Scarcity have not gone away, but they have given ground to the modern preoccupation with a "New Scarcity." The New Scarcity refers to life support assets, such as biological diversity, the global atmosphere, ocean resources, tropical and boreal forests, coral reefs, and wetlands. The shift in concern partly reflects the failure of the Old Scarcity to materialize—we have not, after all, run out of resources—but arguably it is mainly the result of ecological signs that genuine limits to current forms of economic activity have been reached. Dissipation of the stratospheric ozone layer, plausibly linked to increased skin cancer, provides one example. Though the science is still debated, warming of the earth's surface through higher greenhouse gas concentrations in the atmosphere is another. More uncertain still are changes in biological diversity, with at least some eminent scientists speaking of a major human-made extinction comparable to past natural extinctions (Lawton and May 1995).

The New Scarcity embraces resources that take on the nature of "public goods"—assets whose consumption by one person does not restrict consumption by another. The Old Scarcity tended to deal with "private" goods, such as fossil fuels, a unit of which can be consumed by only one person. Public goods usually do not have markets, private goods do. For public goods the only adaptive mechanism tends to be based on policy intervention, which in turn may be prompted by public or scientific perception of scarcity. Markets may be induced by such intervention—for example, the rapidly emerging markets in buying and selling carbon embodied in biomass or in emission-reduction technology—but few would argue that, left to themselves, markets will resolve the New Scarcity problem.

Public goods can be local in scope but may also be global, in that major parts of the world population may benefit from them. These public good and global features mean that damage to these assets is likely to be corrected only by international agreement. Just how much such agreements actually change the course of events is open to debate; for a rigorous and perceptive survey see Barrett (2003). The Montreal Protocol and its amendments almost certainly have changed the fate of the ozone layer. The 1992 United Nations Framework Convention on Climate Change produced targets that no country actually met save by chance, and the Kyoto Protocol of 1997 is likely to have an almost undetectable influence on rates of global warming (Pearce 2002). It is exactly these difficulties of securing sustained international commitment to positive action that make the New Scarcity a challenge to policy.

The severity of this challenge depends on the facts and on perceptions of the facts. For some, the New Scarcity is everywhere evident, and policy action is imperative (e.g., Brown 2001). In the language of economics, the externalities are large and pervasive. For others, the evidence of a serious divergence between private and social well-being is not persuasive (Lomborg 2001), or the social costs of taking action—restrictions on human choice—are not worth the benefits of the reduced environmental externality (Wildavsky 1995). Environmental policy globally exhibits the tensions between these various viewpoints. In the United States, environmental regulations are continually debated in terms of the scientific evidence and the relative balance between the perceived costs of compliance and benefits to health and environment. The debates are perhaps less intense in Europe, but demand for some economic accountability, particularly concerning European Union regulations, is growing. In the international sphere, compliance costs probably underlie the reluctance to sign on to new regulations, as the Kyoto Protocol on climate change control has demonstrated and as the 2002 Earth Summit in Johannesburg further showed.

Whatever the debate about science and cost, few now argue that *nothing* should be done. There is a will to take action, but always restrained by weighing of costs and benefits. This caution is frustrating to those who argue that the environmental challenge is urgent and is best met by remedying the underlying causes of the New Scarcity—economic growth and population growth. They argue that more people consume more resources, and each addition to national output—conventionally measured by GNP—uses more natural resources. Hence the "anti-growth" movement of the 1970s, exemplified by *Limits to Growth* (Meadows et al. 1972), remains a vocal force in the environmental debate. Some of that anti-growth sentiment has found re-expression in the movement, often poorly articulated, against "globalization." Anti-growth protagonists are not always clear about what they are against. If they want a given level of GNP while using lower absolute levels of physical resources, there can be little argument. This is the resource efficiency argument, and it underlies all environmental economics policy. But much of the anti-growth literature is only partially about resource efficiency. Indeed, many see resource efficiency as a short-term measure that simply postpones the inevitable, namely, the need to reduce consumption and incomes themselves.

But abandoning economic growth hardly constitutes a feasible policy stance. It would require substantial shifts to social values of a kind that do not appear to exist anywhere in the world. Value changes could possibly be brought about by *diktat*—limiting working time, setting maximum income limits, and so on—but at the expense of another cherished value, personal freedom. Moreover, it is not clear that any government (or governments, since many nations would have to act in concert) is in control of long-run growth prospects. Even if they were, such determining factors as education and technological change are valued in themselves. While it is possible to make an "anti-growth" case, in the real world context it has little credibility.

Those who remain "anti-growth" usually distinguish measures applicable to rich countries from those applicable to poor countries. The latter need growth to escape often desperate living conditions. The former have "too much" growth. This differentiation is summarized in the notion of "overconsumption," embodied in global wish lists such as *Agenda 21,* issued at the time of the Rio Earth Summit of 1992. But ambiguity rules again: *Agenda 21* refers to the need to change "consumption patterns" rather than the level of consumption. Changing consumption patterns is consistent with maintaining (or increasing) consumption levels while reducing the use of natural resources—resource efficiency again. But the overconsumption argument implies that poor countries need not pursue aggressive environmental policies. What comes first is raising consumption and investment levels, not reducing pollution or conserving resources. It is the rich who need to focus both on consumption levels and on resource efficiency.

While overconsumption is usually charged against the rich as a major cause of environmental degradation, willingness to pay for environmental improvement is a function of per capita income levels, much as we would expect, since willingness to pay is contingent upon ability to pay. The first obstacle to systematic and firm environmental policy in the developing world is therefore income levels; they are low, and hence so is willingness to pay for environmental change. In rich countries, aggressive environmental policies are partly a function of higher income levels.

As the Environmental Kuznets Curve purports to show, environmental degradation worsens at low but rising income levels and improves after some threshold is reached. But higher incomes also bring perceived property rights to those incomes. Even if environmental problems are taken seriously, no government is going to embrace the kind of radical social engineering implied in anti-growth, for reasons previously advanced. The challenge is to balance the competing claims for positive action and comparatively modest action.

The argument that environmental improvement is a low priority in poor countries should not be overstated. The fraction of per capita income spent on clean water, for example, can range up to 10 percent or even 15 percent, indicating that this environmental improvement is highly valued. Many developing countries are recognizing that chronic air and water pollution impose substantial sacrifices of potential economic growth because of impairment of human capital. Nonetheless, it is generally true that public environmental expenditures vary directly with per capita incomes.

These considerations shift the debate to the choice of policy instruments. In rich countries, policy measures that are comparatively cheap in terms of compliance costs provide a compromise between the more radical and expensive demands of environmentalists and the desire of industry and taxpayers to see something done, but not at great expense. The driving force is cost. If action is cheap, there is less likelihood of alliances being formed to prevent it. Environmental policy can appease more radical demands without provoking coalitions that may stifle action. The search for cost-efficient policy makes sense in itself—no rational person should argue for the allocation of more resources to solve a problem than it needs—but it also fulfills a political need. Here lies one clue to the revival of concern with "better regulation": how best to achieve compromise goals at least cost.

The imperative of "efficiency in government" is nothing new. It defined the postwar rise of policy appraisal and the economic analysis that produced such seminal works as McKean (1958), Krutilla and Eckstein (1958), and Eckstein (1958). It tended to show up as a demand for cost–benefit, cost–effectiveness, or multi-criteria approaches to policy appraisal. But in the last 20 years it has also produced a demand for cost-efficient policy instruments. Given that something should be done, the extent of action should be informed by some balancing of costs and benefits, and the mechanism should be some form of market-based instrument, such as a tax or tradable permit. That this approach to policy is not pervasive in rich countries, and that substantial and inefficient anomalies remain, is an issue to be explained later. The textbook promise of cheap and efficient environmental policy has proved to be an illusion. For the moment, however, policy is directed toward market-based approaches.

A similar but perhaps less intensive policy debate concerns developing countries. Even if environmental policy is to be pursued more aggressively once some income target has been reached, what policy there is now should still be at least cost. Indeed, the minimizing cost argument seems all the more powerful for low-income economies. Once again, a market-based approach appears to fit, and the World Bank has persistently advocated just that, as has the United Nations Environment Programme (e.g., Panayotou 1998). But applying it to economies where markets have major imperfections and institutions are weak is even more difficult than applying it in rich countries, and there has been something of a backlash against the approach (Russell and Powell 1996; Bell 2002).

Overall, then, the historical pattern of environmental policy has been limited action while income is low and more pervasive action when incomes are high. Action has traditionally taken the form of technology-based standards for stationary sources of emissions—prescriptions as to what production or abatement technology can be used, plants by plant. Policy has been gradually modified to include emission and concentration standards with choice of technology, but notions of "best available technology" or "integrated pollution control" still pervade most environmental policy. As the demands for cost-efficiency have increased—because of concerns about the scale of public expenditure in rich countries, or the scarcity of funds in poor countries—the role played by market-based instruments has expanded. Interestingly, as global problems—climate change, biodiversity loss, deforestation, and so on—have taken center stage, the call for market-based approaches to solving those issues has strengthened. The Kyoto Protocol on climate change is remarkable for openly embracing the market approach with project-based trades in carbon dioxide ("joint implementation" and the Clean Development Mechanism) up and running in the short term and a regionwide tradable permit scheme in the European Union. Yet, the triumph of the market-based approach should not be exaggerated. Environment-related taxes still comprise only 2 percent of GNP in OECD countries (OECD 2001). The substantial and expanding Europe-wide agreements on acid rain control owe little to market-based controls and much to the extension of technology-based standards. If market-based approaches are so smart, why are they not more widely used worldwide?

The Market-Based Approach to Environmental Policy

Economists have long favored market-based instruments (MBIs). Defining market-based approaches to environmental policy is not straightforward, and many MBIs operate within the context of command-and-control (CAC) regimes. The U.S. sulfur trading schemes under the 1990 Clean Air Act, for example, are embedded in the Act's environmental standards and compliance penalties (Davies and Mazurek 1999). The essence of an MBI, however, is that the policy measure sets a price or a quantity, but leaves the resource user or polluter to respond in a manner largely of their own choice. An environmental tax is the equivalent of a price for use of the environment's role in receiving waste. By raising the polluter's costs of production, it should induce pollution-control activity, which thereby reduces the amount of the tax paid. Provided the tax is paid on all pollution, as opposed to pollution in excess of some agreed standard (a noncompliance fine), there is also an incentive to search continually for less-polluting technology. These two effects constitute the "static" and "dynamic" efficiency features of a tax. Other benefits depend on the way the tax is imposed. Recent experience with environmental taxes, particularly in Europe, has shown the potential benefits of earmarking (or hypothecating) revenues from the tax. Revenues might be used to reduce other taxes thought to have undesirable side effects (e.g., taxes on labor employment—the usual form of the "double dividend" hypothesis) or to finance research into further pollution abatement. Hypothecation may also make the policy more acceptable politically, avoiding costly political lobbying. Finally, a tax should minimize compliance costs—the costs likely to be borne by polluters in complying with any regulation (Baumol and Oates 1988). Apart from the desirable consequence that expenditure is not simply wasted on inefficient policy, keeping compliance costs down should also help to secure future cooperation with new policy measures. High compliance costs simply generate lobbies against future regulation.

In much the same way, a quantity-based system sets a quantity of pollution, say X tonnes of carbon dioxide emissions, and leaves polluters free to adjust to that target. This would also define a traditional environmental standard, but an MBI differs in that the allocated quantity target can then be bought and sold between polluters. Tradable permits, as they are known, have met with considerable success in the United States in the context of controlling acid rain and are now being developed, particularly for greenhouse gas emissions, in Europe. The price of the permit acts just like an environmental tax, so that all the beneficial features of a tax obtain for a tradable permit. Compliance cost minimization is again assured. Permits can also generate revenues if they are allocated through auctions. Political problems have typically prevented this, and some would argue that taxes have an advantage over permits in this respect.

If the new resource scarcities are real and significant, as many argue they are, then it is hard to envisage a "sustainable" future in which people are endowed with many of the environmental assets that contribute to human well-being. While definitions of sustainability abound, most of them center on not shifting environmental burdens forward in time. But it is also difficult to imagine a future in which individuals voluntarily adopt lifestyles that abandon the pursuit of higher real income and consumption levels. Hence, as argued earlier, the real world is far more likely to choose resource efficiency—making each unit of natural resource perform better in terms of producing national output and human well-being. The market-based approach appears to hold out the greatest promise of separating, or "decoupling," economic activity from its environmental impacts. Making environmental inputs into production processes more expensive can be a powerful incentive for economizing on those inputs.

The menu of MBIs is well known, and environmental economics textbooks abound with examples, though not all economists are persuaded of their advantages (for example, Russell and Powell 1996; Bell 2002). Yet in real-world policy, textbook versions are not common. Some tradable permit and tax systems are in place, but they often depart significantly from the "optimal design" that led to their advocacy in the first place. But if MBIs are so theoretically sound, so environmentally effective, so good for business and households, and so attractive as a political compromise between radical demands and environmental inertia, why have they not made more progress in both rich and poor countries? Environmental taxes account for only between 4 percent and 11 percent of total tax revenues in OECD countries, and those figures include vehicle fuel taxes, most of which were introduced solely to raise revenue (OECD 1999). Most environmental policy is still based on setting standards, whether ambient or emission standards, or defining production and environmental technologies that can and cannot be used—in other words, CAC measures. Sometimes the textbook solutions are known, but are not implemented, or at least are not implemented in the form economists recommend (e.g., Hahn 1989). If the emergence of MBIs reflects the demand for efficiency in government, why do CAC measures still dominate environmental policy? Is the failure to implement MBIs symptomatic of a wider failure, namely, to get stricter environmental policy adopted, regardless of the form it takes?

History Matters: Dealing with What We Already Have

That the world of politics differs from the world of textbooks will not occasion any surprise, but it is worth investigating why these differences exist. Some economists have acknowledged that theoretical policy design has been too divorced from real political systems (Cropper and Oates 1992). Political scientists have stressed political complexity (Skou Andersen 2000). What follows is an attempt to uncover the reasons for the divergence between theory and reality.

Textbooks tend to be written as if policy were written on a regulatory *tabula rasa*. No prior legislation is assumed to be in place. But environmental policy in advanced economies has a long history of being based on "best" technology. Essentially, governments or regulators define the kind of technology that may be used in industrial processes, with the best technologies being those that are consistent with some conception of environmental quality. Actual production processes may be defined or, more generally, abatement technologies are mandated to be added to production processes. Whatever the technology specified, the resulting ambient quality effectively defines the target level of quality. Environmental quality is then the outcome of technology, rather than technology being induced by a choice of environmental standard. Moreover, the best technology is determined for each emitting plant rather than for a corporation or a sector as a whole. Not only is the environmental standard implicitly chosen, but so is the means of securing it. This is perhaps the clearest example of CAC, the policy with which MBIs are usually contrasted. "Best technology" takes on various regulatory acronyms. In Europe, "best available technology" (BAT), "best available technology not entailing excessive cost" (BATNEEC—a term peculiar to the United Kingdom), and "best practicable environmental option" (BPEO) have their origins in nineteenth-century air pollution control legislation, where they were referred to as "best practicable means" (BPM).[1] Terms and expressions such as "available," "not entailing excessive cost," and "practicable" convey the idea that the cleanest technology might not be the best because of the expense of installing it. There have long been trade-offs between environmental implications and compliance cost (Pearce 2000). Parallel terms can be found in the United States, such as "reasonably available control technology" (RACT) and "best available control technologies" (BACT).

A century or more of technology-based standards inevitably hinders the transition to MBIs. If each plant is achieving the maximum feasible "end of pipe technology" pollution reduction, no further reductions can be brought about by some MBI. A tax works because the polluter can substitute a more environmentally friendly technological process for the polluting one. If BAT is already employed, that option is not open. Taxes may induce changes in other respects that reduce pollution—changes in management practice, for example. But the principal means of reducing emissions, via technology switches, is not available.

The same limitation applies to tradable permits, whereby firms can buy and sell the rights to emit pollutants. Trading can only take place if those who find it easiest to abate pollution do so and collect credits that they can then sell to those who find abatement most difficult. The levels of abatement will therefore differ between firms. But if all firms' plants are already at their technological maximum in terms of clean technology, and are obliged to stay there because of BAT-type regulations, no trading is possible. To some extent, this explains why Europe had not secured an emissions-trading program for acidic pollutants 10 years after it was enabled in the 1994 Oslo Protocol to the UN ECE Convention on the Long Range Transport of Air Pollution.[2] Because the Protocol's differentiated targets were set according to an aggregate cost-minimizing solution (i.e., the sum of all national compliance costs is approximately minimized), it offers little room for cross-country trading (Klaassen 1996). A second reason is the third-party effects of trading. Trading between any two nations can affect acid depositions in a third country (or more); trades that are beneficial to the trading nations may incur social costs in nontrading countries. In the Oslo Protocol, this effect is constrained by limiting trading to situations in which only minor third-party effects can occur, further limiting the scope for trading to occur at all. How serious these third party effects are in practice is open to question. In the United States, civil action has been brought against some sulfur-trading schemes. However, the United States appears to have accommodated the problem, whereas Europe has effectively strangled the instrument at the outset, perhaps because of a perception issue rather than any real issue.

Other settings often afford some room for maneuver precisely because "best" technology is conditioned by notions of cost. Provided MBIs can be introduced at costs substantially below those of going from "reasonable" control to "best" control, abatement can be improved. But the cost-constrained best-technology policies evolved over time precisely to deal with industry and government concerns about compliance cost. Their perception is that they are doing the best they can subject to the cost constraint. Asking them to do more by way of superimposed taxes or permits systems strains the credibility of regulation and alienates industry. The fact is that the existence of longstanding legislation seriously restricts the usefulness of MBIs.

One other defense of BAT-style regulation can be advanced: BAT confers certainty on the firm having to comply with the regulation. It knows what it has to do and it knows the cost (Weitzman 1974). It is also familiar with the accounting conventions for financing the technology—how it can write off the cost over time, what the tax allowances are, and so on. This contrasts with the uncertainty surrounding MBIs: the tax is unlikely to be fixed over time; the price of permits will vary with market conditions. Certainty brings a premium, uncertainty acts like an additional cost. Moreover, BAT regulations guarantee the market for the technology, enable economies of scale, and provide incentives to innovate, opening the prospect for less-polluting forms of BAT.

Regulatory Capture

One obvious response to the problem of superimposing new legislation is to "wipe the slate clean" and cancel existing legislation in favor of MBIs. This likelihood is remote. First, as will be noted below, the effectiveness of MBIs, and especially taxes, can be uncertain. Technology-based standards offer some assurance that whatever ambient quality they imply will be achieved. Second, large numbers of people have significant personal investments of time and experience in the existing system. The system is "captured" not just by polluters, but by regulators as well. The process of regulatory capture—whereby those who are regulated spend resources influencing the regulator, softening the process, and determining the way regulation is interpreted—means that whatever legislation exists has a political constituency that is not easy to change. The regulators themselves may also see no advantage in change—they have been selected to implement the existing legislation and may well lack the expertise, vision, or will to change it, especially if the alternative threatens their livelihoods.

An environmental tax may be self-defeating if it presents the prospect of tax revenues going to zero. The paradox of an effective environmental tax is that if it is working perfectly no one pays it: if the tax curtails polluting behavior, no one is subject to the tax. But few governments are interested in taxes where revenues shrink to zero. Although they may give credence to the view that the tax is environmental in purpose, political logic requires that taxes yield revenues as well. The experience of the United Kingdom is instructive: the tax on waste going to landfill sites and the tax on the extraction of aggregates materials are taxes on "products" in extremely inelastic demand. Both were introduced as environmental taxes, but most of their revenues are earmarked for reduced labor taxes. If the taxes generated near-zero revenues, the double dividend policy would be moot. As environmental taxes, both are largely ineffective because of the very low demand elasticity.

Sorrell (1999) analyses the failure to introduce tradable sulfur dioxide permits in the United Kingdom, despite favorable experience in the United States (EPA 2001). Some of the reasons were "exogenous," that is, factors distinct from environmental policy. Developments in energy markets resulted in a switch to low-sulfur fuels, reducing the need for controls on sulfur emissions. But the conflict with existing policy was also relevant. Trading simply did not fit with Integrated Pollution Control (IPC), a BATNEEC-oriented policy of emission limits at individual sources.

Sorrell notes conflicts of regulatory culture. There was (and is) a tight-knit "policy community" largely responsible for implementation. Mostly operators and pollution inspectors, and excluding the public, central government, and environmental groups, they have a belief system into which MBIs simply do not fit. They did not initiate tradable permits, nor would they have, since these would have reduced their discretionary power. Sorrell also argues that tradable permits had limited political backing. Support came from the Department of the Environment, but not from other government departments and certainly not from the regulators. Even if the "exogenous" factors had not been present, Sorrell doubts that tradable sulfur permits would have been introduced.

Other conditions present at that time still obtain in the United Kingdom. The Environment Agency, formed in 1996 from various separate agencies, is responsible for the implementation of environmental policy. Its large staff are almost entirely scientists and engineers, despite its own constitution requiring it to embrace economic approaches. Helm (2000, *26*) notes:

> The Environment Agency's reluctance to embrace CBA [cost–benefit analysis] and economic instruments is not an accident. It reflects the incentives of the Environment Agency's management and employees. Scientists and engineers are unlikely to welcome the idea that pollution licences should be subject to detailed scrutiny of costs and benefits, and that economic valuations might help to determine the optimum level of pollution. Many current activities and decisions might be consequentially questioned. More threatening still are economic instruments, since their application removes much of the role of the experts in fixing and revising regulation, and would make the activities of some employees redundant.

Others argue that, because lawyers write the legislation, the problem lies with their training, which tends not to include an appreciation of economic instruments (Sprenger 2000). Whichever the "guilty party," the result is that the same lengthy history of traditional control that hamstrings MBIs has vested power in those who have the least understanding of MBIs and the least incentive to embrace them.

What of the wider question of limits to environmental policy? Technology-based standards may seem the obvious way to solve an environmental problem. Choice of technology is causing the problem in the first place, so it is natural to see technology as the solution. But the attenuation of the policy with notions of "practicable," "available," and "excessive cost" is also an effective loophole that limits the strictness of environmental policy. "Best" technology has to be affordable to polluters, so environmental quality is determined by the technology rather than an environmental standard.

Inconsistent Policies

Contrary to the textbook assumption that governments maximize the achievement of some well-ordered set of policy goals, all regulatory regimes tend to adopt inconsistent political objectives. These inconsistencies take various forms.

Governmental goals will often conflict. Government is about compromise between lobbies with divergent goals: environmental improvement may be inconsistent with the need to keep production costs down to ensure international competitiveness. Protecting vulnerable income groups may be inconsistent with charging desirable prices for basic resources, such as energy and water. In the United Kingdom, the first responsibility of energy regulators is keeping consumer prices down; honoring social and environmental goals is secondary. As yet, no one has provided rigorous guidance on what to do if the goals conflict, as they obviously do. High prices encourage conservation, lower prices encourage wasteful use.

"Political capture" of different government departments can also cause inconsistency. A department primarily responsible for agriculture is almost certain to be swayed by the farming lobby, a defense department reflects the views of the armed forces and arms manufacturers, and so on. Since the interests of these groups are not the same, government departments are unlikely to act in unison. The result is usually some form of compromise that draws the instrument far from the text-book solution. The United Kingdom's "climate change levy" is a tax on energy designed to help comply with obligations under the Kyoto Protocol. But the tax is unrelated to the carbon content of the different fuels, exempts electricity generators and householders, and gives elaborate tax discounts to energy-intensive industries. The result is a hybrid tax with little of the environmental effectiveness of a proper carbon-based tax.

Even within one policy sector, the uncoordinated growth of legal provisions results in inconsistent goals. Perhaps the most startling example of economic irrationality in environmental legislation occurs in the United States. Costs are not a consideration in the Clean Air Act, are implicitly a factor in the Clean Water Act, and are explicitly traded against benefits in the Toxic Substances Control Act (Davies and Mazurek 1999, 19). The requirement that the U.S. EPA should *not* consider costs dates from a lower court ruling in 1980 (*Lead Industries Association v. EPA*, 499 U.S. 1042). Studies in preparation for revised ozone standards in the mid-1990s suggested that the costs of control exceeded the benefits, but if costs are deemed inadmissible as evidence for or against a regulation, the only issue is whether it provides health benefits. A D.C. Circuit Court, and later the Supreme Court, upheld the ruling: costs do not matter (Lutter 2001). As noted earlier, the principal argument for MBIs is that they minimize compliance costs; this ruling implies that MBIs are invalid. The function of "regulatory impact" assessments—statements justifying regulation on the basis of gains and losses—should ensure that such inconsistencies do not arise, but in practice the evolution of regulations tends to override such assessments. As Hahn (2000) notes, the quality of such assessments is also often poor.

Explaining why costs should be ignored in promulgating standards is not easy. In the United States the emergence (or re-emergence) of the "public trust" doctrine in the context of damage liability has reinforced the neglect of cost. The standard economic approach to regulation is to evaluate it in terms of the costs and benefits the regulation produces, costs and benefits being losses and gains in human well-being. Applications of the cost–benefit approach are extensive in the United States (Hahn 2000), but some regulatory developments have downplayed cost considerations. By 1996 two sets of regulations regarding economic valuation had been issued for liability cases. Department of the Interior guidelines for damage assessment, under the 1994 Comprehensive Environmental Response and Liability Act (CERCLA) stressed restoration costs as the primary component of damage but allowed any valuation technique to be used to value interim losses. On the other hand, the National Oceanic and Atmospheric Administration (NOAA) 1996 regulations for damage assessment under the Oil Pollution Act (OPA) effectively exclude economic valuation and require that the cost of restoration be the measure of damage done.

The precise reasons for this shift of emphasis are not easy to discern. Hanemann (1999) offers an explanation derived from the legal interpretation of the public trust doctrine. Under this doctrine, natural resources are considered to be held in trust by the state and federal governments for existing and future generations. It could be inferred that damage to natural resources must be negated, that is, the natural environment must be restored to its pre-damage situation. Moreover, the public trust doctrine allows those who act as trustees to use any money recovered from actions against liable parties only for enhancing or creating natural resources (Jones and Pease 1997). Monetary compensation, actual or hypothetical, would then have no role to play because, of itself, compensation does not restore the status quo. This contrasts with the standard economic view: to the economist the status quo refers to the well-being of the individual. If individuals are compensated so as to be as well off—in their own judgment—as they were before the damage occurred, then compensation is efficient and just. So long as the individual regards the compensation as a substitute for the environmental resources that were destroyed, it is not necessary for the damage itself to be undone.

To date, the distinction between public trust and cost–benefit appears not to have influenced the European regulatory environment, although the same issues are debated in academic and policy circles. There are also signs that something akin to the public trust doctrine is beginning to have influence. The European Union Habitats Directive, for example, requires that member states indicate, by 2000, an entire network of conservation areas. The presumption in the directive is that no form of economic development may take place in those areas, save for "overriding public interest." Various European Court of Justice rulings have indicated that such overriding interest must be at least national in scope and that, in general, ecological integrity is more important than economic considerations. Moreover, even where public interest prevails and development is permitted, a compensatory project must be introduced, of equal "ecological" value, to compensate for the loss of the named site. In effect, the Directive embodies a public trust approach only partly attenuated by cost-benefit considerations.

To these forms of policy inconsistency Davies and Mazurek (1998) add "fragmentation" (in the United States), by which they mean controlling pollution without considering the effects of the controls on other forms of pollution. Wastewater treatment may reduce water pollution, but contaminate land if sludge is diverted. In Europe this problem is purportedly addressed by the notion of "integrated pollution prevention and control" (IPPC). But IPPC is still rooted in technology-based standards, although in principle it should favor MBIs aimed at pollution prevention.

Policy inconsistencies limit not just the diffusion of MBIs, but environmental policy generally. Different government departments have different goals, with some effectively acting as agents for economic sectors. Agricultural departments have been, and in some countries still are, the voice of farmers' interests. The United Kingdom is one of very few countries that have abolished their agricultural ministry on precisely these grounds. Significantly, the UK agricultural industry has escaped many of the environmental controls common to other sectors, a situation that is beginning to change.

Subsidies and Corruption

Environmental policy is invariably formulated in surroundings where generally unjustified subsidies to polluting and resource depleting activities prevail. Subsidies either lower costs or take the form of price guarantees. Price guarantees are common in agriculture and effectively guarantee a price for the product regardless of how much is produced. The incentive in either case is to overproduce potentially polluting products. Worldwide, subsidies to natural resource sectors and to energy and industry sectors are estimated to total more than one trillion dollars (van Beers and de Moor 2001), equivalent to around 4 percent of the entire world's gross product. Nearly 70 percent of global subsidies are in OECD countries, and 40 percent of all subsidies occur in agriculture, nearly all them in OECD countries. The highest per capita agricultural subsidies are in Japan, but Europe and the United States are close behind. Some subsidies have valid social goals, but most reflect long histories of protecting vocal lobbies. Subsidies lower the chances of an effective environmental policy being introduced in several ways.

The "culture of subsidies" is inimical to environmental conservation: subsidies primarily reward environmentally damaging activities. They could, of course, be reformed to pay for environmentally beneficial activities if it were thought that property rights rest with polluters (an issue that is not always clear—Who owns the rural countryside?). The 2003 reforms to the European Common Agricultural Policy signal the first real shift away from production-related subsidies to subsidies contingent upon environmental performance.

Subsidies create "rents"—opportunities to secure funds in return for lobbying and exerting political pressure rather than producing a product or service (Krueger 1974). Rents attract "rent-seekers"—individuals and corporations seeking to maximize their share of the rents. Their efforts are essentially unproductive in the social sense—they merely lessen the chances of introducing environmental policy and spawn powerful groups who campaign for the retention of the subsidies. Environmental policy will generally threaten rent-owners, especially where the rents become capitalized in asset values. Farmland prices, for example, are known to incorporate expectations about continuing agricultural subsidies. This capitalization process makes it all the more difficult to remove subsidies without overt compensation for the asset value loss.

The processes of rent-seeking and capture may be carried to extremes, and corrupt practices ensue. Typically, these will involve some form of payment, in money or in kind, to officials or politicians in an effort to secure a share of the available rents. The sources of corruption are many and varied (for an extensive discussion see Rose-Ackerman 1999), but our focus is on the role of corruption in inhibiting environmental policy and market-based approaches in particular. Environmental policy can reduce rents. Forestry restrictions on allowable cutting areas reduce short-run profits, as do requirements for sustainable management of forests. It is well documented that clear-cutting or felling without concern for damage to noncommercial species is more profitable than sustainable forest practices (Pearce et al. 2001). The rents associated with illegal forestry are therefore higher than those for environmentally managed forest practices. The resulting corruption thwarts environmental policy and sound resource management.

By threatening rents, MBIs threaten the power of the corrupt, and hence attract even stronger lobbies. The clearest examples come from resource-rich developing countries, but it would be a mistake to suppose that corruption is not endemic elsewhere. Returning to illegal forestry, the resulting deforestation could be partially avoided by the introduction of forest laws that tax away part of the substantial rents residing in the forest sector. Self-evidently, some measures would achieve little on their own, and deforestation cannot be solved by the forest sector alone. The widespread political malaise of corruption at all levels of governance must be dealt with.

Who Lobbies for MBIs?

It is easy to identify the lobbies against MBIs, but who argues for them? Sustained pressure to introduce MBIs can be identified in the U.S. context, where the World Resources Institute and the American Enterprise Institute have argued for various forms of MBIs, and particularly tradable permit systems. Resources for the Future has rehearsed the arguments for and against MBIs and sought to have them heard in policy circles. By contrast, few "think tank"–style lobbies arguing for MBIs exist in Europe, and not many undertake the presentation of the pros and cons. Most advocacy and debate has come from individual academics rarely acting in concert. The Organisation for Economic Co-operation and Development has been instrumental in promoting MBIs and was the first to formulate international guidance on the "polluter pays principle" (OECD 1975). It has continued its work with repeated assessments of progress in introducing economic policy instruments and with its Environmental Performance Reviews carried out on each member country. As a government-financed institution, the OECD plays an interesting role, because it exists to reflect members' views while at the same time coaxing governments to take more action. Politicians are not prominent in the OECD: delegates are civil servants who may well seek a sounding board for their own ideas as well as a forum for the exchange of information.

Sweden (Brännlund 1999), Denmark (Mortensen and Hauch 1999), and Norway (Christiansen and Gren 1999) have had positive experiences with so-called green tax commissions. Such parliamentary and expert groups have contributed to arguments about the efficient design of policy, usually in the wider context of "ecological tax reform," that is, the shifting of the tax base away from personal income and toward "bads," such as pollution. Taking this concept one step further would mean turning such commissions into more inclusive bodies, widening membership to all relevant stakeholders or their representatives. Elsewhere, such commissions have been strongly resisted as encroaching on the power of government finance departments. The UK Treasury, for example, has no interest in any independent body discussing tax affairs, which it sees as exclusively its own concern. The variations in democratic traditions and internal power politics help to explain differing attitudes toward MBIs.

The role of individual advocacy and leadership should not be overlooked. It is no accident that countries such as Chile, Colombia, and Costa Rica have been leaders in the developing world in the introduction of various market approaches to environmental policy: tradable water rights, carbon offsets, payments for watershed protection, and so on. Individuals or small groups of educated economists have been instrumental in arguing for and designing such measures. Moreover, the closer the links between government and academic economists, the more likely it is that MBIs will be introduced. This holds for Norway, Sweden, Italy, and the United Kingdom. The role of ministries of finance is generally, but not always, crucial. Ministerial support for environmental taxation was important in Scandinavian countries, but the absence of such support did not prevent the introduction of various MBIs in Germany, where there is a strong "green" political party.

Convery (2000) points to some cultural factors that make MBIs difficult, and sometimes impossible, to introduce. Ireland, for example, does not have any charges for water use and is a frequent but lone voice opposing harmonized water charges in the European Union. One political party actually advocated water metering in the 1997 Irish general election and suffered accordingly. Even the terminology matters. The term "tax" usually excites more opposition than the emotionally more neutral "charge." The opposition of the American public to environmental taxes on energy is well known. President Bill Clinton's "Btu tax" was quickly abandoned, American car owners pay only a fraction of the fuel tax levied in European countries, and President George W. Bush's opposition to the Kyoto Protocol on climate change has its foundation in concerns about the cost implications for energy. This "anti-energy-tax" culture has, of course, furthered emissions trading in the United States.

Finally, environmental pressure groups can play an important role in MBI-based policy. The (misguided) view that tradable permits contravene a basic moral precept because they make emissions "legal" has certainly prevailed among European environmental groups and may help to explain the limited role that tradable permits have played in Europe until recently. Even in 2004, environmentalists protested at meetings on nascent emissions trading schemes in the United Kingdom. This contrasts with the role that prominent environmental groups played in the United States in securing tradable permits, although the environmental debate there has had its fair share of moralists opposing such measures.

How far can this analysis be extended to environmental policy generally? Rich countries have no shortage of environmental advocates. Some nongovernmental organizations (NGOs), however, pursue policies of radical lifestyle change that, as we have seen, have negligible chances of success. Their efforts can be counterproductive because they channel energy into campaigns that governments know will not appeal to the public. Efforts advocating environmental policy based on rational arguments for conservation would provide governments with more critical opposition and hence more incentive to act.

In poor countries, NGO movements can also be strong, but again may be misdirecting efforts at overconsumption rather than the immediate causes of degradation. Environmentalism is often seen as tantamount to crime because it challenges the interests of the corrupt. The absence of voices for stronger environmental policy correlates well with indices of corruption and lack of freedoms.

The Baseline Issue

In much of the debate about the virtues and shortcomings of MBIs, their adversaries use different "rules of the game." The issue in most unresolvable debates is the "baseline." The baseline can be defined as what will happen if a given policy instrument is not adopted. If the parties to the debate make different assumptions about what the baseline is, then they will differ on the desirability of a given regulatory instrument.

Consider a hypothetical tax on pesticides: The alternatives are (a) no change in the status quo, that is, no tax but no other regulation either; and (b) some different regulation. Each stakeholder will measure or judge their position toward the tax (as they perceive it) relative to the baseline they assume. Farmers may assume that the only alternative to the tax is no regulatory action at all, but regulators may assume that if there is no tax, there will have to be quantity restrictions, perhaps bans on some pesticides. The farmers measure their loss as the cost of the tax. For the regulators, the benefit is the environmental and health gains from pesticide reduction due to the tax compared to the gains from the alternative quantity control. If, instead, the farmers calculate their costs as the losses from the tax *minus* the losses from the restrictions they would otherwise have to tolerate, it is possible for both parties to gain from the tax.

The framing of policy decisions can seriously affect the chances of adoption. Of course, in practice those opposed to the tax are likely to ensure that the baseline is the one they favor. In other words, they campaign against the tax but also, implicitly, against any other form of regulation. Manipulating the baseline is important in the context of MBIs because, if they are less expensive in terms of compliance than some CAC alternative, regulators can present MBIs as the cheapest option, which should encourage acceptance. But this strategy will fail if polluters either believe they can influence the choice of baseline so that it effectively becomes one of the status quo, or acknowledge that the MBI solution is initially the cheapest solution but suggest that it will become more expensive in the future.

Polluters may legitimately feel that MBIs are unacceptable in the long run, even if they acknowledge that they are acceptably cheap in the short run. They may fear that an MBI will be manipulated to their detriment in the future. For example, a tax may be raised beyond the level initially justified in environmental terms and be used to raise revenue. More subtly, it is more difficult for polluters to "capture the regulatory process" when MBIs are used. Polluters use their lobbying activities to seek favor with the immediate regulators (an inspectorate, for example) and to soften the actual implementation of regulations. Consequently, while MBIs may be cheaper than alternative regulation, even in the long run, the potential for regulatory capture could lead polluters to favor the alternative measure. Governments often fail to stress the baseline issue. The carbon/energy tax proposed by the European Commission as a European Union–wide tax was subject to a massive industrial lobby that succeeded because the instrument was presented as the sole choice, making it comparatively easy for the lobby to "select" the baseline, that is, no policy at all.

A further example of failed baseline tactics is the fuel duty escalator (FDE) in the United Kingdom. The FDE was originally introduced under a Conservative administration as a "perpetual" surcharge on gasoline and diesel fuels, originally at 3 percent per annum on fuel prices (1993), then 5 percent (late 1993), then 6 percent (1997). As an annual escalator, the effect was potentially formidable: 6 percent compounded over five years would have been a 34 percent increase in the real price (assuming fuel prices were unchanged). In 1999, under a Labor administration, the FDE was changed from an *automatic* annual price increase to discretionary. In November 2000 any increase at all was cancelled, which was tantamount to a real price decrease.

The political background to these changes of intent included early pressure from the freight transport sector against rising diesel prices and direct action by an unofficial group of lorry drivers. Many factors contributed to the success of the protest, but an important one was the failure of the Labor government to make clear why the FDE existed. At various times it was justified as a carbon tax designed to help achieve the United Kingdom's 2000 target under the Framework Convention on Climate Change; as a general pollution tax; as a tax to generate revenues needed to pay off debt inherited from the Conservative government; and as a tax needed to pay for social programs, such as hospital building. Public confusion increased when newspapers claimed that cuts in FDE were cost-free because the government was "awash" with money. As a result, there was an unsurprising wide level of public support for cuts in fuel taxation. Convery (2002) notes a similar public belief in Germany and Ireland that environmental taxes are about raising revenues rather than correcting an environmental problem, and if governments are perceived not to need the money, the tax cannot be sustained. Even where the tax is understood as an environmental measure, there is often a widespread belief that taxes have little or no impact on environmental performance. This perception is easier to understand as a further manifestation of the baseline problem. Vehicle and fuel taxation may increase, but vehicle kilometers may also increase, fostering the view that the tax changes have had no effect. But relative to the true baseline of what the traffic volume would have been, the tax may well have been effective.

The media also help to determine the perception of the baseline through their preoccupation with losers and indifference to winners. Finding "losers" is part of the culture of journalism. Price rises constitute bad news, even if expenditures may decline as a result of such rises. Whereas conventional regulation may cost more than a policy based on MBIs, the CAC measures are "invisible" with respect to their effect on costs, MBIs are transparent—they show up in a market price. MBIs raise costs so there will always be losers, and the media can always exploit that fact.

One final factor determining the baseline is countervailing power. Polluters are less likely to strive for a "no-action" baseline if they believe that in so doing they will lose the respect of the public or their own employees. Many also pay close attention to NGOs. Governments may therefore influence the choice of baseline by encouraging countervailing power. One effective measure is the public right to know, for instance, pollution-release inventories. Coalitions of "gainers" should be encouraged to counteract the influence of vocal "losers." As noted earlier, the case for MBIs has lacked a pressure group, and countervailing power has not been present.

The baseline issue clearly has many facets, but it does suggest that governments that cultivate media and public opinion are more likely to succeed in introducing MBIs than those with more distant relationships with the media and the public.

Competitiveness

Probably the major factor in government reluctance to extend environmental policy and to introduce MBIs is the fear that such measures will harm international competitiveness. By and large, industry is well organized to lobby against measures it does not like. Individuals tend to be relatively poorly organized. This may explain why individuals pay environmental taxes that are higher than those on industry for comparable pollutants, notably CO_2 (Svendsen et al. 2001). The usual industrial rationale for opposing stricter environmental policy and taxes in particular is based on the effects on MBIs on competitiveness. The "climate change levy" in the United Kingdom met with substantial industrial opposition; in return for agreements to reduce emissions, energy-intensive industries secured large discounts against the tax, along with reductions in the absolute rate of tax. In Norway, efforts to extend to all industry a CO_2 tax originally introduced in 1991 met with formidable industrial opposition that has so far prevented the tax from being extended (Svendsen et al. 2001).

Competitiveness also involves the baseline. If, as most economic analysis argues, the aggregate compliance cost of MBIs is less than that of alternative CAC measures, then MBIs should impose less of a competitive burden than the baseline alternative. It is not always clear that politicians and civil servants perceive the issue this way. They may believe that MBIs impose a greater cost burden: a tax, for example, is levied on all emissions and not just on emissions over and above some prescribed standard. This seems unlikely, however. It is more probable that the greater cost burden of MBIs is an illusion arising from (a) the transparency of the measure and (b) the lobbying that tends to accompany such measures. In short, polluters react to a measure they can see and understand, and decisionmakers respond to the ensuing lobby.

It is not always clear what competitiveness means, or how the impact of environmental regulation should be measured. Various tests of the impact on competitiveness can be imagined:

- The extent to which net exports of environmentally regulated goods change with regulations or the extent to which net exports of environmentally regulated goods fall behind less regulated goods;
- The extent to which firms facing heavy regulation relocate outside the regulating country (the "pollution haven hypothesis");
- The extent to which investment occurs away from strictly regulating countries;
- The extent to which labor or total factor productivity is affected by regulation.

Net exports have not been found to be significantly affected by regulations (Jaffe et al. 1995; Sorsa 1994). Corporations' location decisions are generally unaffected by environmental costs, primarily because they tend to be a small fraction of total costs (Jaffe et al., 1995; Eskeland and Harrison 1997). Nor is there any evidence that firms invest more abroad in pollution-intensive industries to compensate for higher environmental costs at home (Eskeland and Harrison 1997; World Bank 1999).

Most studies of productivity have found modest adverse effects, but even those effects tend to reflect the adoption of traditional regulatory policies rather than MBIs. Moreover, several commentators have noted a methodological problem. The effect of environmental regulation on productivity must be negative, almost by definition. Productivity is usually measured as the ratio of output (say, GNP) to labor inputs or to some aggregate of all factors of production (in practice, this "total factor productivity" measure is calculated as a residual of unexplained GNP change, i.e., growth that is not due to changes in capital or labor). Most environmental regulation in advanced economies has been based on technological standards such as BAT and BATNEEC. Hence any regulation forces firms to purchase abatement technology that is not productive in the sense of contributing to the firm's output. Costs rise, and there is no offsetting increase in output.

But measures to reduce pollution may themselves contribute to productivity. The business and environment literature is replete with examples, unfortunately not always rigorously investigated, of firms introducing environmental measures and securing significant financial returns on those measures. Obvious examples are energy efficiency measures prompted by regulation. While purists may doubt such self-financing gains, it seems fair to say that they are possible. Firms do not necessarily minimize costs in the manner suggested by textbooks, and they may well ignore productivity gains from controlling activities, such as energy consumption, that contribute to environmental pollution. So the view that regulations necessarily reduce productivity can be challenged.

The negative effects of regulation on productivity shed little light on the desirability of MBIs as a policy instrument. None of the studies involves significant MBIs. Productivity must be correctly measured. Repetto et al. (1996) measure the damages of the environmental impacts arising from economic activity and then deduct them from the output measure. Conversely, regulation has environmental benefits that should be *added* to the conventional productivity measure. Regulation can then enhance productivity: even if conventional GNP falls because of regulation, the value of other things that people care about increases. Studies of U.S. electricity, the pulp and paper industry, and farming find that conventional measures of productivity change for 1970 to early 1990s were −.35 percent, +0.16 percent, and +2.3 percent, respectively. But the revised productivity measures, allowing for the benefits of environmental improvement, are +0.68 percent, +0.44 percent, and +2.41 percent. For electricity and paper, then, the proper measurement of productivity makes a stark difference.

The idea that regulation may actually improve competitiveness is associated with Michael Porter and the "Porter hypothesis." It is not completely clear what the hypothesis is:

- Self-evidently, more regulation benefits firms that manufacture compliance equipment. Markets for pollution-control technology and services are projected to rise well into the hundreds of billions of dollars in the next decade.
- It may be that firms that can easily comply with regulations squeeze out those who find it harder to comply, increasing the market share of the lower cost firms. Those who anticipate market changes, for example, to smaller more fuel-efficient vehicles, may gain.
- Perhaps stricter regulation forces a closer scrutiny of costs generally, "shaking out" waste and making firms more competitive. Very inefficient plants may finally be forced to close.

Most economists have been skeptical of the Porter hypothesis. If it were true it would mean that corporations are ignorant of the potential for cost reductions and require the stimulus of regulation to recognize those opportunities. This seems fairly unlikely (Jaffe et al. 1995; Oates et al. 1994). Sorsa (1994) finds no evidence to suggest that rising standards improve competitiveness. But there may be other benefits—as environmental concerns become "globalized," the green image of corporations is becoming internationally important. Environmental standards in the so-called lax environmental standard countries are in fact rising rapidly.

This brief overview of the literature suggests that environmental policy, whatever form it takes, has a negligible effect on competitiveness. Indirect impacts via productivity seem credible at first, but the use of traditional measures of output, ignoring the environmental benefit of the regulation, makes these findings suspect. Nonetheless, what matters for policy is the political perception of these issues. Whatever the empirical studies suggest, business is frequently and firmly of the view that regulations divert significant productive resources away from "output" and toward "environment." The resulting lobbies reflect this view and, as noted earlier, lobbies in favor of MBIs are nonexistent or poorly coordinated. Politicians remain comfortable with indicators of economic activity, such as GDP or labor productivity, and uncomfortable with ideas like "environmental gain" that cannot be readily measured or attributed to policy. Convery (2002) notes that the evidence may be perceived as immaterial anyway. The focus is on the future rather than the past, and the future appears to be one of more stringent environmental regulation rather than less. Past financial burdens may be no guide to future financial burdens. For these reasons there is an asymmetry between what the detailed studies find and what politicians believe. MBIs remain problematic.

Institutions and Environmental Policy

By and large, the institutions to effect environmental policy exist in rich countries. Laws can be promulgated and enacted by one or more regulatory agencies, usually an environmental agency, and backed by a centralized government department. Even rich countries need flexibility in changing institutions to meet new challenges and ensure environmental protection. The creation of agencies either dedicated to particular functions, like radiation control, or embracing interconnected environmental concerns is a case in point. In quasi-federal contexts, such as the European Union, compliance with environmental law is varied due to the substantial variation in institutional strengths across the member states.

In poor countries, the problems are obviously far more serious. Transitional and developing countries may lack an understanding of market principles. Firms have not been obliged to make profits, stock exchanges may not exist, and formal accounting may not be practiced. Asking these countries to take on systems like permit trading that require a sophisticated understanding of the workings of markets is asking too much. Emission trades need to be monitored, and any transgressions must be legally pursued. Poor countries simply lack this legal and institutional infrastructure. Trades also require considerable transparency—documents must be public and accessible, trades must be published and open to inspection by the public. Again, infrastructures simply do not exist. Trading schemes and taxes only work if polluters are aware of the costs of abatement, otherwise they cannot react to the signal being sent out. In most developing countries this requirement is simply not met. The culture of minimizing costs may not exist, especially in transitional countries, so that the very basis of MBIs, that they are cheaper than CAC, will not register with polluters as a reason to accept them.

Institutional weakness in developing countries thus inhibits the advance of environmental policy and the introduction of MBIs in particular. There are excellent examples of institutions emerging to tackle rising environmental problems—Thailand and China come to mind—while Chile has tradable water rights and has considered tradable permits for air quality.

The Social Incidence of MBIs

If preserving competitiveness is the main reason for the reluctance of rich governments to extend environmental policy, effects on socially disadvantaged groups are probably the second. Curiously, this concern becomes stronger when MBIs are proposed. Some MBIs, notably environmental taxes, are widely perceived as being regressive—the burden on the poor is greater than on the rich. This is obviously the case if the tax is on a commodity expenditure that consumes a higher proportion of low incomes than high incomes, like energy. Regressivity usually occurs when the object of taxation is a product or an input rather than the emission or environmental damage. Thus, a tax on gasoline appears to be regressive because it affects those who have to rely on motor vehicles most and who spend more of their income on private transport. Those who cannot substitute public transport for private transport—e.g., most rural dwellers—and those who earn a low income are the most affected by gasoline taxes.

As with competitiveness, the political perception of this issue is often at odds with the available evidence. Most of the evidence relates to energy and carbon taxes. It is difficult to generalize beyond these measures, but the shortage of analyses of other taxes may reflect the judgment that they are unlikely to figure prominently in household budgets and regression is unlikely. This suggests that the most important taxes for distributional assessment are those on "necessities" (i.e., goods with low price elasticity of demand). Nonetheless, the relevant information is not available, so the incidence of nonenergy taxes is not generally known.

OECD has conducted several reviews of the distributive incidence of economic instruments, with the main focus being on energy taxes (Harrison 1994; Smith 1995; OECD 1997a). The general finding is that, while taxes on household energy are regressive (although even here the evidence is not conclusive), the limited number of studies on gasoline taxes suggest that they are not regressive if the total sample of households is considered. Gasoline taxes are regressive across vehicle-owning households. This conclusion has also been reached in other assessments of vehicle fuel taxes (Barker and Kohler 1998; Speck 1999).

Popular discussion of the regressivity of environmental taxes tends to dwell on the cost of the tax and not on the benefits. The technically correct measure of "incidence" of the tax is the cost to an individual minus any benefit that he or she secures from increased environmental improvement. Thus, a tax might be regressive, and the benefits of reduced pollution might also be regressive in the sense that the rich gain more than the poor from reduced pollution. If so, the overall effect of the tax would be regressive. It is clearly important to know if this is the case with any specific MBI and with the overall structure of MBIs in general. Thus, if the overall impact of MBIs is regressive in terms of costs and benefits, some readjustment of the general tax system might be warranted in order to ensure that the advantages of MBIs are secured with overall equity.

Surprisingly little is known about the net incidence effects of environmental policy, mainly because virtually no research has taken place on the issue (beyond the energy tax incidence). While the evidence on the social incidence of MBIs is not conclusive, the gap between political perception and the available evidence is substantial. This is explained by (a) the limited evidence on incidence, (b) the complexity of thinking in terms of the "net" incidence of a tax (cost minus benefit), and (c) the "baseline" issue (comparison between the incidence of MBIs and that of the alternative policy). Policy initiatives that do not win "media acceptance" face serious obstacles. If the media portray few "losers," or if there are credible mitigation or compensation measures for losers, the policy may come across as acceptable. Illustrations of losers offered by journalists to one study (Pearce et al. 2000) included "the motorist as victim" of petroleum fuel taxes and "mothers unable to bathe children" following the introduction of domestic water metering.

There might be several justifications for ignoring equity considerations in MBI policy:

- the existence of winners and losers is not specific to MBIs—all economic adjustments, policy induced or otherwise, create losers;
- the distributional effects may be small and the costs of addressing them may greatly outweigh the benefits;
- distributional impacts may be better addressed with income distribution or spatial distribution policies than by adjusting environmental policy; and
- compensation may open the way to rent-seeking by lobbies.

Moreover, if CAC really is more expensive in terms of compliance costs than MBIs, and if the choice is between MBIs and a baseline CAC policy, one could argue that the social incidence of CAC costs will be worse than that of MBI costs. That the issue is not perceived that way has a lot to do with the transparency of MBIs and the concealed nature of price effects arising from CAC policy.

Nonetheless, equity considerations clearly do worry governments, and it may be difficult to advance the case for MBIs without a policy that avoids significant impacts on the vulnerable or includes compensation for such groups. The OECD's conclusion is,

> The distributional issues and concerns that arise with environmental taxes are greater than with economic instruments which do not raise revenues and have no fiscal character, but they are nevertheless largely overstated. Distributional issues should not, however, be disregarded, especially if the number and level of eco-taxes increase in the future. (OECD 1997a)

Summary and Conclusions

The question that prompted this essay was: Does society have the political will to face what we have termed the New Scarcity? If, by and large, the Old Scarcity has been taken care of by natural resource markets, how is the New Scarcity of essentially nonmarketed resources going to be solved? The economist's answer is approaches that essentially mimic markets: ideally, find the price that the nonmarketed resource would have if only there were a market, and then artificially introduce that price through some market-based mechanism—taxes, tradable permits, and the like. The resulting prices will be high if the environmental threat is severe, and low if the threat is a modest one. Above all, the use of market-based instruments will ensure that the requirements for efficiency in government are met. Efficiency matters because high-cost solutions divert resources away from better uses and also stimulate coalitions against further measures. Efficient solutions also aid the political compromise between those who seek radical but politically infeasible solutions and those who advise caution in the interest of employment, profits, and competitiveness. So the political economy outcome points to solutions that use market-based approaches to minimize compliance costs while achieving compromise environmental goals.

The factors inhibiting stricter environmental policy can largely be explained by differences in willingness to pay for environmental quality, a feature central to the Environmental Kuznets Curve depiction of the relationship between incomes and environmental quality. We expect poor countries to set lower environmental standards than rich ones, and this is generally what we find. Natural resources are sources of rents, rents attract rent-seekers, and corruption almost inevitably follows. Introducing environmental regulations of any kind into this setting is fraught with difficulty, and what is introduced on paper may never be enacted in practice. Institutional weakness reflects other resource endowments—levels of education, for example—but also reflects low comparative willingness to pay. Institutions tend to emerge when there is a demand for them.

Environmental policy has evolved some way toward the textbook vision of cost-minimizing, net benefit–maximizing measures. Market-based instruments do exist on a bigger scale and across more sectors than a decade ago and have arguably kept the costs of policy below what they otherwise would have been. More attention is also paid to cost–benefit comparisons in the design of policy. But technology-based and other standards still define most policy; the principle that cost matters—fundamental to MBIs—has penetrated only some areas of policy. Otherwise cost receives passing acknowledgment, and public trust doctrine declares cost irrelevant to rational policymaking.

MBIs do have genuine problems, and these have not been discussed here. Any environmental tax has uncertain effects, and if policy needs to be precisely targeted on a quantity change, a tax may be ineffective. The argument that taxes can be shifted if targets are missed is theoretically correct, but ignores the effects of uncertainty about tax rates in corporate abatement investment decisions. It is possible to argue that some MBIs have not been introduced simply because they are not as efficient, in some respects anyway, as their alternatives. In most cases, however, it is likely that political problems explain their limited progress.

The discussion has highlighted some of the difficulties of introducing environmental policy generally, and MBIs in their "optimal" form in particular. These may be classified as institutional obstacles, inconsistent policy, constituencies, and perception issues.

Institutional Obstacles

- MBIs have to coexist with the prevailing regulatory framework. This cannot be eliminated in the short run and is not inconsistent with "full-blown" MBIs.
- Regulatory capture, the "comfort" that polluters and regulators enjoy with a system they can understand and influence, hampers the transition to MBIs.
- In developing countries, institutions may simply not exist to enable the introduction of MBIs or, for that matter, environmental policy generally.

Inconsistent Policy

- Competing political goals may be inconsistent with the environmental goals of MBIs. From a textbook standpoint, securing competitive prices and then taxing the externality (the environmental impact) makes sense. From an environmental point of view, however, monopolistic prices above competitive prices are more oriented toward environmental conservation.
- Pervasive subsidies foster rent-seeking and ultimately corruption.
- More subtle forms of inconsistency arise from the uncoordinated evolution of legal precedent. "Public trust" doctrine in the United States dismisses the relevance of compliance cost to the setting of environmental standards. If costs are irrelevant, then so is regulation based on MBIs.
- Inconsistent policy also inhibits environmental policy in general. Government departments speaking for industry and agriculture hold views different from those who speak for the environment.

Constituencies

- The lobbies against MBIs are conspicuous. Regulatory capture suggests that they will comprise not only the polluter or resource degrader, but also government departments acting as the voice of those lobbies, and sometimes even the regulatory agencies themselves. Lobbies for MBIs are likely to reside in academic and "think tank" institutions.
- Rich countries are well endowed with nongovernmental "voices" speaking for the environment, but that is not true of many developing countries. This may reflect low public concern for the environment, but it may also reflect organized suppression of environmental groups in countries where they are seen as a challenge to authority and to rent appropriation.

Perception Issues

- How MBI proposals are presented matters a great deal; how they are received by the media matters also. All too often, an MBI appears to be contrasted with

doing nothing at all, when the alternative is actually some other form of regulation. Finding problems with MBIs without seeing if the alternative fares any better is self-evidently illogical. Yet that is how the issue is often perceived.
- Politicians are sensitive to competitiveness and equity (social incidence). The evidence that MBIs perform badly in these terms is limited; the evidence that they perform worse than the alternative is negligible. But the evidence is not easy to assimilate. Measurement issues are complex and should involve evaluating the net impacts—costs and benefits—rather than the cost impacts only.
- Culture contributes to the perception problem. No one likes taxes, but it is not uncommon in public debates on MBIs to find the view that charging for "basic needs" is wrong per se (the Irish water example). If incidence issues are a political concern in rich countries, naturally they are significant in poor countries. Raising energy or water prices could well harm the poor, although vocal opposition to such policies tends to be engineered by those better off.

Does all this mean that the pursuit of stronger environmental policy and the ideal of an MBI-based policy is misplaced? The answer has to be "no." Arguments about the desirability of environmental policy are changing. More people recognize that, while Nature matters for itself, for future generations, and for amenity and aesthetic purposes, the environment matters for human health and for the traditional policy goals of economic growth. Nowhere is this argument more powerful than in the developing world. But if there is to be even stricter environmental control, it is also vital to adopt efficient environmental policy with the promise of keeping compliance costs low. That still favors a market-based approach. What matters is not just the efficient design of policy, but a more detailed understanding of the political economy of the choice of environmental instruments.

Notes

1. In radiation protection, the equivalent is "as low as is reasonably achievable" (ALARA), and in electromagnetic radiation it is "prudent avoidance" (PA).
2. UN ECE is the United Nations Economic Commission for Europe, based in Geneva.

References

Barker, T., and J. Kohler. 1998. Equity and Ecotax Reform in the EU: Achieving a 10 percent Reduction in CO_2 Emissions Using Excise Duties. *Environmental Fiscal Reform Working Paper 10*. Cambridge: Cambridge University.

Barnett H., and C. Morse. 1963. *Scarcity and Growth: The Economics of Natural Resource Availability*. Baltimore: Johns Hopkins University Press for Resources for the Future.

Barrett, S. 2003. *Environment and Statecraft: The Strategy of Environmental Treaty-Making*. Oxford: Oxford University Press.

Baumol, W., and W. Oates. 1988. *The Theory of Environmental Policy*. 2nd ed., Cambridge: Cambridge University Press.

Bell, R.G. 2002. Are Market-Based Instruments the Right Choice for Countries in Transition? *Resources* 146 (Winter): 10–14

Brännlund, R. 1999. The Swedish Green Tax Commission. In *Green Taxes: Economic Theory and Empirical Evidence from Scandinavia*, edited by R. Brännlund and I.-M. Green. Cheltenham, UK: Edward Elgar, 23–32.

Brown, L. 2001. *State of the World 2000*. New York: W.W. Norton.

Christiansen, V., and I.-G. Gren. 1999. The Governmental Commission on Green Taxes in Norway. In *Green Taxes: Economic Theory and Empirical Evidence from Scandinavia*, edited by R. Brännlund and I.-M. Gren. Cheltenham, UK: Edward Elgar, 13–22

Convery, F. 2002. Acceptability and Implementation Problems. In *Green and Bear It? Implementing Market-Based Policies for Ireland's Environment*, edited by D. McCoy and S. Scott. Dublin: Economic and Social Research Institute, 85–102.

Cropper, M., and W.E. Oates. 1992. Environmental Economics: A Survey. *Journal of Economic Literature* 30 (June): 675–740.

Davies, C., and J. Mazurek. 1998. *Pollution Control in the United States: Evaluating the System.* Washington, DC: Resources for the Future.

Eckstein, O. 1958. *Water Resource Development: The Economics of Project Evaluation.* Cambridge, MA: Harvard University Press.

Eskeland, G., and A. Harrison. 1997. *Moving to Greener Pastures? Multinationals and the Pollution Haven Hypothesis.* Policy Research Working Paper 1744. Washington, DC: World Bank.

Gray, L. 1914. Rent under the Assumption of Exhaustibility. *Quarterly Journal of Economics* 28: 466–489

Hahn, R. 1989. Economic Prescriptions for Environmental Problems: How the Patient Followed the Doctor's Orders. *Journal of Economic Perspectives* 3(2): 95–114.

Hahn, R. 2000. *Reviving Regulatory Reform: A Global Perspective.* Washington, DC: American Enterprise Institute–Brookings Joint Center for Regulatory Studies.

Hanemann, M. 1999. Water Resources and Non-Market Valuation in the USA. Paper presented to the Chartered Institution of Water and Environmental Management (CIWEM) conference, "Valuing the Environment Beyond 2000," London.

Harrison, D. 1994. *The Distributive Effects of Economic Instruments for Environmental Policy.* Paris: Organisation for Economic Co-operation and Development.

Helm, D. 2000. Objectives, Instruments and Implementation. In *Environmental Policy: Objectives, Instruments and Implementation*, edited by D. Helm. Oxford: Oxford University Press, 1–28.

Hotelling, H. 1931. The Economics of Exhaustible Resources. *Journal of Political Economy* 39 (April): 137–175

Jaffe, A., S. Peterson, P. Portney, and R. Stavins. 1995. Environmental Regulation and the Competitiveness of US Manufacturing: What Does the Evidence Tell Us? *Journal of Economic Literature* 33: 132–163.

Jevons, W.S. 1865. *The Coal Question: An Inquiry concerning the Progress of the Nation, and the Probable Exhaustion of Our Coal-Mines.* London: Macmillan.

Jones, C., and K. Pease. 1997. Restoration-Based Compensation Measures in Natural Resource Liability Statutes. *Contemporary Economic Policy* 15(4): 111–122.

Klaassen, G. 1996. *Acid Rain and Environmental Degradation: The Economics of Emission Trading.* Cheltenham, UK: Edward Elgar.

Krueger, A. 1974. The Political Economy of a Rent-Seeking Society. *American Economic Review* 64: 291–303.

Krutilla, J., and O. Eckstein. 1958. *Multipurpose River Development.* Baltimore: Johns Hopkins University Press.

Lawton, J.H., and R.M. May. 1995. *Extinction Rates.* Oxford: Oxford University Press.

Lomborg, B. 2001. *The Skeptical Environmentalist: Measuring the Real State of the World.* Cambridge: Cambridge University Press.

Lutter, R. 2001. *Ignoring All Costs Won't Clean Our Air.* Policy Matters 01-05. Washington, DC: American Enterprise Institute–Brookings Joint Center for Regulatory Studies.

McKean, R. 1958. *Efficiency in Government through Systems Analysis.* New York: Wiley.

Meadows, D., D.L. Meadows, J. Randers, and W.W. Behrens III. 1972. *The Limits to Growth: A Report for the Club of Rome's Project on the Predicament of Mankind.* Washington, DC: Earth Island.

Mortensen, J.B., and J. Hauch. 1999. Governmental Commissions on Green Taxes in Denmark. In *Green Taxes: Economic Theory and Empirical Evidence from Scandinavia*, edited by R. Brännlund and I.-M. Gren. Cheltenham, UK: Edward Elgar, 1–12.

Oates, W., K. Palmer, and P. Portney. 1994. *Environmental Regulation and International Competitiveness: Thinking about the Porter Hypothesis.* Discussion Paper 94–02. Washington, DC: Resources for the Future.

OECD (Organisation for Economic Co-operation and Development). 1975. *The Polluter Pays Principle: Definition, Analysis, Implementation.* Paris: OECD.

———. 1997. *Environmental Taxes and Green Tax Reform.* Paris: OECD.

———. 1999. *OECD Environmental Data Compendium 1999.* Paris: OECD.

———. 2001. *Environmentally Related Taxes in OECD Countries.* Paris: OECD. www.oecd.org/env/policies/taxes/index.htm.

Panayotou, T. 1998. *Instruments of Change: Motivating and Financing Sustainable Development.* London: Earthscan.

Pearce, D.W. 2000. The Economics of Technology-Based Environmental Standards. In *Environmental Policy: Objectives, Instruments and Implementation,* edited by D. Helm. Oxford: Oxford University Press, 75–90.

———. 2002. Will Global Warming Be Controlled? Reflections on the Irresolution of Humankind. In *Challenges to the World Economy: Festschrift for Horst Siebert,* edited by R.Pethig and M.Rauscher. Berlin: SpringerVerlag, 367–382.

Pearce, D.W., F. Putz, and J.Vanclay. 2001. Sustainable Forestry in the Tropics: Panacea or Folly? *Forest Ecology and Management* 172: 229–247.

Repetto, R., D. Rothman, P. Faeth, and D.Austin. 1996. *Has Environmental Protection Really Reduced Productivity Growth? We Need New Measures.* Washington, DC: World Resources Institute.

Rose-Ackerman, S. 1999. *Corruption and Government: Causes, Consequences and Reform.* Cambridge: Cambridge University Press.

Russell, C.S., and P.T. Powell. 1996. *Choosing Environmental Policy Tools: Theoretical Cautions and Practical Considerations.* Social Programs and Sustainable Development Department. Washington, DC: Inter-American Development Bank.

Skou Andersen, M. 2000. Designing and Introducing Green Taxes: Institutional Dimensions. In *Market-Based Instruments for Environmental Management: Politics and Institutions,* edited by R.-U. Sprenger and M. Skou Andersen. Cheltenham, UK: Edward Elgar, 27–48.

Smith, S. 1995. *Review of Empirical Evidence of Distributional Effects of Environmental Taxes and Compensation Measures.* Paris: Organisation for Economic Co-operation and Development.

Smith, V.K. (ed.). 1979. *Scarcity and Growth Reconsidered.* Baltimore: Johns Hopkins University Press for Resources for the Future.

Sorrell, S. 1999. Why Sulphur Trading Failed in the UK. In *Pollution for Sale: Emissions Trading and Joint Implementation,* edited by S. Sorrell and J. Skea. Cheltenham, UK: Edward Elgar, 170–207.

Sorsa, P. 1994. *Competitiveness and Environmental Standards.* Policy Research Working Paper 1249. Washington, DC: World Bank.

Speck, S. 1999. Energy and Carbon Taxes and Their Distributional Implications. *Energy Policy* 27: 659–667.

Sprenger, R.-U. 2000. Market-Based Instruments in Environmental Policies: The Lessons of Experience. In *Market-Based Instruments for Environmental Management: Politics and Institutions,* edited by R.-U. Sprenger and M. Skou Andersen. Cheltenham, UK: Edward Elgar, 3–26.

Svendsen, G.T., C. Daugbjerg, L. Hjollund, and A. Branth Pedersen. 2001. Consumers, Industrialists and the Political Economy of Green Taxation: CO_2 Taxation in the OECD. *Energy Policy* 29: 489–97.

EPA (U.S. Environmental Protection Agency). 2001. *The United States' Experience with Economic Incentives for Protecting the Environment.* Report No. EPA-240-R-01-001. Washington, DC: EPA.

van Beers, C., and A. de Moor. 2001. *Public Subsidies and Policy Failures: How Subsidies Distort the Natural Environment, Equity and Trade, and How to Reform Them.* Cheltenham, UK: Edward Elgar.

Weitzman, M. 1974. Prices versus Quantities. *Review of Economic Studies* 41: 477–491.

Wildavsky, A. 1995. *But Is It True? A Citizen's Guide to Environmental Health and Safety Issues.* Cambridge, MA: Harvard University Press.

World Bank. 1999. *Greening Industry: New Roles for Communities, Markets and Governments.* Washington, DC: World Bank.

CHAPTER 11

Public Policy

Inducing Investment in Innovation

Molly K. Macauley

T HE GOVERNMENTS OF MOST industrial nations wield influence over their countries' research and development (R&D) activities in natural resources and environmental management. Government involvement assumes many forms, including protecting intellectual property; carrying out research at government laboratories or other facilities, often in partnership with the private and academic sectors; and financially supporting research and development. But government's influence is in fact more widespread: a host of other policies, although not directed toward R&D, significantly affect the rate and direction of innovation. Examples include rationing access to and use of natural and environmental resources, controlling pollution emissions, and setting energy efficiency standards.

What effect, if any, has government influence on R&D had in alleviating resource scarcity? Previous case studies reveal a mixed record. Ineffective government management of fundamental property rights appears to have biased the direction and rate of innovation away from a socially desirable path in some natural resource industries. Government sponsorship of R&D through joint endeavors with industry or by way of a variety of financial incentives, including direct subsidies and tax provisions, has also brought uneven results. But tradable permit regimes for environmental management seem to have brought about an appropriate amount and type of innovation.

Innovations largely undertaken independently by the private sector have alleviated much of the effect of any increasing scarcity of natural resources during recent decades. Simpson (1999) discusses in detail an array of new technologies that have made significant productivity improvements in coal mining, petroleum exploration and development, the copper industry, and forestry. The authors in this edited volume scarcely mention any government actions.

The first section of this chapter summarizes the conceptual literature on the relationship between government policy and innovation, including rationales for policy intervention. It discusses the evolution in analysis of "induced innovation," starting with the original concept enunciated in Hicks (1932)—that price changes induce innovation as firms seek to economize on inputs that have become relatively expensive. Extensions of Hicks's concept consider explicitly the role of government policy in influencing price changes—and thus inducing innovation by the actions of public policy.

Next we review rationale for government intervention in innovation. Policy-induced innovation may derive from policy explicitly directed toward R&D, or it may occur by a more circuitous route from public intervention not overtly directed toward R&D.

The next section looks at case studies of the effects of government policy on technological change in several natural resource and environmental industries, including energy, fishing, water, forestry, the airwaves, and pollution abatement and control. These case studies illustrate the gamut of policy-induced innovation. A much larger empirical literature addresses the role of government in the diffusion of new technology—through adoption, imitation, or licensing—for managing environmental quality. A paper by Jaffe et al. (2003) summarizes this literature; therefore, technology diffusion is given less coverage here.

The last section offers suggestions for further research and some concluding observations.

Prices and Innovation

Public policy can influence innovation by directly or indirectly affecting the prices of inputs and outputs of firms as they produce goods and services. In the literature on the economics of technological change, these influences of policy on the behavior of firms are loosely referred to as a form of induced innovation.

Induced Innovation

Before analysts began to model the role of government in technology and innovation policy, the concept of induced innovation merely described how firms alter their inputs in response to changes in the prices of inputs—whether or not government intervened. It signified that firms may respond to an increase in prices by reducing quantities of inputs, substituting among inputs, or perhaps changing a production process or technique to use less of a more expensive input. The induced innovation hypothesis as initially proposed by Hicks (1932, *124*) holds that

> a change in the relative prices of the factors of production is itself a spur to invention, and to invention of a particular kind—directed to economizing the use of a factor which has become relatively expensive.

Hicks did not mathematically formulate his hypothesis; a variety of formal specifications were developed later (summarized in Binswanger and Ruttan 1978).

These models have in turn led to a variety of empirical tests. But direct measurement of the relationship has been problematic, even in the absence of the confounding influence of policy intervention. Early research—generally without detailed formal models—focused largely on time trends in aggregate price data. For instance, Barnett and Morse (1963), in their discussion of natural resource scarcity, traced inflation-adjusted prices from about 1900 to 1957. They pointed out that prices had gone up and down, but that the general trend for most resources was a decline in prices. Hence they rejected much of the then-prevalent thinking that resources were becoming increasingly scarce.

Underlying determinants of these price trends, including the role of innovation, are less obvious. Barnett and Morse conceded that it was difficult to disentangle the effects of demand and supply merely by observing prices. Smith and Krutilla (1979, *279*) comment,

> The literature on induced technical change was not widely accepted at the time Barnett and Morse prepared their book. Their discussion of technical change was more in terms of a description of the past patterns of technological change rather than a specific generating mechanism. Nonetheless, it seems clear from their explanations that they did consider the increasing scarcity of natural resources among the motivations for developing new technologies.

The Barnett and Morse research thus introduces a host of questions about the contribution of technology to the trends they observed. Were downward trends in the prices of resources an outcome of innovation, in that as prices rose, industries using the resources found innovative ways to economize on inputs, thus lessening pressure on supply and leading to lower prices? Were extractive industries finding new extraction techniques that increased supply? Were there other trends in the economy at large that led to price changes—for example, slowing productivity, recession, technical change, complementary or substitute goods and services? Another question raised by recent research is the role of not only trends in price levels but price variability (Carey and Zilberman 2001; and Moreno 2001). Does innovation offer a hedge against this type of uncertainty? What part, if any, does government play?

"Policy-Induced" Innovation

Researchers have extended the relationship depicted by Hicks between prices and innovation to include innovation in response to price changes that in turn were induced by government policy intervention. If a firm's production process is thought of as combining inputs to produce outputs, and government policy affects input or output decisions, then the firm may respond with innovations in its mix of inputs, outputs, or the production process itself. Even the firm's own internal plans for innovation—itself an input into the production process—may be influenced through policy. An example is a tax credit for expenses incurred in conducting R&D. In the 1960s and 1970s, government regulation triggered a spate of theoretical research and innovation (see Kamien and Schwartz 1969; Smith 1974, 1975; Okuguchi 1975; Magat 1976). More recent empirical research has modeled innovation specifically in response to environmental regulation (e.g., Newell et al. 1999).

The case studies surveyed here exemplify the myriad policy-induced innovations, ranging from firms' response to governmental allocation of property rights (airwaves and fish resources) and financial payments and tax credits for new investment (renewable energy technology) to innovation in response to mandatory standards (energy efficiency in appliances and autos). The case studies also include examples of technological change in response to R&D conducted by government alone or in partnership with universities or the private sector (synthetic fuels and forestry). They reveal a wide range of reasons for government intervention.

Rationale for Public Policy Intervention in Innovation

Before turning to reasons for public policy intervention in innovation, a few definitions are in order. This chapter interchangeably uses "innovation," technological change," and "R&D." Imprecision in this choice may be excused by what engineers and others who actually carry out R&D themselves admit is an artificial distinction among these activities in practice. Freeman and Soete (1997) offer insight on the nonlinearity of the innovation process; see also Stokes (1997). Any or all stages of the innovation process can be carried out by the private or public sectors, separately or jointly. By and large, basic research—that aimed at deriving fundamental knowledge—is thought to be the purview of government, for reasons discussed further below, and applied research—engineering, testing, and prototyping—is considered to be the role of the private sector. But the line separating these domains—who should do the research, and who does—is blurred.

Scholars have long known a variety of reasons why the private sector may fail to provide enough innovation—that is, a socially optimal amount of R&D. They have also prescribed ways for government involvement to mitigate this shortfall—that is, improve economic efficiency. Related studies rationalize government involvement on the basis of the political attractiveness of R&D spending. One reason is that it provides a "pork barrel" from which politicians may draw benefits for their constituents. It may also garner votes across party lines based on a conviction that sponsorship and support of R&D demonstrates national leadership in technological prowess.

Concerns about Economic Efficiency

Three sources of failure of private markets are typically invoked to justify government intervention in R&D:

- nonappropriability of the benefits of R&D or innovations—that is, the difficulty in securing ownership of and profit from these activities;
- unwillingness of markets to finance R&D, given its uncertain financial return; and
- distortions caused by existing government policies or regulations that negate incentives to the private sector to pursue R&D.

Additional rationales for government intervention include issues involving the market structure of an industry and the influence of political institutions.

Nonappropriability. Firms tend to underinvest in R&D (Arrow 1962). R&D can be, like environmental improvement, a public good. Underinvestment—the failure to engage in the socially optimal level of R&D, where its incremental benefits and costs just offset each other—can result for several reasons. Use of the new knowledge brought about by the R&D is both nonrival—others can share it without reducing its inherent value—and nonexclusive—it is difficult to keep others from using it, absent enforceable property rights. If a firm can improve its production technology or lower its costs by observing another firm's experience, it has less incentive to invest in acquiring experience itself. If industry-wide learning economies are large relative to firm-specific learning economies, this nonappropriability may discourage private development of the technology. If innovations diffuse rapidly through an industry, then each firm has less incentive to invest in innovations because the firm can "ride free" off others' investments. If a firm expects to bear all the costs but only some of the benefits, it will tend to invest less than the socially optimal amount in the project.

Patents and other "elaborate distinctions of partial property rights of all degrees" (Arrow 1962) are a possible solution to the externality. But nonappropriability can also be a consequence of spillovers of knowledge and technology from nonpatentable results of basic R&D or from industry-wide learning by doing. The results of basic research are not typically patentable because they are directed toward general use rather than a specific application. For this reason, much basic research takes place in universities and government laboratories, supported by, among other agencies, the National Science Foundation, the high-energy physics programs at the Department of Energy, and space science at the National Aeronautics and Space Administration.

Would a monopolistic firm be more inclined to undertake R&D? One camp says yes, because a monopolist does not have competitors ready to imitate an innovation or to circumvent a patent. Others conclude that a monopolist earns monopoly profit without the innovation and therefore is in less of a hurry to innovate than a competitive firm.

Firms are less likely to underinvest in R&D if they are competing for patents. By increasing its R&D effort, a firm reduces the probability that a rival will obtain the patent. But firms engaged in a patent race may overinvest in R&D and duplicate too much of their research effort. This phenomenon is analogous to rent dissipation. Each time the patent process engenders a rent, there is competition for it, and the rent tends to be partially dissipated by the additional costs incurred in an attempt to appropriate it. (See Reinganum 1982.) Firms in a patent race may wastefully differentiate their R&D activities, say, by making a small change in a formula or algorithm to distinguish their research from that of competitors, but with no benefit to consumers. Critics of this patent-race model argue that it makes the unrealistic assumption that firms do not learn during the R&D process. The model ignores strategic behavior—for instance, the advantage of being the first to innovate may be so large that being second or third, or concentrating on product differentiation, is of less value. Reasoning that "two heads are better than one," government often establishes several independent research programs. If these are conducted simultaneously, they could slow a patent race. Firms would not have as much opportunity to internalize the loss of patent revenues incurred by (sequentially) preempted rivals. (Tirole 1988 offers a useful overview of much of this literature.)

Financial markets—uncertain returns. Underinvestment in R&D may also result from inadequate spreading of risk. The return to investment in research is inherently speculative. Hall (2002) discusses the conceptual and empirical research on venture capital for funding R&D, particularly the gap between the private rate of return and the cost of capital when the innovation investor and the financier are different entities and have different information about the technical risk of the R&D. This gap is not the same thing as the gap between private and social returns arising from the nonexclusivity of R&D. In addition, R&D sometimes may have large fixed costs and long lead times for project development or payback. In these circumstances, capital markets may be unwilling to finance R&D.

Arguments that capital markets are limited in their capacity to absorb risk can be overstated. Rose (1986) offers several examples. Pharmaceutical firms invest roughly $1 billion annually in R&D on new drugs, despite highly uncertain and distant returns. Development and production of IBM's System/360 involved nearly half a billion dollars in R&D and risked $5 billion, a "bet the company" gamble on a highly innovative computer design. The rapid development in the 1980s and 1990s of biotechnology and genetic engineering firms attests to the willingness of private investors to finance high-risk ventures in R&D areas (even if the experience of the late 1990s proves that not all of these investments were wise).

Even if barriers to R&D financing loom large, a lack of private capital alone may be an insufficient rationale for government intervention. If private investors decline to finance a project because there are more productive uses of their capital, then, unless government is better at "picking winners" than the private sector, it reduces total social product if it intervenes to finance the project. The extensive case studies in Cohen and Noll (1991) on supersonic transport, the Clinch River breeder reactor, synthetic fuels from coal, and photovoltaics commercialization programs show a relatively poor track record of public sector financing. Other studies report beneficial results of government investment. The National Research Council (NRC) (2001) asked whether fossil energy research supported by the U.S. Department of Energy has been worthwhile. In the council's review of 39 R&D initiatives ranging from electronic ballasts in compact fluorescent tubes to atmospheric fluidized-bed coal combustion, at least a handful of projects were deemed successful in that the estimated net realized economic benefits were positive. A long series of studies initiated by Griliches in 1979, revisited in Griliches 1998, and continued by other researchers, finds that the social rate of return to *all* government spending on innovation is positive. This argues that the sum of government's investments in R&D may be considered a portfolio with gains and losses that whether by skill or by luck yield a good net return.

In many industries, government has encouraged research joint ventures (RJVs) in which firms agree to share the expenditures and the benefits associated with a given research project. RJVs may increase R&D activity by correcting the externality of nonappropriability or enable firms to share R&D costs to undertake innovation that a single firm may not be willing to support. Such ventures may decrease R&D activity, however, if they simply redistribute rather than increase industry profit. This may occur if the firms are in a concentrated industry and the RJV helps rivals to avoid competing in the R&D market. The ventures may also facilitate horizontal collusion in preventing the firms from competing on the output market.[1]

Government-induced distortions. Existing government policies and programs may also influence private sector R&D. A case study discussed later in this chapter is government regulation of the electromagnetic spectrum, or airwaves. The manner in which the government allocates this resource, and in turn influences its shadow value as an input into telecommunications services, appears to have biased the direction of private R&D in some communications technologies. Another case discussed below is regulation of the fishing industry. Analysts have found that by affecting the relative profitability of fishing, some government regulations have affected R&D in that industry. Analysts have also concluded that natural gas wellhead price regulation discouraged producers from investing in natural gas exploration during the 1960s and early 1970s (Stauffer 1975). Emissions limits specified as engineering standards can discourage investment in emissions control technology.

Uncertainty about future legislation and administrative actions can affect either the expected return from a particular technology or the risk of that return. Tax credits provided for R&D have always been a temporary provision in the U.S. tax code. Even though it has been extended repeatedly, most public finance economists agree that its provisional nature, together with frequent changes in its tax rate, exemptions, and other provisions, has greatly weakened its potential to stimulate R&D.

Innovation and the Environment

Innovation in environment-related science and technology is the object of even greater apprehension. The environment itself is a public good and may not be "priced" correctly if at all. The fundamental catalyst for Hicks's induced innovation process—responses to price changes—may be missing if prices of environmental inputs (air, water, and so forth) are missing.

In the case of open-access natural resources, like most fishing resources, the same lack of property rights can prevent price signals from conveying the desirability of innovating to conserve or better manage outputs, like fish stocks. In an overview of historical trends in renewable resource markets, Frederick and Sedjo (1991, *17*) state,

> Institutional factors such as the laws, policies, programs, and administrative arrangements directed to managing resource use have had major influences on the changing status of the resources. The institutions themselves have evolved over time as resource demands and conditions have changed. Major institutional shifts, however, have generally come only after resource conditions had deteriorated to a state that attracted widespread attention and concern. Frequently, the institutions encouraged use, tolerated questionable practices, and as with water rights, limited the incentives and opportunities

for adjusting resource use and management efficiently to changes in the underlying condition and availability of the resources.

For many natural resources, incentives to innovate depend in large part on the establishment of clear property rights. Rivers, streams, and groundwater resources that flow from one property to another are common property resources; in the absence of property rights, they are un-owned until captured for use and in the interim can be overused. Improved environmental performance itself is a public good, and incentives to invest in such improvements may be blunted.

In the case of natural resources and the environment, two types of resource misallocation may occur. Use of the resources and environmental pollution levels may be too high, and incentives to engage in innovation to improve resource use or protect the environment may be too low. These mismatches occur because the resources are not priced correctly and because appropriating returns to R&D is difficult.

In theoretical research at the forefront of addressing both misallocations, Parry et al. (2000) set up a dynamic social planning model to compare the effects of innovation to reduce future costs of pollution control with the effects of controlling pollution now, at a constant state of technology. They found that controlling pollution now is likely to be more attractive, largely because innovation is costly and takes time. Their results hold for stock and flow pollutants, different environmental damage functions, and short and long planning horizons. Their results are derived from a "first-best" case, in which both pollution control and innovation take place at the socially optimal amount. Nonetheless, such research could go far in setting priorities for government policy concerning the twin issues of intervening to control pollution and supporting innovation for pollution control.

The public goods nature of natural resources and the environment challenges not only theoretical research on innovation but empirical research as well. Researchers cannot easily observe the shadow prices of natural resources and environmental goods. For this reason, empirically testing Hicks's price-induced link between government policy and technological change—that is, discovering when prices signal scarcity and lead to innovation—relies heavily on proxies for shadow prices.

Political Economy of Innovation

Efficiency goals are not the only motivation for government intervention. In R&D policy, as in many other areas, political objectives like demonstrating technological leadership, furthering international competitiveness, and enhancing national prestige often dominate policy debates. In their extensive study of the role that government has played in a variety of commercial R&D programs, Banks et al. (1991) find that political institutions introduce systematic biases in incentives that lead to cost overruns and poor performance.

A potentially large bias toward higher costs and low performance stems from the distributive politics of R&D programs. In most cases, the costs and benefits of programs are geographically unevenly distributed. Renewable energy projects, in which the natural endowments of an energy resource such as wind, geothermal, hydropower, or solar thermal vary markedly among regions of the country, can arouse contention among states in supporting federal outlays for renewable energy

R&D. Politicians subject to electoral cycles operate within short time frames. This limits willingness to fund projects that have only long-run results, like R&D. It also means that policymakers may initiate projects in response to short-term crises.

Technology policy in the form of mandating pollution control technology or energy efficiency standards, for instance, may be more attractive to policymakers than are other strategies to encourage innovation. In the case of environmental regulation, Parry et al. (2000) suggest that mandating stringent emissions reductions may be politically difficult when the costs are concentrated in one industry while the benefits are widely diffused. Governments may consequently turn to promulgation of technology policy as a less controversial approach. Perry and Landsberg (1981) point to the proclivity of government intervention in R&D to tend toward supply rather than end-use problems. They reason that it is easier to construct and run programs involving only a small number of sophisticated performers than programs that involve millions of consumers.

Although there may be imperfections in the operation of the private market supplying R&D, this does not imply that market imperfections necessarily warrant government intervention. Government policy is seldom perfectly constructed or implemented, it typically introduces a new set of distortions into market operations, and it offers "pork" opportunities.

Some researchers maintain that new scientific knowledge generated by government-financed R&D expands firms' basic knowledge and stimulates their own R&D. Others argue that publicly financed and company-financed R&D are substitutes, because either the output of the public R&D activity is internalized by the firms, or the publicly financed R&D performed by the firms causes the firms to reach their full R&D capacity in lieu of internal effort. Tests of these theories are ambiguous and appear to depend on how "R&D intensive" an industry is; public and private R&D appear to be somewhat complementary in low R&D-intensive industries and apparently weak substitutes in high R&D-intensive industries (Mamuneas and Nadiri 1996).

Taken together, these considerations urge a higher standard for government intervention than a simple "market failure" test and reinforce the conclusion that imperfect markets must be balanced against imperfect government in designing policy.

Case Studies: Mixed Results from Policy-Induced Innovation

This section illustrates results from case studies of the effects of government policies on innovation in natural resources and environmental management. It also briefly reviews the empirical literature on the effectiveness of broad-based R&D tax credits not specifically targeted at natural resources and the environment, but nonetheless potentially effective.

Synthetic Fuels from Coal—Demonstration Projects and the Politics of "Eastern Coal"

The idea of a large synthetic fuels (synfuels) industry in the United States began in the 1960s with a handful of pilot plants undertaking research sponsored by the newly established Office of Coal Research at the U.S. Department of Energy. Synfuels gained attention after the Arab oil embargo in 1973 and as part of the Ford administration's energy policy announced in 1975. In early 1979, oil import prices began to rise sharply, and in June 1980, Congress approved legislation designed to make the synthetic fuel industry a dominant feature of U.S. energy policy, with the U.S. government as the linchpin in the endeavor. Justification for government intervention was the slow pace at which industry seemed to be developing synthetic fuels, the potentially severe consequences of oil import dependence, and the worldwide implications of a further tightening of the liquid fuel market through increased supply manipulation by major exporters.

The legislation provided for establishment and government funding of a variety of R&D activities, including a Synthetic Fuels Corporation (SFC) to promote a domestic synfuels industry marketing solids, liquids, and gaseous fuels derived from coal, oil shale, peat, and tar sands. The government would make available some $88 billion in the form of price guarantees, purchase commitments, loan guarantees, and direct loans; the government could also enter joint ventures by obtaining minority equity interests in small-scale projects.

The idea of creating a synfuels industry was long debated and by no means universally accepted. Enthusiasts pointed to German success with synfuels in World War II, the South African record of producing liquids from coal, and the synthetic rubber production undertaken by the U.S. government as a wartime emergency measure in the early 1940s. Yet these projects produced a small fraction of the expected daily output of the planned new U.S. synfuels industry, and scaling up to its anticipated size would require output more than 30 times greater than the largest pilot plant then under testing. Supplying 10 percent of the country's liquid fuel requirements would call for the production of 230 million tons of coal, or nearly one-third of total U.S. production in 1979 (Perry and Landsberg 1981).

In their analysis of the synfuels program, Cohen and Noll (1991) comment that at first glance, the government's overall approach to synfuels R&D seemed archetypical as a government R&D effort. It was reasonable to develop a backup technology that would hold down imported oil prices and protect the United States against oil supply disruptions. It offered a mixed strategy or portfolio of small- and large-scale demonstration projects in partnership with government to test very speculative research of a wide variety of technological options—not picking "winners" or "losers." The approach also avoided large projects with "concentrated pork barrel benefits." Yet they wrote,

> The entire synfuels program had a quality of madness to it. Project after project failed. Cost estimates were connected to the price of substitutes rather than to the program itself. Goals were unattainable from the start. Official cost–benefit studies estimated net benefits in the minus billions of dollars. Even apart from the Synthetic Fuels Corporation, the dogged continuation of the research and development program seems incredible. However, despite the rhetoric, it is clear that the wildest scenarios never came close to happening. The nation's entire construction industry was not pressed into service to build plants that would not work in order to impress the Organization of Petroleum Exporting Countries (OPEC) of the United States'

commitment to energy independence. Nevertheless, substantial funds were expended, and several patterns emerge. (*297*)

In a study of fossil energy R&D, the National Research Council (2001, *66*) concludes in the case of synthetic fuels programs during the 1970s and early 1980s,

> In retrospect, technology development in direct coal liquefaction and other synthetic fuels programs … was not handled well by the government or industry. Technologies were targeted for major demonstration expenditures before they were well understood.

Cohen and Noll take the diagnosis a few steps farther. The patterns they observe are "dramatic expansions of the size of the program after the energy crises of 1973 and 1979" and "hasty" project choices in "knee-jerk responses to the crises." The project's use of eastern coal led to votes in Congress dominated by westerners, who saw their states bearing the costs but not enjoying any benefits of the program. Political opposition and improvements in oil markets eventually led to termination of the program.

Energy Conservation—The Checkered Record of Standard Setting and Tax Credits

Like the synfuels program, government involvement in R&D to improve energy efficiency was also initiated in response to the first oil embargo in 1973. And like sponsorship of synfuels demonstration projects, programs centered on applied product and process research in joint endeavors with industry to develop more efficient heating, lighting, and refrigeration, and new multifuel transportation technologies. Other aspects of federal efforts to improving energy efficiency were very different from the synfuels program. New legislation established automotive corporate average fuel economy (CAFE) standards, the labeling of appliances to state their energy consumption, tax credits for energy-efficient retrofitting of residential buildings, low-income weatherization grants, and standards for the energy efficiency of new buildings.

One of the standard-setting programs was the National Appliance Energy Conservation Act, passed in 1987 to set minimum efficiency standards for household air conditioners and gas water heaters. In sharp contrast to the synfuels experience, some standard-setting activities indeed brought about a modicum of innovation. Newell et al. (1999) extend the Hicks concept of price-induced innovation to include the price effects brought about by these regulatory standards. Specifically, their econometric study measures the extent to which the energy efficiency of home appliances changed in response to the promulgation of new efficiency standards between 1958 and 1993. They find that a sizeable portion of efficiency improvements was autonomous, and the rate of overall innovation was independent of energy prices and regulation. However, some innovations did appear to be the result of energy price changes for some products, and manufacturers modified models that were offered for sale after the energy-efficiency product labeling was required.

Standards for automobile fuel efficiency yielded ambiguous results. In 1975, in the wake of the 1973 oil crisis, Congress passed the Energy Policy and Conservation Act to reduce dependence on foreign oil. Among other provisions, the act established the corporate average fuel efficiency program to require automobile manufacturers to increase the sales-weighted average fuel economy of the passenger car and light-duty truck fleets sold in the United States.

An extensive review of the program by the National Research Council (2002) finds that the standards have played a leading role in preventing fuel economy levels from dropping since the early 1980s, even as inflation-adjusted fuel prices continued to decline and fuel consumption for auto transportation continued to increase. To improve fuel economy, manufacturers downsized and lightened cars, using alternative materials (aluminum and plastic rather than steel), fewer drive train components, and better fuel metering (fuel injection). Smaller, lighter autos led to a higher highway fatality rate (2 of the 13 authors of the study disagreed with this finding).

The NRC study speculates that improvements in fuel economy over the next 15 years or so will likely include greater use of existing technologies throughout the auto fleet rather than spur radical innovation. Fuel efficiency improvements may be less likely to come at the expense of safety if consumers insist on safety, but could lead to two other effects that would "undo" some of the fuel savings. Reducing the cost of driving could lead to a "rebound" effect in which consumers drive more (about 1 percent to 2 percent for each 10 percent reduction in the cost of driving, according to research). Consumers may also retain their old, less fuel-efficient cars longer to the extent that more fuel-efficient autos become more expensive.

Another prominent instance of policy intervention to promote technology has been tax credits and other financial incentives for renewable energy investment. These subsidies have supported installation of renewable energy generation capacity rather than innovation per se, but many observers say that the credits have gone far in propelling suppliers along a "progress curve" or "learning curve." The learning curve phenomenon is not understood in detail, but is generally thought to be influenced by scale and time: improvements in knowledge may come from manufacturing large quantities, for instance, and with time may come improved knowledge and capability in operating and maintaining facilities. Some of the learning may be internal to the firm; some may come from external activities of other firms or government R&D (Zimmerman 1982). In any case, learning curve effects have often been offered as justification for government efforts to underwrite production output.

Tax provisions for renewable energy include the Renewable Energy Production Incentive (REPI) in the 1992 Energy. Policy Act (EPACT) to promote use and development of renewable energy technologies. REPI provided payments of 1.5 cents per kilowatt hour (in 1993 dollars and indexed for inflation) to new qualifying renewable energy production facilities owned by state and local governments and not-for-profit electricity cooperatives starting operations between 1993 and 2003. The net impact of REPI on production has been minor; production has increased from 42 million kilowatt-hours to about 529 million kilowatt-hours, which represents less than .02 percent of total recent U.S. energy production (U.S. DOE 1999).

The 1992 act also extended a 10 percent investment tax credit for businesses investing in or purchasing qualifying solar or geothermal energy systems. Also known as the business energy tax credit, the credit was originally part of the Energy Tax Act of 1978, which established the credit for photovoltaic projects and added an additional 15 percent credit to an existing 10 percent credit for solar, geothermal, and wind generation facilities. The Tax Reform Act of 1986 eliminated the 10 percent credit, reduced the 15 percent credit to 10 percent, extended it to 1988, and eliminated credit for wind. The credit was then extended year to year until 1992, when EPACT made the credit permanent. Even with this incentive, the higher generation costs of renewable energy compared with conventional fuels have limited commercial application of renewables (U.S. DOE 1999).

Another provision of the 1992 act was a production tax credit of 1.5 cents per kilowatt hour for electricity produced from wind and biomass electricity generation from facilities placed in service in the United States between 1993 and 1999 for the first 10 years of the facility's operation. In 2000, the credit was increased to 1.8 cents per kilowatt hour. New wind facilities were installed in several states and some of them embodied technical improvements in turbine fabrication and operation. Analysts have concluded that the federal credit alone was unlikely to have spurred this growth, however. A favorable depreciation life for wind (five years—much shorter than the life of conventional power supply investments) in the Tax Reform Act of 1986 also contributed. In addition, state incentives and mandates also probably played a large role, particularly establishment of renewable portfolio standards by a handful of states. The standards required that increasing percentages of their electricity supply be provided from a menu of eligible renewable energy resources.

The government has also supported research on renewable energy through cost-sharing partnerships with industry, which typically contributes 20 percent to 50 percent of a project's funding. The U.S. General Accounting Office (GAO) (1999) reported that from 1978 to 1998, federal spending on renewable energy R&D totaled about $10.3 billion (in current dollars). The GAO observed that spending had shifted from fundamental research to enhanced market opportunities, for instance, in cost sharing with utilities to deploy wind turbines through the Department of Energy's Turbine Verification Program (U.S. DOE 1997).

The generation cost of renewable power today is even lower than was once projected. The cost of wind generation was expected to decline 64 percent between the 1980s and 1990s, and it actually fell 67 percent (McVeigh et al. 2000). The authors of this study were not able to distinguish the effects of the federal, state, and local government subsidies and other incentives for development of renewable energy from improvement in generation costs that would have occurred without these financial incentives. But technological innovation may have played a major role, as evidenced in the improvements in the engineering specifications of technologies now operating. For example, some of the innovation leading to improvement in the cost of wind power included larger diameter wind blades, enabled by lighter, stronger, stiffer materials; new transmission technology; special airfoils; and improved drive trains (U.S. DOE 1997).

Renewable energy continued to supply only a small share of power generation—about 10 percent of total energy production in the United States in 2000.

The major reason is a concurrent decline over the past decades in the cost of conventional power—a decline also due to continued innovation in fossil fuel technology. Nationwide electric generation costs of conventional power declined from 6.3 cents per kilowatt hour in 1984 to 3.6 cents per kilowatt hour in 1995. This 40 percent decline, although less than the percentage decline in the cost of the most promising of the renewable technologies—wind power—was large enough for conventional power to retain the lion's share of generation capacity. The factors leading to lower costs of conventional power include the emergence of more competitive supply markets, productivity improvements in oil and gas exploration and coal production, deregulation of railroads (the major factor in the cost of shipping coal), and progress in conventional technology (McVeigh et al. 1999; Darmstadter 2001). Conventional technology also benefited from federal funding and favorable tax provisions. Because of continued technical advances expected in conventional power (centered around advanced combined-cycle gas turbines), and given assumptions about future fuel prices, the Department of Energy projects that renewable energy will be a decreasing share of U.S. electricity generation, declining to about 8.5 percent in 2020 (U.S. DOE 2000).[2]

Pollution Abatement—Complementarity of Permit Trading and New Technology Development

Government intervention in pollution abatement has a better track record than it does in energy. This success has largely represented technology adoption rather than innovation, however. Two of the most prominent examples are the elimination of lead from gasoline and reduced emissions of sulfur dioxide by electric utilities. In both cases, emissions trading systems delivered on the predictions of theorists who argued that if firms were allowed to choose among control strategies, they could make economically efficient use of new technology and achieve policy goals.

The U.S. phase-out of leaded gasoline is generally thought to be one of the most successful market-based environmental policies to control auto pollution. Kerr and Newell (2001) econometrically relate the timing of the policy intervention to changes in operating characteristics of lead refineries. They find that a tradable permit system implemented during the later years of the phase-out provided greater incentives for refineries to adopt lead-reducing technology than the requirement to use unleaded gasoline that was imposed in the early years. Their study focuses on technology adoption rather than innovation per se, and it may suggest that government's influence in pushing new technology is greater if the technology is already waiting in the wings. Kerr and Newell do not address the factors that led to innovation.

Title IV of the 1990 U.S. Clean Air Act Amendments instituted an emission allowance-trading program to regulate emission of sulfur dioxide from electricity-generating facilities. Burtraw (2000) finds that rather than patentable discoveries, innovation meant organizational innovation at the firm and market level. The act also led to process innovation by electricity generators and upstream fuel suppliers. Much of this change was already in the works independent of the trading program, but the program encouraged its exploitation.

Allocation of the Electromagnetic Spectrum and Innovation in Satellite Technology—Policy-Induced Scarcity

The electromagnetic spectrum, or the airwaves, is the medium over which communications signals travel—including those for television, cell phones, radar, even garage door openers. Like other public goods, the spectrum is considered to be "the common heritage of mankind." But this line of thinking, after much debate, has led to scarcity of spectrum regions, centralized government allocation of the spectrum, and the imposition of technical operating standards on the industries using it. The spectrum is a highly differentiated resource because different frequencies have distinct carrying capacities. Some frequencies are attenuated by rain and other atmospheric conditions, and others are blocked by physical structures like buildings. In this sense, spectrum is exactly like land, with its extensive and intensive margins determined by fertility and other attributes.

Government regulation of frequencies around the world has consisted of administratively allocating its use among competing services—for instance, the assignment of radio stations to unique regions along the AM and FM dials. That government fails to allocate the spectrum efficiently is attested to, and has been the subject of proposed reformation in an historic and growing economics literature (inefficiencies in radio spectrum regulation inspired Coase's "theorem"). Researchers generally agree that inefficiencies have persisted (Levin 1971; Hazlett 1998; Cramton 1998; Macauley 1998). This static resource misallocation has occasioned study of the dynamic effects on the pace and direction of innovation in satellite technology.

Using a model based on theoretical research of regulation-induced innovation (Kamien and Schwartz 1969; Smith 1974, 1975; Okuguchi 1975; Magat 1976), empirical research finds that the allocation regime used by government up to 1999 incorrectly signaled the true scarcity of spectrum frequencies (Macauley 1986a, 1986b, 1998). As a result, the shadow values of different regions of the spectrum were distorted, and innovation by telecommunications companies to adjust the mix of spectrum and other inputs (in the form of the hardware and software used in the communications devices) on the basis of relative scarcity were misdirected. Telecommunications equipment makers and service operators, for example, tended to substitute away from innovation to make better use of capital inputs that would enable more intensive spectrum use (more communications capacity per unit of spectrum), such as more powerful transmitters and receivers or differently polarized signals, when spectrum was "free." More flexible allocation would have done a better job of signaling the pace and direction of efficient innovation. It would also have enabled realization of gains from innovation that would have been economically feasible and socially desirable if spectrum were "priced." Such opportunities to innovate were discussed in engineering meetings and technical publications. By 1999, the federal government had begun to experiment with spectrum markets and lotteries with resale opportunities for some regions of the spectrum, and many of the new spectrum-efficient technologies that had been envisioned were eventually developed and adopted.

Although the model does not rigorously test political economy influences as did Cohen and Noll in their study of synfuels, the study found a pattern of interests opposing and supporting auctions. Incumbent holders of spectrum, including well-organized interests like the National Association of Broadcasters, sought to maintain administrative assignment and claimed that innovation to improve efficiency of spectrum use was unattainable any time soon. Less well organized newcomer companies urged opening spectrum through market procedures like auctions and supported their argument with prototypes of new, more efficient spectrum-using technology.

Fishing—Innovation to Circumvent Government Quotas

Governments support innovation in fisheries management in many of the world's major fishing nations. These approaches include funding and testing technology in order to encourage conservation, modernization of fleet vessels, innovation in fishing gear, development of aquaculture, and in some countries, exploration of new regulatory enforcement techniques (through use of satellite data, for example) and research grants to use e-commerce and information technology in fleet and fisheries management (Iudicello et al. 1999; Irish Sea Fisheries Board n.d.; National Marine Fisheries Service 2002).

One of the most noted effects of government on innovation has been the industry's response to regulation of access to fishing resources. Iudicello et al. point out that until the 1960s, almost all the marine fisheries of the world were open—there was no owner and no regime to create property rights. Over time, regulated open access or controlled open access evolved and included catch limits, gear restrictions, and time and area closures. For example, as New England fish stocks declined, regional government managers limited the size of landed fish and the size of the mesh, shortened fishing seasons, and closed some areas to fishing. In response, Iudicello concludes, "with each regulatory stricture, skippers found a new way to change technology or improve their catching prowess." Stocks continued to decline.

Although conventional technology does not preclude overfishing, new technology implemented in response to regulation—such as synthetic line rather than cotton or hemp—can enable even larger catches per unit of fishing effort and aggravate overfishing. Iudicello et al. also point out that policy-restricting technology can lead to innovation with undesirable effects. They cite the development of "twisted twine," a net-tying technique that creates shrinking mesh, that came about in response to an increase in mesh size mandated for conservation purposes in the Atlantic Ocean.

In the late 1970s, New Zealand and Iceland began establishing individual transferable quotas (ITQs) to manage some fish species. ITQ programs operate like tradable permit programs. Fishermen are allocated ITQs based on government forecasts of seasonal fish populations and estimates of sustainable yield. Improving incentives through property rights, such as use of ITQs, may offer a solution for some stocks, provided populations can be forecast accurately and landings can be monitored (Newell et al. 2002).[3]

Some researchers are skeptical of the likelihood of improving control of access and catch rates by way of ITQs or other management change. Christy (1997) remains concerned about long-run fish stocks in light of heavy subsidies paid to many of the world's fishing fleets, including in Russia, Japan, and Europe; largely ineffective decisionmaking authority within national economic zones; and a lack of institutions and authority over stocks that are shared, straddle economic zones, or are located on the high seas.

Water—Innovation in Management

New management techniques largely organized under government auspices, rather than changes in physical inputs or production processes, characterize innovation in water resources. The road to these changes has been difficult. In his detailed study of the history of water resource policy, Frederick (1991, *66*) states,

> it is not scarcity *per se* that justifies the fear of costly water shortages.... The roots of the concerns of future shortages lie in the laws, administrative practices, and other institutions that create uncertainty over water rights, pose obstacles to developing new supplies or reallocating existing supplies to new uses, and provide little incentive for conservation. No one would argue that the role of "getting prices right" has been tested in water markets, let alone the opportunity to observe the effect of policy to influence innovation through the price system.

In their historical study of natural resources, Frederick and Sedjo (1991) observe that, steered by the unique circumstances of the American experience, states created innovative institutions. In the East, where water was plentiful, the English system of riparian rights (which gives owners of land adjacent to a water body the right to use the water) was adopted. In the West, however, the riparian system was soon deemed inadequate and a system of appropriative rights (which gives priority to the earliest users) evolved and was enacted into law in the nineteenth and early twentieth centuries. During the 1970s and 1980s, western water law evolved to protect instream flows and to facilitate transfers of water rights. Virtually all western states now have some provision for instream water rights. Farmers who continued to irrigate successfully in the face of much higher water costs in the 1970s and 1980s adopted a variety of water-conserving techniques, including more efficient water-management practices and less-water-intensive crops and seed varieties.

Another example of institutional innovation in the early 1980s was integrated management of some water supply systems. The storage and distribution facilities of the three principal water-supply agencies servicing the Washington, D.C., area initiated joint operations in 1982. With relatively little new investment in infrastructure, the new managerial arrangements increased drought-condition water yields in the region by nearly one-third. Frederick (1991) considers this type of innovation promising for other regions of the country, but also notes that there are substantial institutional obstacles, including separate ownership of facilities, multistate jurisdictions, and state water laws.

Beginning with the Federal Water Pollution Control Act in 1972, Congress specified technology-based effluent standards based on the control achievable using the kinds of pollution abatement equipment then available. By basing the standards on technological measures rather than water quality objectives, the act did away with the need for regulators to estimate the assimilative capacity of water bodies and the relationship between individual discharges and water quality. Assessment of the act by the General Accounting Office found that dischargers had installed the modern equipment, but noted widespread failure to comply with allowable discharges. Overall, national water quality by some measures has improved since 1972, but some water bodies have declined. Many analysts now argue that effluent charges approximating damages or marketable discharge permits could be used to provide incentives for developing and adopting more cost-effective technologies for reducing pollution discharges (Frederick 1991; Freeman 2000).

Forestry—Different Outcomes from Government Product Research and Management Innovation

One of the most comprehensive analyses of the effects of government policy on innovation in the forestry sector finds little direct evidence that government programs improve overall performance. Hyde et al. (1992) conducted detailed econometric studies modeling industry production functions and using data on forest industry input prices (capital, labor, and materials) and public and private expenditures on research from 1950 to 1980. They measured the effect of publicly funded research in contributing to measures of net present value of increased output, benefit–cost ratios, and internal rates of return.

Part of the study looked at innovation in forest products in the sawmill, wood pulp, and wood preservative industries. In the sawmill industry, publicly funded research at the Forest Products Laboratory (FPL) of the U.S. Forest Service was instrumental in the development of both structural particleboard and truss framing. These developments substitute for softwood plywood and offset restrictions on herbicide usage in timber production and caps on harvests from designated wilderness lands. Other research at the FPL led to computerized programs for operating saw cutting equipment, leading in turn to more complete use of the wood resource. The software was then made available at no cost to all firms that chose to computerize their production. Hyde and his colleagues found that all of these innovations improved industry performance.

FPL research on low-quality wood species, such as the use of short-fiber woods to improve the productivity of wood pulp in paper manufacturing, was less productive. Hyde et al. (1992, 99) wrote that "some research investment was perhaps a misguided political imposition" by congressional representatives from states with short-fiber wood species.

Research on wood preservatives also appears to have been less effective. Much of the research focused on reducing levels of petroleum-based, environmentally harmful residuals. In this case, Hyde and coauthors note that their modeling approach does not capture potential environmental quality improvements.

According to the same study, public research on timber growth and management has been less successful. This included projects at U.S. Forest Service experiment stations, state forest research organizations, and forestry extension organizations. The federal research focused largely on management, including forest inventories and surveys, regenerating unproductive cutover pine forestland, and controlling insects, diseases, and fire. This part of the study is limited to four or five surveys of forest production, and gaps weakened the data on public research expenditures during the 30 years they analyze. With this caveat in mind, the authors found negative or at best, under some model specifications, very small net present value benefits of public investment. They surmised that this may be largely because forestland and timber resources have been in plentiful supply relative to labor and other forms of capital, and timber prices have been relatively low. Land and capital have not been scarce because old fields have been converting naturally to timber while timber still remains elsewhere, if on less accessible sites. They predicted that research can contribute to timber production only if prices rise to exceed the cost of growing stumpage.

R&D Tax Credits—Ambiguous Results from a Broad-Based Fiscal Approach

The numerous energy-related tax provisions discussed earlier typify direct subsidies to resource industries. Another broader-based subsidy in the form of a research and experimentation tax credit extends to R&D in any industry. The credit has always been a "temporary" provision of the tax code; lawmakers have extended it nine times and substantively changed it five times since first enacting the credit in the Economic Recovery Tax Act of 1981.

The current version of the credit reduces the marginal cost of qualifying R&D activities by allowing firms either a percentage credit tied to their increase in research intensity (expressed as a percentage of revenues) compared with a fixed historic base or a smaller percent credit for all research expenses exceeding 1 percent of sales. A special provision, the basic research credit, is intended to stimulate research partnerships between universities and private firms; it applies to basic research performed under contract by educational institutions and certain other nonprofit organizations. This credit applies to incremental expenses (over an inflation-adjusted fixed base period from 1981 to 1983) for contract research that is undertaken without any "specific commercial objective."

Critics argue that the credit is a weak stimulus of R&D. The provision is not permanent. Its marginal effective rate is below its statutory rate because the credit interacts with other tax provisions including the investment tax credit, depreciation, and capital gains (see Guenther 1999). If plant and equipment investment is a major part of conducting R&D and plant and equipment are less expensive because of an investment tax credit, then the marginal effective rate of the credit for R&D is smaller. The tax code also allows expensing of R&D costs incurred in the "experimental or laboratory sense" (not including spending on structures and capital equipment). Because of limitations on the research expenditures that can be expensed, the value is only a fraction of the costs that are expensed, and an even smaller fraction of their total R&D spending. In addition, the credit treats investment preferentially through expensing, so there is an incentive to "re-label" as R&D activities that would be undertaken anyway.

Analyses of the credit have been inconclusive. Its effectiveness in closing the gap between private and social returns to R&D is "almost impossible" to evaluate, in part because it changes so often and the credit is not permanent. Evidence supports the criticism that firms redefined some activities as R&D after the introduction of the credit. Studies have also generally assessed the net additional R&D

spending brought about by the credit (after deducting the revenue loss to the government) or the resulting elasticity of R&D spending (the percentage increase in R&D spending arising from the percentage decrease in the cost to the firm of this spending). Studies in the 1980s, when the credit structure was substantially more generous, showed that the credit stimulated as much as two dollars of additional R&D for every dollar of tax expenditure (Hall 2002). Not all recent studies agree. Some find that the credit stimulates additional research with a social value larger than the foregone government revenues but smaller than the effect found in earlier research, whereas other studies find it much smaller (Mamuneas and Nadiri 1996).

With the large amount of research on R&D tax credits, it is surprising and disappointing that natural resources have been largely overlooked. What has been the effect of the credit on firms engaged in R&D related specifically to natural resources and the environment? Research using grossly aggregated industry classifications suggests that the vast share of credits were claimed by large manufacturing firms with assets in excess of $250 million; of these, nearly 60 percent of the credits claimed are by companies making computers, electronic components, drugs, and motor vehicles. This suggests that effects on natural resources industries may be small, but future research might shed better light.

Conclusions

The case studies reviewed here illustrate the varied effects and uneven performance of public policy–induced innovation. In some it yielded sufficient quantities and qualities of innovation, whether intended or not, to alleviate nature resource scarcity—forestry, the airwaves, some energy markets. In some cases—allocation of the airwaves and the use of permit trading for pollution control—policy shifted toward more "innovation friendly" strategies and management regimes. Portions of the spectrum are now auctioned, and some pollution control is now attained by way of permit trading. Policy still needs to make rapid progress in the cases of fishing and water.

Unfortunately, it is impossible to say why some strategies work and others do not; some studies surveyed here analyze what went right and what went wrong, but no one has yet discovered the overall pattern. The potential usefulness of a more systematic assessment is obvious. A more detailed understanding of the conditions for success could go far in answering the question If government had an additional one million dollars to spend on innovation in natural resources and environmental management, what should it do?

- Set up a government laboratory?
- Provide a tax credit for R&D that would be the financial equivalent of a million dollars of additional spending?
- Underwrite a joint research venture among all the firms in a particular industry?
- Fund the establishment, administration, and enforcement of standards mandating the use of new technology?

As Jaffe et al. (2003) note of the relationship between policy and innovation undertaken to control pollution (as distinguished from innovation to alleviate resource scarcity), ideally studies would trace the relationship among innovation, the shadow or implied price of pollution or environmental inputs, and the policy intervention. In practice, proxies for hard-to-observe shadow prices would include prices of polluting inputs or expenditures on pollution. In part because of these measurement problems, the size of the effects of induced innovation remains uncertain. Studies of the effect of environmental regulation (measured by assessment of relative regulatory stringency, number of enforcement actions) on productivity and investment in R&D for abatement offer some indirect evidence that induced innovation can be small or overshadowed by other costs of regulation (see references in Jaffe et al. 2003).

Linking government intervention to the alleviation of resource scarcity is no easier. Some of the most detailed econometric studies among the cases reviewed here are those of government product and management research in forestry, but still the authors lament data gaps and the difficulty of modeling the "black box" of R&D effort. More intellectual innovation—that is, research that in itself is creative in modeling R&D and policy—could help decide whether to target policies on technology or on other regulatory schemes that may or may not lead to innovation.

Most of the examples in this chapter are drawn from the U.S. experience. It is only fair to point out that New Zealand was the first country to implement tradable fishing quotas and to auction electromagnetic spectrum. Other countries, the United Nations, nongovernmental organizations, and international scientific bodies have begun to consider alternative water policies that rely on new management regimes, much like those discussed above (Boxer 2002). Chile has established a comprehensive water allocation system that establishes tradable property rights (Rosegrant and Binswanger 1993). R&D programs in industrialized countries are aggressive. Finland, Canada, Australia, Hong Kong, the United Kingdom, most European countries, Japan, and Israel all have major government programs in basic research (including nondefense science and technology) and, in the past two decades, a growing number of programs to advance technology for commercial development. The Organisation for Economic Co-operation and Development routinely surveys R&D activities by government and industry in member countries.

Another set of innovations not addressed here could be collectively called "improved scientific understanding." We are acquiring a deeper understanding of ecosystems and a greater ability to detect and monitor changes in the physical and biological world. This research is aided by heightened public interest, scientific interest, and improved remote measuring capability (for instance, sensors on satellites) for resources that are hard to access in situ.

Finally, an epilogue of sorts is in order about the double-edged sword of technology (Austin and Macauley 2001).

Aquaculture—the farming of fish, shellfish, and aquatic plants—supplied one-third of worldwide seafood consumption as of 2000. Aquaculture research can introduce unwanted seaweeds, fish, invertebrates, parasites, and pathogens that can harm wild species. Some experts are calling for closer government supervision of aquaculture research, arguing that existing controls are inconsistent and in some cases too lax (Naylor et al. 2001; Christy 1997).

Biotechnology in forestry is another frontier R&D area encountering the two-edged sword. In forestry, efforts have moved beyond traditional hybridization and cell and tissue cultures to genetic engineering to increase timber stocks and improve disease and pest resistance. Concern about the long-term effects of biotechnology, in forestry and other sectors (most notably, agriculture) is widespread. The Cartagena Protocol on Biosafety was signed by 130 governments in January 2000 to counter adverse effects of the international movement of genetically modified organisms. Many European consumers have shunned genetically modified foods. In 2002, European governments began requiring that they be labeled. As a result, field trials in Europe fell from 239 in 1997 to about 35 in 2002 (Mitchener 2002). As with aquaculture, some experts are calling for more systematic government control of these new developments.

These examples bring the role of government-induced policy innovation full circle, from efforts to conduct, fund, or otherwise directly support innovation, and allocation and regulatory approaches that indirectly influence innovation, to the role of government in shepherding new technology to avoid untoward effects. Future research at the nexus of innovation, natural resources, and government will continue to build on the seminal concepts set forth by Barnett and Morse.

Acknowledgments

I thank for their comments Toni Marechaux and participants in the workshop "Scarcity and Growth in the New Millennium," held at Resources for the Future in November 2002.

Notes

1. The National Cooperative Research and Production Act alleviates some of the antitrust concerns that might be associated with joint ventures. More than 800 research joint ventures (RJVs) were formed in the United States from 1985 to 2000 according to filings required by the National Cooperative Research and Production Act. Half of the RJVs involved companies in electronic and electrical equipment, communications, and transportation equipment (National Science Foundation 2002).

2. This discussion has focused on R&D in primary energy. The U.S. Department of Energy has also funded R&D in energy end-use by residential consumers and industry, including home weatherization, improved efficiency of industrial energy use, and improved gas turbines. Of these, the largest effort has been in the transportation sector to encourage development of energy efficient advanced vehicles (U.S. DOE 1999).

3. Problems can arise from the role of government as gatekeeper of stock data as well. In Iceland, the fishing authority, the Marine Research Institute, seriously overestimated the size of the cod stock in the late 1990s. This led to catch allocations that were too high under Iceland's quota system, which limits the total harvest each season and makes available a catch percentage of the total harvest based on an amount "pre-agreed" with industry. Weaknesses of the original system, such as exclusions for small boats and poor enforcement, had been remedied, but the problem of overestimation led to overfishing (George 2002).

References

Arrow, K. J. 1962. Economic Welfare and the Allocation of Resources for Invention. In National Bureau of Economic Research, *The Rate and Direction of Inventive Activity: Economic and Social Factors.* Princeton: Princeton University Press.

Austin, D., and M.K. Macauley. 2001. Cutting through Environmental Issues: Technology as a Double-Edged Sword. *Brookings Review* 19(1): 24–27.

Banks, J.S., L.R. Cohen, and R.G. Noll. 1991. The Politics of Commercial R&D Programs. In *The Technology Pork Barrel,* edited by L.R. Cohen and R.G. Noll. Washington, DC: Brookings Institution, 53–76.

Barnett, H.J., and C. Morse. 1963. *Scarcity and Growth: The Economics of Natural Resource Availability.* Baltimore: Johns Hopkins University Press for Resources for the Future.

Binswanger, H.P., and V.W. Ruttan. 1978. *Induced Innovation: Technology, Institutions, and Development.* Baltimore: Johns Hopkins University Press.

Boxer, B. 2002. Global Water Management Dilemmas. *Resources* 146(Winter): 5–9.

Burtraw, D. 2000. Innovation under the Tradable Sulfur Dioxide Emission Permits Program in the U.S. Electricity Sector. Discussion Paper 00-38. Washington, DC: Resources for the Future.

Carey, J., and D. Zilberman. 2001. Irrigation Technology Adoption with Stochastic Prices. *American Journal of Agricultural Economics* forthcoming.

Christy, F.T. 1997. Economic Waste in Fisheries: Impediments to Change and Conditions for Improvement. In *Global Trends in Fisheries Management,* edited by E.K. Pikitch, D.D. Huppert, and M.P. Sissenwine. Bethesda, MD: American Fisheries Society, 28–39.

Cohen, L.R., and R.G. Noll. 1991. *The Technology Pork Barrel.* Washington, DC: Brookings Institution.

Cramton, P. 1998. The Efficiency of FCC Spectrum Auctions. *Journal of Law and Economics* 41(2): 727–736.

Darmstadter, J. 2001. The Role of Renewable Resources in U.S. Electricity Generation: Experience and Prospects. Testimony before the Committee on Science, U.S. House of Representatives, February 28. http://www.rff.org (accessed Fall 2002).

Frederick, K.D. 1991. Water Resources: Increasing Demand and Scarce Supplies. In *America's Renewable Resources: Historical Trends and Current Challenges,* edited by K.D. Frederick and R.A. Sedjo. Washington, DC: Resources for the Future, 23–71.

Frederick, K.D., and R.A. Sedjo (eds.). 1991. *America's Renewable Resources: Historical Trends and Current Challenges.* Washington, DC: Resources for the Future.

Freeman, A. M. 2000. Water Pollution Policy. In *Public Policies for Environmental Protection,* edited by P.R. Portney and R.N. Stavins. Washington, DC: Resources for the Future.

Freeman, C., and L. Soete. 1997. *The Economics of Industrial Innovation.* Cambridge, MA: MIT Press.

George, N. 2002. Strong Cod Stocks Are Proving Elusive. *Financial Times,* Jan. 25, 27.

Guenther, G. 1999. The Research and Experimentation Tax Credit. Congressional Research Service Report 92039. Washington, DC: Congressional Research Service.

Hall, B.H. 2002. The Financing of Research and Development. National Bureau of Economic Research Working Paper W8773. http://papers.ssrn.com/paper (accessed Fall 2002).

Hazlett, T.W. 1998. Assigning Property Rights to Radio Spectrum Users: Why Did FCC License Auctions Take 67 Years? *Journal of Law and Economics* 41(2): 529–576.

Hicks, J. 1932. *The Theory of Wages.* London: Macmillan.

Hyde, W.F., D.H. Newman, and B.J. Seldon. 1992. *The Economic Benefits of Forestry Research.* Ames: Iowa University Press.

Irish Sea Fisheries Board. No date. Towards Innovation and Sustainability in the Fisheries Sector, National Development Plan. http://www.isfb.org (accessed Fall 2002).

Iudicello, S., M. Weber, and R. Wieland. 1999. *Fish, Markets, and Fishermen: The Economics of Overfishing.* Washington, DC: Island Press.

Jaffe, A.B., R.G. Newell, and R.N. Stavins. 2003. Technological Change and the Environment. In *Handbook of Environmental Economics,* vol. 1, edited by K.-G. Maler and J. Vincent. Amsterdam: North-Holland/Elsevier Science, 461–516.

Kamien, M.I., and N.L. Schwartz. 1969. Induced Factor Augmenting Technical Progress from a Microeconomic Viewpoint. *Econometrica* 37(October): 668–684.

Kerr, S., and R. Newell. 2001. Policy-Induced Technology Adoption: Evidence from the U.S. Lead Phasedown. Discussion Paper 01-14. Washington, DC: Resources for the Future.

Levin, H.J. 1971. *The Invisible Resource: Use and Regulation of the Radio Spectrum.* Baltimore: Johns Hopkins University Press for Resources for the Future.

Macauley, M.K. 1986a. Out of Space? Regulation and Technical Change in Communications Satellites. *American Economic Review* 76(2): 280–284.

———. 1986b. The Contribution of a Partnership between Economics and Technology. In *Economics and Technology in U.S. Space Policy,* edited by Molly K. Macauley. Washington, DC: Resources for the Future, 3–24.

———. 1998. Allocation of Orbit and Spectrum Resources for Regional Communications: What's at Stake? *Journal of Law and Economics* 41(2): 737–764.

Magat, W. 1976. Regulation and the Rate and Direction of Induced Technical Change: Comment. *Bell Journal of Economics* 6(Autumn): 703–705.

Mamuneas, T.P., and M.I. Nadiri. 1996. Public R&D Policies and Cost Behavior of the US Manufacturing Industries. *Journal of Public Economics* 63: 57–81.

McVeigh, J., D. Burtraw, J. Darmstadter, and K. Palmer. 2000. Winner, Loser, or Innocent Victim? Has Renewable Energy Performed as Expected? *Solar Energy* 68(3): 237–255.

Mitchener, B. 2002. Europe Has No Appetite for Modified Food. *Wall Street Journal,* Sept. 27, B5.

Moreno, G. 2001. Factor Price Risk and the Diffusion of Conservation Technology: Evidence from the Water Industry. Working Papers in Economics, Dec. 19. Claremont, California: Claremont College.

National Marine Fisheries Service. 2002. U.S. Department of Commerce: Aquaculture Policy. In *Mission Statement and Vision for U.S. Aquaculture.* http://www.nmfs.noaa.gov/trade/DOCAQpolicy.htm (accessed Fall 2002).

National Research Council. 2001. *Energy Research at DOE: Was It Worth It?* Washington, DC: National Academy Press.

———. 2002. *Effectiveness and Impact of Corporate Average Fuel Economy (CAFE) Standards.* Washington, DC: National Academy Press.

National Science Foundation. 2002. *Science and Engineering Indicators.* Arlington, VA: National Science Foundation.

Naylor, R.L., S.L. Williams, and D.R. Strong. 2001. Aquaculture—A Gateway for Exotic Species. *Science* 294(Nov. 23): 1655–1656.

Newell, R.G., A.B. Jaffe, and R.N. Stavins. 1999. The Induced Innovation Hypothesis and Energy-Saving Technological Change. *Quarterly Journal of Economics* 114: 941–975.

Newell, R.G., J.N. Sanchirico, and S. Kerr. 2002. Fishing Quota Markets. Discussion Paper 02-20. Washington, DC: Resources for the Future.

Okuguchi, K. 1975. The Implications of Regulation for Induced Technical Change: Comment. *Bell Journal of Economics* 6(Autumn): 703–705.

Parry, I.W.H., W.A. Pizer, and C. Fischer. 2000. How Important Is Technological Innovation in Protecting the Environment? Discussion Paper 00-15. Washington, DC: Resources for the Future.

Perry, H., and H.H. Landsberg. 1981. Factors in the Development of a Major US Synthetic Fuels Industry. *Annual Review of Energy* 6: 233–266.

Reinganum, J. 1982. A Dynamic Game of R&D: Patent Protection and Competitive Behavior. *Econometrica* 50: 671–688.

Rose, N.L. 1986. The Government's Role in the Commercialization of New Technologies: Lessons for Space Policy. In *Economics and Technology in U.S. Space Policy,* edited by M.K. Macauley. Washington, DC: Resources for the Future, 97–126.

Rosegrant, M.W., and H.P. Binswanger. 1993. *Markets in Tradable Water Rights: Potential for Efficiency Gains in Developing Country Irrigation.* Washington, DC: International Food Policy Research Institute.

Schurr, S.H., J. Darmstadter, H. Perry, W. Ramsay, and M. Russell. 1979. *Energy in America's Future.* Baltimore: Johns Hopkins University Press for Resources for the Future.

Simpson, R.D. (ed.). 1999. *Productivity in Natural Resource Industries.* Washington, DC: Resources for the Future.

Smith, V.K. 1974. The Implications of Regulation for Induced Technical Change. *Bell Journal of Economics* 5(Autumn): 623–632.

———. 1975. The Implications of Regulation for Induced Technical Change: Reply. *Bell Journal of Economics* 6 (Autumn): 706–707.

Smith, V.K., and J.V. Krutilla. 1979. Summary and Research Issues. In *Scarcity and Growth Reconsidered*, edited by V.K. Smith. Baltimore: Johns Hopkins Press for Resources for the Future, 276–290.

Stauffer, T. 1975. Liquified and Synthetic Natural Gas—Regulation Chooses the Expensive Solutions. In *Regulating the Product: Quality and Variety*, edited by R.E. Caves and M.J. Roberts. Cambridge, MA: Ballinger Publishing Company, 171–198.

Stokes, D.E. 1997. *Pasteur's Quadrant.* Washington, DC: Brookings Institution.

Tirole, J. 1988. *The Theory of Industrial Organization.* Cambridge, MA: MIT Press.

U.S. Department of Energy. 1997. Renewable Energy Technology Characterizations. TR-109496. Washington, DC: Department of Energy.

———. 1999. *Federal Financial Interventions and Subsidies in Energy Markets 1999: Primary Energy.* SR/OIAF/999-03. Washington, DC: Energy Information Administration.

———. 2000. *Annual Energy Outlook 2001.* DOE/EIA-0383. Washington, DC: Energy Information Administration.

U.S. General Accounting Office. 1999. *Renewable Energy: DOE's Funding and Markets for Wind Energy and Solar Cell Technologies.* Report GAO/RCED-99-130. Washington, DC: U.S. General Accounting Office.

Zimmerman, M.B. 1982. Learning Effects and the Commercialization of New Energy Technologies: The Case of Nuclear Power. *Bell Journal of Economics* 13(Autumn): 297–310.

CHAPTER 12

The Marvels and Perils of Modernity

A Comment

Sylvie Faucheux

T HE PURPOSE OF THIS chapter is to highlight the importance of cultivating a critical capacity for reflection about underlying philosophical, moral, and cultural presuppositions of arguments put forward with respect to resource scarcity and technological innovation. According to a well-known characterization of science and technological innovation, progress in knowledge (primarily or essentially, depending on one's persuasion) offers humanity the benefits of new discoveries and applications. These benefits are the normal or natural profit from investment in new knowledge. The destructive sides of technology are not intrinsic. They are effects of either inexperience and inadequate mastery (accidents, pollution, etc.) or human perversity. In the latter case, the knowledge may be "abused," that is, put to wanton, perverted, or destructive uses. Science and technology are marvelous, but society may be delinquent; this potential delinquency is the source of perils, not science and technology themselves.

I think that we all know and understand this argument quite well. It suggests that knowledge and technology are only a means, to be employed or not, depending on the goals and purposes of society. It has been widely discussed and criticized for many years. We are confronted with the question, how can we avoid, or at least reduce, the outbreaks of this delinquency with their sometimes terrible consequences? The problem of science and technology governance is likened to keeping matches out of the reach of small children and vandals.

What I want to suggest is that we can draw different insights, and perhaps some very important ones, from insisting that scientific practice and technology innovation practices are quite inseparable from their social and cultural contexts, including sets and types of societal value systems. The production of knowledge is not really dissociable from particular circumstances and practices that make the knowledge possible. The ways in which science and technology may take on a benign, marvelous, irksome, or diabolical character depend on complex amalgams of social and cultural values.

Do we have reasons to believe that the "keep the matches away from children" strategy is a viable risk management policy? Can we be sure that the specification of the roles, that is, the wardens and the delinquents, is satisfactory? Maybe the specification is a part of the problem. Lessons from a wide range of contemporary situations suggest that this way of framing the governance of technology risk is likely to produce perverse results. The "delinquents," feeling shabbily treated, will become irritated and resentful and will continue to find ways to be delinquent.

What conclusions should we draw from this? One outlook contends, it's sad, we just have to live with delinquency, and the legitimate authorities (whoever they are) should invest more in various police and security services, and so on. Alternatively, we could have second thoughts about the conventional wisdom that (a) technology is basically good (or at least value free), and (b) geopolitics can and should be built on simple categories of Good (morally responsible) and Bad (delinquent).

Technical Marvels ... and Their "Dark Side"

Relatively few people in the Western world have doubts about the great benefits that technology and scientific progress can bring. Many sections of the so-called developing countries are clamoring to "catch up." Yet, increasingly large fractions of the world's populations express anxiety about the double face—the potential for Good and for Bad—of scientific knowledge and technological innovations. This ambivalent potential of science and innovation is a cause of widespread alarm in both North and South (mad cows, radioactive waste, cloning, etc.).

It has been relatively commonplace to acknowledge the existence of risks from failures (a machine breaks down, a bridge collapses, and so on). It has also been commonplace to highlight risks associated with uncertainty and ignorance about the "world out there." Increasingly, however, these risks are seen to relate to complex systems whose changes are partly induced by our own actions. We throw stones in the pond, but the full effects are unknown; the ripples travel out and down with repercussions farther than we can see. A desirable result in the short term and on a small scale can be a disaster in its long-term indirect consequences or if replicated on a larger scale. Ecosystem perturbations and toxic chemicals, which nature or society can cope with on small scales, can accumulate and grow into serious threats.

Suggestions that might be made about good governance of innovation, including management of the risks associated with progress and innovations in technology, will depend strongly on the interpretive framework—including underlying cultural and moral premises—used to analyze this ambivalence. The challenges of governance have several facets:

- They are not only the already vexing ones of transferring benefits and know-how and building capacity in the South for poverty relief and sustainable development (affordable, reliable equipment for basic health care, water quality control, and refrigeration).
- They are not only the empirical and methodological challenges of working on the frontier between knowledge and ignorance in systems complexity, irrevers-

ibility, and long-term uncertainty. These issues must be faced to achieve better understanding of our environment and the planet's life-support systems.

- They are also the (now highly visible) political, military, moral, and economic challenges of reducing the risk of deliberate—not just accidental—abuse of technological_prowess to provoke major disruptions in human society, including possible long-term impacts on health, organism integrity, and ecological stability.

Advances in science and technology greatly increase our capacity to create new services and products and to exploit and transform our physical surroundings. Innovation promises to continually extend this capacity. Yet, almost everybody seems to agree, the pursuit of societal goals through science and technology innovation is not without risks. What is less universally accepted, yet seems quite strongly supported by the scientific record itself, is that some of these risks are inherent in the potentialities of science and technology themselves.[1]

- The permanent process of pushing back the frontiers of knowledge and scientific intervention also confronts us, in new ways, with the limits both to our knowledge and to our intervention capacity.
- Advances in our knowledge permit more and more sophisticated interventions in ecosystem functioning and in the components of life itself; yet our scientific understanding of the physical environment and of the impacts of human activity on life processes and ecosystems remains very incomplete and in many cases lags far behind our interventions.
- Science-based innovation has, in the past, contributed to industrialization processes that have proven highly disruptive to ecosystems at local and global levels. Some of the new commercially attractive technologies may also be incompatible with ecological stability and environmental quality goals.
- Some forms of commercially driven innovation and technology transfer may intensify socioeconomic stratification, and perhaps even worsen poverty for disadvantaged populations rather than reduce it.

Social choice, in the sense employed by economists and political theorists, is concerned with the distribution not only of material wealth, entitlements, economic goods and opportunities, but also of risks, burdens, and damages. Some of the new forms of knowledge increase the potential for mass terrorism, and some increase the potential for highly authoritarian surveillance and coercion. The scope for this terrorism and surveillance is not limited to a few peculiar combinations of circumstances. On the contrary, there are a huge variety of options for "bringing down the house" (or threatening to) by allying a few well-focused perturbations (notably chemical, biological, and informational) to techniques of electronic communication and rapid mobility; and the forces of surveillance have a wide spectrum of tools for tracking the movements of persons, networks, and populations.

It is uncertain where these innovation paths will lead us. Some work in industrial economics, concerned with quantifying or estimating trends of productivity in an econometric framework, seeks to define measures for "accumulation of knowledge" (as a capital stock of some sort) or for the manifestation of knowledge growth in a tangible secular trend of "productivity growth." The problem lies deeper than the need to come to an accurate econometric estimate of a "residual." Moses Abramowitz once suggested that the "residual" of an econometric estimation problem is "a measure of our ignorance." This is apt enough, but it raises the paradoxical question, What do we know (or suspect) in our ignorance? More precisely, What is hiding behind the residual? Is progress a secular trend, or can there be changes of sign (ups and downs) or changes in qualitative meaning and signification?

Two hundred years ago the Marquis de Condorcet, a contemporary of Ricardo and Malthus, wrote a tract called the *Sketch for a Historical Picture of the Progress of the Human Mind* (1795), where he set out a fine dialectical argument:

> by appeal to reason and fact that nature has set no term to the perfection of human faculties; that the perfectibility of man is truly indefinite; and that the progress of this perfectibility, from now onwards independent of any power that might wish to halt it, has no other limit than the duration of the globe upon which nature has cast us. This progress will doubtless vary in speed, but it will never be reversed as long as the earth occupies its present place in the system of the universe, and as long as the general laws of this system produce neither a general cataclysm nor such changes as will deprive the human race of its present faculties and its present resources.

In this argument, everything is captured in the *residual* that, in this formulation, is the repeated phrase "as long as." This is echoed currently in quite a large literature on "external effects" suggesting that the problem of social/environmental costs can be considered as a "residual indeterminacy" whose magnitude (and uncertain sign) may—if the prophets of catastrophe are to be believed—dominate any attempted corrections to market prices and national accounting aggregates.

To tease out the complexities of technological progress that are tucked within such terms as "residual" and "externality," it is useful to portray innovation explicitly as a two-faced affair: on the one side the production of significant (but difficult to quantify) opportunities for benefits along with, on the other side, significant (but indeterminate) nuisance values. To explain, let's list some examples of things about which we know our ignorance, but which include joint benefits and nuisances that are hard to evaluate and compare:

- Toxic waste, including the management of accidentally or willfully contaminated sites.
- Old mine tailing and mines, some posing the risk of large-scale water and ecological contamination (while the location of the mine is lost from human memory).
- Dilemmas in the disposal of spent nuclear fuel and radioactive waste. These are now the object of scientific reflection and societal enquiry in several nations (notably in Europe, but also Canada). The debate about the "reversibility" of radioactive waste storage reveals the underlying anxiety about guaranteeing—at the level of societal assurance of stewardship as well as technological mastery of containment—a safe disposal process for waste whose nuisance lifetime is hundreds or thousands of years.

- Soil, water, and ecosystem changes. These are brought on by human agency, with consequences now difficult to master or prevent. Many of the "miracles" of increased productivity within the agro-food industry depend on permanent use of pest-control chemicals, fertilizers, hybrid or genetically modified stock, and other capital inputs. This technology can heighten the vulnerability of food production systems to technological, economic, or natural disruption. Intensive production is also, in many regions, seriously damaging soil and water quality, which will undermine productivity in the long term.
- The international agreements at Kyoto (in December 1997) for worldwide stabilization and eventual reductions of greenhouse gases. These have been widely heralded as a step toward sustainability. At the same time they sharpen some other development–environment tensions, including (a) risks associated with the use of nuclear power and (b) historically rooted North–South asymmetries in the distribution of benefits from industrialization processes.
- The risks of biotechnology. These include those not quantifiable, yet potentially serious, with irreversible health or ecosystem consequences (such as genetically modified crop plants). There is considerable skepticism among populations of many developing countries concerning the equitable distribution of benefits.
- Fishing industry. When mechanization, allied with sophisticated scanning technologies, can dramatically increase catch effectiveness, yet the catch volumes (including discarded by-catch) in many of the world's seas jeopardize the sustainability of fisheries as a food resource. Aquaculture using genetic recombination may alleviate some productivity constraints while introducing new indeterminacies for species viability and stability, marine ecosystems, and consumer health.

The Complex Societal Demand for Technology Governance

Many new forms of science-based innovation intervene in complex biological, ecosystem, and sociocultural processes where assurance of desired outcomes is almost impossible. It has long been recognized that industrial production activities, mass consumption, and intensive agriculture can have unwanted effects on ecosystems and environmental quality. Now critics emphasize that some of the adverse consequences can be long term and difficult to control.

We must now recognize that technological interventions in complex natural processes can constitute, in themselves, a self-renewing source of problems that jeopardize community livelihoods, health, and economic prospects. This is almost universally conceded for the risks in the electronuclear industry and in biotechnology applications based on genetic engineering. It is also true for the complicated yet fragile systems of food production and communication upon which modern societies depend. But now we must see that it is equally true for technologically mediated interventions in politics and human relations, whether at local or international scales.

Current environmental costs and risks push us to perceive joint benefit and nuisance outcomes as inseparable but also incommensurable (and, in various ways, also indeterminate). In the face of such technological uncertainties, some risk analysts and economists will adopt the stance nonetheless that we should make an estimate of the *net* result in order to estimate net productivity—if necessary introducing probabilities, Bayesian learning, and sensitivity analyses in order to get a handle on the uncertainties. This is part of the conventional wisdom (e.g., estimates of probabilities of different types and scales of nuclear reactor accidents), and it has its place in some problems. However, such expert procedures have been the object of much controversy in Europe and elsewhere, and their use as justifications of public and enterprise policy has uncertain legitimacy.

In the face of this uncertain legitimacy for "technocratic" risk assessment, it is socially much more satisfactory *not* to aggregate goods and bads too quickly. One option for the appraisal of the potential of science and of technological innovations for human development is to evaluate them (and the perceived risks and uncertainties) against a spectrum of criteria for societal acceptability and well-being. In other words, adopt a perspective on science and technology governance that confronts the deep ambiguities of technological innovation by proposing the inseparable "good and bad" to be evaluated in a societal context, for example, through multistakeholder deliberative processes.

This is, in fact, the direction in which science and technology evaluation practices are evolving in Europe at the present time. Developing the necessary awareness, outlooks, and habits for such evaluation practices is a major challenge associated with governance capability.[2] This new governance adopts the attitude that scientific practice is not fundamentally value free and that technological innovation is not automatically a boon for society. Rather, science and innovation practices have to find their justifications explicitly with reference to prevailing social concerns.

If science and innovation are not universally beneficial or welcome, then how are they perceived? Empirically, we see that science may be cast, depending on the social and cultural context, in a variety of roles that depend on social values and cultural significations. We give three topical examples.

Diversity and Coexistence Ideals—Science for a Common Future?

The argument goes as follows: Short-sighted exploitations of technological prowess, even when well-intentioned, have often led to ecological or social destruction. If sustainability goals are to be achieved, science and technological development as potential forces for public good will have to be guided by ethical, political, and epistemological reflection. New notions of social responsibility in public policy and science will need to be explored.

The old conception of scientific communication as a one-way traffic of information from the experts to the public would need to be replaced by a notion of partnership through reciprocal learning. If this is to happen, policymakers and the public must be involved in the quality assurance of innovations in science and technology. Scientists must learn as well as teach, policymakers must specify their needs and accept uncertainty as well. The general public, grounded in cultural and community values, must use their discrimination on scientific questions as on all others of public concern. A major challenge for science for sustainable development is, indeed, bridging the communication gaps so that mutual learning and trust can be established among all the parties.

The agenda of sustainable development would mean, in this regard, guiding scientific work and technology applications toward innovations that respect fundamental sustainability values, such as local ecosystem resiliency, mitigation of global climate change impacts, energy efficiency, food security, and enhanced problem-solving capacities of local populations. The process must include the design and implementation of agreed social processes for quality assurance in technological implementations. This will entail the emergence of new social institutions to perform the quality assurance function. In this style of science, place-specific knowledge and resources of local communities should be integrated as a complement to the universal knowledge of traditional scientific practice.

Historical Liability and "Science against Unsustainability"

The argument goes as follows: The scientific community finds itself placed, more and more, in a reactive role of trying to fill a "knowledge deficit," driven by growing awareness of problems like hazardous waste, water contamination, renewable resource depletion, climate change, atmospheric pollution, and disruption of aquatic and terrestrial habitats.

This position can be considered a variant of the first one. We hear a call for scientific activity to be designed around goals of sustainable development (Lubchenco 1998). However, this science for sustainability would be fundamentally *issue driven*—not just curiosity generated and not just "mission oriented," but also prepared to take up problems created by past science and technology, which for society are like sunk costs.

Jane Lubchenco (1998), in her Presidential Address to the AAAS, made the following statement about environmental change:

> Between one-third and one-half of the land surface has been transformed by human action; the carbon dioxide concentration of the atmosphere has increased by nearly 30 percent since the beginning of the industrial revolution; more atmospheric nitrogen is fixed by humanity than by all natural terrestrial sources combined; more than half of all accessible surface fresh water is put to use by humanity; about one quarter of the bird species on earth have been driven to extinction; and approximately two thirds of major marine fisheries are fully exploited, overexploited or depleted....
>
> The current and growing extent of human dominance of the planet will require new kinds of knowledge and applications from science—knowledge to reduce the rate at which we alter the earth's systems, knowledge to understand earth's ecosystems and how they interact with the numerous components of human-caused global change, and knowledge to manage the planet.

Then she called upon the scientific community to "formulate a new social contract for science" with the view that scientists should

> address the most urgent needs of society, in proportion to their importance; communicate their knowledge and understanding widely in order to inform

decisions of individuals and institutions; and exercise good judgment, wisdom and humility.

This "science for sustainability" would, by necessity, address salient problems of unsustainability, regardless of the prospects for a traditional solution. These would include complex and difficult issues, where sometimes our knowledge is swamped by uncertainty, ignorance, and conflicting values, and where dilemmas about noncontrollability of dangerous processes cannot be overcome. We have already reviewed such problems. Lasting effects include land degradation, saltwater intrusion in aquifers, pesticide residues and durable toxic wastes that may accumulate in ecosystems and food chains, radioactive waste from nuclear reactors, and climate change triggered by greenhouse gases. Interventions in social, economic, and ecosystem processes that, once initiated, cannot easily be mastered include changes or increased variability in hydrological and regional climate patterns due to the enhanced greenhouse effect, the environmental release of "transgenic" organisms for food production or other purposes, the cloning of animals (and perhaps humans), the presence of BSE (mad cow disease) in cow and, perhaps, human populations.

The Diabolization of Science

The two agendas, science in the service of sustainability and science as a component of culturally sensitive (and sustainable) development, push for scientific work and technology applications to be oriented toward innovations that respect fundamental societal values (local community integrity, ecosystem resiliency, mitigation of global climate change impacts, energy efficiency, food security, enhanced problem-solving capacities of local populations). Explicitly stating these agendas makes it the presumption that the desired orientation is not inherent in science practice, but that it can, with effort, be put on the right track.

However, there are some people who, having survived for some years the "counterproductivities" of technology and having viewed a wide spectrum of real and imagined technological catastrophes,[3] have decided that science and innovation themselves are the real source of the problem. Science and technological innovation are then cast in a diabolical role of being historically, or indeed, intrinsically associated with various categories of error and evil—an agent of domination, exploitation, and aggression, and social, cultural, and ecological destruction.[4]

Summary

Notwithstanding the rhetoric of modernization and "development for all" over the last half-century, the prevailing social concerns of different societies around the world are *not* converging. Opinions about what can be hoped and expected from science and technological innovation are not necessarily converging either.

Not only do clear divergences separate various factions of South and North ("development first" in environmental debates, tension between Islamic and

Christian traditions, etc.), rifts are also opening up (or becoming visible) between the Old World and the New World within the Organisation for Economic Co-operation and Development. Europe and the United States were already diverging on issues such as genetically modified organisms (GMOs) and genetic intrusion into foods before the latest round of geopolitical and military tensions.

What are the consequences of this divergence of societal perspectives for technology risk governance? To answer this question we should consider a range of plausible models in the "marketplace for ideas." Exploring different perspectives on governance can stimulate dialogue between intellectual and social factions.

In Europe at least, many people are proposing that the inherent complexity, high stakes, and urgency of technological risk, international security, and sustainable development cannot be left to technological advances in the service of self-serving doctrines. Hope for civilized coexistence on our small planet requires concerted efforts for affirming solidarities at different scales.[5]

Public policy toward technological development and risk can seek simply to mitigate problems by controlling the delinquency levels of populations in conditions of alienation and fatalism. Or it might seek to repair, revitalize and rebuild damaged communities, and to engage people in dialogues about their own interests and futures. In either case, technology and (fossil) energy are only about 1 percent of the resource scarcity problem. In times of stress, for evaluating the prospects for a society, the symbolic and emotional "residual" dominates everything.

Trust is the willingness of a person, group, or community, to make themselves vulnerable in the expectation or hope of a benefit from association with others. Trust in governments, as in commercial enterprise, as in science and technology, depends on hopes for meaningful benefits and on confidence in the capacity and will of society's leaders and innovators to assure the sharing—symbolic as well as material—of those benefits.

People do not only seek material comfort and security; they also pursue intangibles, like status, honor, and dignity. It is hardly surprising to find doubt about the benefits of new technologies, to see people using technology in a "delinquent" way to show alienation, to see technology used to challenge and disrupt the structures of authority, if they believe they are held in contempt or disdain.

It is therefore rare that new technology solves the problems that old technology has created. For the most part, technology alone is not what creates the problems; the focus should be on the social conditions and purposes underlying the mobilization of technology. The resource scarcity is not only material; it is just as much imaginative, moral, and symbolic. Many anthropologists would suggest that, even if conflict in various forms is inevitable, the specificity of human culture is the emergence of conventions and symbolic systems that regulate the forms that are acceptable in a given society. If an objective in public policy is to mitigate violence within and between societies, a big part of the needed imaginative capacity would be to develop new means to interpret "technological risks" in terms of the symbolic, cultural, and emotional dimensions of people's relations to each other.

There might, in this respect, be a high return on investment in investigating and remedying some of the social and economic factors—commonly called intolerance, prejudice, and injustice—that contribute to social delinquency.

People become desperate, mistrustful, aggressive, and the like partly in a complex cumulative causation process within society. The emergence of "rogue states" might well be, as John Kenneth Galbraith (1968) pointed out (in his little book *The Triumph*), at least partly a dialectical outcome of other states' shortsightedness and obsession with hegemony. None of the great economists—such as Adam Smith, John Stuart Mill, J.M. Keynes—ever pictured people as self-interested automata. Smith referred to moral sentiments that, in his perception, revolved around a form of vanity. Mill affirmed a type of sympathy as the affective basis of civilized liberal society. Keynes referred to the animal passions. At the macro scale, identity and relations are not summed up in aggregate demand for goods. They are codified in complex societal institutions and symbols, including religions, conventions of justice, tribal identities, informal networks of exploitation and solidarity, community belonging, and so on. These are all important terrain for economists who want to contribute to environmental and technology policy issues.

Notes

1. This formulation comes from some of my risk governance colleagues (Funtowicz et al. 1998; Gallopín et al. (2001).

2. Multistakeholder evaluation through deliberative processes is a complex task that must involve policymakers and the public at large as well as the scientific community, and that is not without pitfalls and risks. Some starting points in the European literature include Bailly (1998); De Marchi & Ravetz (1999); Faucheux & Hue (2000, 2001); Funtowicz & Ravetz (1990, 1993), O'Connor (1999, 2002, 2003).

3. For example, the possible use of a new generation of "mini" nuclear bombs in the war of Good against Evil. Whatever the direct health and economic consequences of the immediate radioactive fallout of these weapons, it seems speculative to try to "calculate" the consequences (the net costs or benefits) in terms of geopolitics, human relations, governance regimes, and so on.

4. The international catalog of environmental side effects and "historical liability" problems has a set of "cumulative causes," including, not least, material affluence and a belief in the benefits of "technological progress" led by the rich, developed, industrialized countries. The ordinary consumer and the onlookers in the South (hoping to catch up) are merely conforming to the roles offered to them in the Western cultural project. But it is not totally surprising that some of these consumers or onlookers decide to repudiate the Western solution in favor of other visions of a possible humanity.

5. The spirit of "multistakeholder deliberation," as also of cross-cultural dialogue, is not just to share existing wisdom but also to let wisdom and purposefulness emerge and be renewed in the course of the dialogue process itself. There are, evidently, situations where people, or different cultures, simply cannot or do not want to find a basis for a durable coexistence. Reflective deliberation as advocated here, may thus work to highlight appreciation of tensions, but it does not necessarily find a way to put an end to them.

References

Bailly, J.-P. 1998. *Prospective, Débat, Décision Publique. Avis du Conseil Economique et Social.* Published as vol.16 (July 17) of the *Journal Officiel de la République Française.*

De Condorcet, A.-N. 1795. *Sketch for a Historical Picture of the Progress of the Human Mind.* Translated from the French by June Barraclough (1955). New York: Noonday Press.

De Marchi, B., and J. Ravetz. 1999. Risk Management and Governance: A Post-Normal Science Approach. *Futures* 31(7): 743–757.

Faucheux, S., and C. Hue. 2000. Politique Environnementale et Politique Technologique: Vers une Prospective Concertative. *Nature Sciences Sociétés* 8(3): 31–44.

———. 2001. From Irreversibility to Participation: Towards a Participatory Foresight for the Governance of Collective Environmental Risks. *Journal of Hazardous Materials* 86: 223–243.

Funtowicz, S.O., and J.R. Ravetz. 1990. *Uncertainty and Quality in Science for Policy.* Dordrecht: Kluwer Academic Press.

———. 1993. Science for the Post-Normal Age. *Futures* 25(7): 735–755.

Funtowicz, S., J. Ravetz, and M. O'Connor. 1998. Challenges in the Use of Science for Sustainable Development. *International Journal of Sustainable Development* 1(1): 99–107.

Galbraith, J.K. 1968. *The Triumph: A Novel of Modern Diplomacy.* Boston: Houghton Mifflin.

Gallopín, G., S. Funtowicz, M. O'Connor, and J. Ravetz. 2001. Science for the 21st Century: From Social Contract to the Scientific Core. *International Journal of Social Science* 168: 209–229.

Lubchenco, J. 1998. Entering the Century of the Environment: A New Social Contract for Science. Presidential Address to the American Association for the Advancement of Science, February 15, 1997. *Science* 279 (January 23): 491–497.

O'Connor, M. 1999. Dialogue and Debate in a Post-Normal Practice of Science: A Reflection. *Futures* 31: 671–687.

———. 2002. Social Costs and Sustainability. In *Economics, Ethics and Environmental Policy: Contested Choices,* edited by D.H. Bromley and J. Paavola. Oxford: Blackwell Publishing, 181–202.

———. 2003. Building Relationships with the Waste. In *Public Confidence in the Management of Radioactive Waste: The Canadian Context.* NEA Forum on Stakeholder Confidence, Workshop Proceedings, Ottawa, Canada, 14–18 October 2002. Paris: OECD, 177–190.

CHAPTER 13

Intragenerational versus Intergenerational Equity

Views from the South

Ramón López

ENVIRONMENTAL SCARCITY IS CONVENTIONALLY analyzed as an issue concerning intergenerational equity. If the present generation is leaving so little environmental resources for future generations that their welfare growth and even the preservation of current standards of living are threatened, then we say that current growth is unsustainable. No one can prove or disprove this with certainty. We may be able to find out, however, whether current growth patterns are reducing certain environmental resources at a rapid pace. This increased scarcity probably means a greater risk of unsustainable growth. As discussed elsewhere in this volume, it appears that scarcity created by current (and past) generations mostly affects so-called environmental amenities (ecosystem health, air, atmospheric carbon concentration, etc.). Scarcity of environmental commodities (foods, minerals, energy, etc.), on the other hand, is not imminent. The consensus is that given current trends, there is a significant risk that scarcity of environmental amenities may affect the welfare and economic growth of future generations within the present century.

Separability

Analysts have commonly separated intergenerational equity from intragenerational equity. Given the complexity of the question of intergenerational equity and sustainability, there is much to be said for this separation as a simplifying device. Most contributions to this volume follow this tradition. Separability does entail, however, considerable disadvantages, especially in assessing the role of public policy in affecting sustainability. It is particularly inappropriate for developing countries, where public policies are frequently responsible for large wastes of natural resources for the sake of benefiting politically powerful economic elites (Ascher

1999; Myers and Kent 2001; López 2003). As Dasgupta points outs elsewhere, governments have often allowed or even facilitated the disenfranchisement of the poor through usurpation of local resources.

Equity and the Control of the State

Integration of intragenerational and intergenerational analyses is preferable for developing countries because intergenerational socioenvironmental inequities (unsustainable growth) may to a significant extent be grounded in the existence of intragenerational inequities. A common denominator of both types of inequities is the dominance of the state by a small fraction of the population, which is able to direct public policies for their own benefit rather than maximal social welfare (López 2003; Van Beers and de Moor 2001). Will governments that systematically neglect the welfare of the vast majority of the current population follow public policies that consider the interests of those not yet born? The answer is not obvious. It is in principle possible that if the elites that dominate the state are concerned about their descendants, this concern will be reflected in current public policies. It depends on the political economy mechanisms they use to influence public policies.

Public Policy and the Economic Elites

Let us look at certain aspects of the mechanics of public policy generation in developing countries. First we need to introduce some greater precision in terminology. The "elites" are those who are able to influence public policy through influence peddling, corrupting government officials, political campaign contributions, and so on. Their political power is derived from their economic power. But these are only the most obvious ways by which these groups influence public policy. There are more subtle means that are likely to become more important as democracy and democratic institutions evolve. "Think tanks" and other organizations funded by organized power groups, and media controlled and funded by the same groups, are important instruments in promoting a certain ideology: namely, whatever is good for the elites is also good for economic development.

Regardless of the means used, influencing public policy is costly; given the usual imperfections in capital markets, mainly wealthy individuals who can access capital markets or have enough wealth to directly finance such investment can do it. Buying public policy is an investment like any other: to be attractive, it needs a rate of return equal to or greater than that of other investments. The potential rate of return on investing in influencing public policy is presumably much greater for wealthier individuals than for poor individuals. Collectively the poor may have even greater stakes in public policy than the rich, but as Olson (1965) pointed out, collective action to influence government is greatly facilitated when the lobby groups are small and homogenous. The fact that the poor are many and the wealthy are few means that collective action on the part of the poor to influence public policy is rare, and collective action by the wealthy is the norm.

Control by the elites often induces governments to undersupply public goods, including environmental protection, for the sake of providing private goods or subsidies to the elites (López 2004).[1] Moreover, beyond direct budgetary subsidies—credit subsidies, irrigation subsidies, outright financial transfers, and others—another important subsidy mechanism is the giveaway to the wealthy of the usufruct of fishing, water use, and other resources often not amenable to private property rights. Almost unlimited polluting rights may be given away to powerful economic interests, especially in rural areas where monitoring by civil authorities is more difficult (Ascher 1999; Myers and Kent 2001).[2]

Policy Traps

The greater the concentration of wealth, other things being equal, the more concentrated the dominance of the state and the more biased the policies in favor of a smaller elite. Biased policies, by increasing economic benefits to the wealthy, perpetuate the historical concentration of wealth, which, in turn, causes the policy biases to remain in place. Thus a *policy trap* arises. Even governments that seem ideologically committed to wealth redistribution have not been able to significantly change the old policies, much less to alter historical wealth distribution. Chile and Brazil over the last decade constitute important examples of the policy trap in resource-rich countries: democratic and seemingly "progressive" regimes have consistently failed to significantly alter the old policies, originated under dictatorship, while at the same time they have not been able to improve their wealth distribution, among the worst in the world.

On Natural Resource Endowments and Policies

Wealth tends to be more concentrated in natural resource–rich countries than in resource-poor countries. Resource-rich countries in Latin America and Africa, oil-rich nations, and the United States generally have much greater concentration of income than the resource-poor countries of Asia, Europe, or Japan.[3] Rents for resources affected by ill-defined property rights (which the poor can have access to) are either greatly dissipated or are fully captured by the few who control highly capital-intensive resource extraction activities, such as oil, minerals, large forests, and agricultural plantations, where property rights often become well defined. Thus natural resource–rich countries, which also tend to be the depository of most significant natural habitats around the world, are precisely those that are prone to have the most egregious anti-environment (and generally anti-poor) policies. As a consequence of the policy trap discussed above, they are also the least likely to reform them.[4]

Sustainability and the Economic Elites: The Role of Substitution

A public policy agenda dictated by the economic elites affects the likelihood of environmental sustainability and, therefore, of intergenerational equity. The (short-run) benefits of environmentally unsustainable growth disproportionably go to the economic elites, while the (medium- and long-term) burden of environmental degradation is mostly shouldered by everyone else. In general the ability of an individual to pollute and to extract natural resources is directly proportional to their capital endowment; therefore, charging user costs or pollution taxes is proportionally more expensive to those with larger capital endowment. Equivalently, the benefits of allowing greater environmental degradation accrue to wealthy individuals. The cost of environmental degradation is lower for wealthy individuals because they can afford to substitute natural environmental services with human-made services (through, for example, defensive expenditures) while the poor generally cannot (Antoci and Bartolini 2002).[5] Moreover, the cost of substitution is lower when only a small segment of society attempts it than when a larger portion makes efforts to substitute environmental services with human-made services. If a large segment of society demands more human-made services that substitute for lost natural services, the prices of such human-made services increase. So, in countries with a concentrated wealth distribution, a majority of the population remains excluded from the market for defensive goods. This means that the costs of substitution for those that can afford it tend to remain low.

The net implication of all this is that the (private) "optimum" pollution and environmental degradation is likely to be higher for the elite households than for everyone else. The degree of environmental "unsustainability," therefore, is likely to be higher the more the economic elites are able to dictate public policy. To the extent that the descendants of the wealthy have a high probability of also being wealthy (especially when "death taxes" are nonexistent or easily bypassed, as in most developing countries and in at least one prominent developed one), intergenerational considerations are not likely to alter this conclusion. The rich know that their children will have the same advantages that they have, or perhaps more because of technological change, and will be able to continue substituting the loss of the services produced by the natural environment with more human-made services.

Governance and Democracy

The new as well as the old literature on endogenous government policies emphasizes competition among groups to influence public policy (Grossman and Helpman 1994). Although subgroups of the elites unquestionably have conflicting interests that often trigger competition among them to steer public policy in different directions, they also have certain common goals on which they are able to act with a significant degree of coherence. The real conflict is between the wealthy elites and the rest of society, not among subgroups within the elites. Well-established democratic rules, the influence of that part of the intellectual class not contaminated by the ideology of the elites (writers, scientists, and social scientists, but generally few economists), and the political organization of the nonelites are the only possible mechanisms that can at least in part offset the enormous power of the economic elites (who usually count orthodox economists among their most valuable intellectual allies).

Respect for democratic institutions is all-important. Only if broader civil society participates in electing and monitoring political authorities is it possible to counter at least the most primitive forms of lobby, especially those influencing public policy through corruption and other illicit tactics. Unfortunately, the more evolved mechanisms—those adapted to more perfected forms of democracy—used by the elites to influence public opinion are harder to counteract through civil participation. Control of the media and other important institutions is a particularly effective mechanism used by the economic elites to continue exerting their political power even under relatively perfected forms of democracy. The breaking of the policy trap needs not only perfected democratic institutions but also outside help or a great deal of luck (in the form of chance conditions that dramatically incline the pendulum in favor of policy reform).

The economically powerful groups generally win, although, depending on the strength of the offsetting democratic conditions, they are occasionally forced to make concessions. The resulting policies in most developing countries, and even some industrial economies are, however, likely to be closer to those demanded by the economic elites than those that benefit the large majorities.

Implications

Because policies are endogenous, what is the point of bringing up these issues in this forum, especially when the ideas presented here are fairly simple, perhaps obvious? The point I hope to make is that we economists have ignored these issues for too long and, indeed, more often than not have been part of the problem. By empirically understanding the political motivation of environmental and certain economy-wide policies we (the economic profession) could incline the balance a bit more (perhaps a lot more) in favor of the vast majorities instead of continuing to tacitly support the public policy dominance of the powerful elites. We could contribute to both intra- and intergenerational equity and increase the likelihood of environmental sustainability. It seems absurd to study sustainability in a social and political vacuum. It is also absurd to attribute government policy failures to "mistakes" and ignorance instead of recognizing that most such failures are the logical outcome when governments serve the economic elites instead of all society.

Notes

1. López (2004) presents empirical evidence that governments in several Latin American countries allocate more than 50% of their total expenditures in rural areas to nonsocial subsidies. He shows that such subsidies not only promote environmental destruction but also reduce the per capita income of the rural population and increase rural poverty. That is, they virtually define *perverse subsidies*.

2. Some have argued that giving away the usufruct rights of natural resources to the wealthy may have equity connotations, but it does not necessarily undermine environmental sustainability. However, when resource usufruct is allocated to individuals who have the means to influence governments, they are more likely to pressure governments to relax environmental regulation of such resources, or worse, to evade legal controls through bribery and other means. The result? Greater environmental destruction and worsening social equity.

3. Of 16 developed countries that have data on income distribution that is comparable across countries, the unweighted average income Gini coefficient for the late 1990s was 30.7. These countries include Australia, Austria, Belgium, Canada, Denmark, France, Germany, Italy, Japan, the Netherlands, Norway, Spain, Sweden, Switzerland, the United Kingdom, and the United States. The majority of these countries are clearly resource poor with the exceptions of Australia, Canada, the United States, and possibly Norway. The average Gini for the 12 resource-poor countries was 27.6, compared to 33.3 for the four resource-rich developed countries. Though the United States (40.8) significantly influences this average, Australia (35.2) and Canada (31.5) each has a Gini significantly worse than the average of the resource-poor developed countries and also worse than the average for all developed countries. Only Norway (25.8) escapes this norm. A similar exercise for LDCs yielded similar results; given the much greater data and methodological heterogeneity in the Gini estimates, the numbers for LDCs are less reliable.

4. The United States is an interesting example of a resource-rich country, with relatively high degree of wealth concentration and consistently having policies that are much less pro-distribution and more anti-environment than most other industrial countries. It is no coincidence that it refused to sign the Kyoto Protocol.

5. A common argument is that the environment is a normal or even a "luxury" good, so the wealthy tend to demand a better environment than the poor (McConnell 1997 rejects the premise that the environment is a luxury good). However, the fact that the wealthy can protect themselves from many effects of environmental degradation (for example, by choosing where to live), while the poor cannot, considerably weakens this argument. The demand for environmental quality is just one side of the equation. The wealthy can increase their income at the cost of environmental degradation to a much greater extent than can the poor. The empirical evidence of the so-called Environmental Kuznets Curve—pollution first increases and then declines as income exceeds a certain threshold—may be taken as supporting the luxury good hypothesis. However, apart from the many objections to the way these curves are derived and interpreted, hardly any developing country has a per capita income above the estimated threshold for any significant pollutant. See the various articles in the 1997 survey issue of *Environment and Development Economics*.

References

Antoci, A., and S. Bartolini. 2002. Defensive Expenditures and Economic Growth in an Evolutionary Model. Unpublished paper. University of Florence.
Ascher, W. 1999. *Why Governments Waste Natural Resources—Policy Failures in Developing Countries*. Baltimore: Johns Hopkins University Press.
Grossman, G., and E. Helpman. 1994. Protection for Sale. *American Economic Review* 84: 833–850.
López, R. 2003. The Policy Roots of Socio-Economic Stagnation and Environmental Implosion: Latin America 1950–2000. *World Development* (forthcoming).
———. 2004. *The Structure of Public Expenditures, Agricultural Income and Rural Poverty: Evidence for Ten Latin American Countries*. Washington, DC: World Bank.
McConnell, K. 1997. Income and the Demand for Environmental Quality. *Environment and Development Economics* 2: 383–e99.
Myers, N., and J. Kent. 2001. *Perverse Subsidies: How Tax Dollars Can Undercut the Environment and the Economy*. Washington, DC: Island Press.
Olson, M. 1965. *The Logic of Collective Action: Public Goods and the Theory of Groups*. Cambridge, MA: Harvard University Press.
Van Beers, C., and A. de Moor. 2001. *Public Subsidies and Policy Failures*. Northampton, MA: Edward Elgar.

CHAPTER 14

Sustainable Economic Development in the World of Today's Poor

Partha Dasgupta

*E*NVIRONMENTAL AND RESOURCE ECONOMICS, as practiced in the United States, has not shown much interest in economic stress and population growth in poor countries. In reviews of environmental and natural resource economics, Kneese and Sweeney (1985–1993), Cropper and Oates (1992), and Oates (1992) altogether bypass the links between abject poverty, population growth, and degradation of the natural resource base. They were right to do so, for it has been common practice among economists in the North (I am using the term in its contemporary geopolitical sense) to regard Nature as an amenity. The strong emphasis on valuation exercises in environmental economics is indicative of that viewpoint. At an extreme it is even believed that countries need to grow out of poverty before they can begin to care for their environment, and "trade improves the environment, because it raises incomes, and the richer people are, the more willing they are to devote resources to cleaning up their living space" (*The Economist* 1999).

These passages reflect the view that the natural environment is a luxury. It is a detached view.[1] Closer to home, matters look different. Producing as it does a multitude of ecosystem services, a large part of what the natural environment offers us consists of necessities. A number of services reach the global context, but many are local. When wetlands, forests, and woodlands in poor countries are destroyed (for agriculture, urban extension, or commercial timber), traditional cultures suffer. For them—and they are among the poorest in society—there are no substitutes. For others (e.g., ecotourists), there is something else, often somewhere else, which means there are substitutes. Moreover, the range separating a need and a luxury is enormous and tied to context. What is a luxury for one may be a need for another. Income matters, because income in part determines opportunities to employ substitutes. Microeconomic reasoning has identified a bewildering variety of harmful spillovers associated with the use of natural resources in poor countries.

Such spillovers, symptomatic as they are of institutional failure, are mechanisms by which official economic activity, for example, gross national product (GNP), can grow at the expense of the lives of the poorest.[2]

Growth Accounting and Technical Change

Macroeconomic reasoning glosses over the heterogeneity of the earth's resources and the diverse uses to which they are put—by people at the source or elsewhere. Modern theories of economic growth stress new ideas as a source of progress and suppose that the growth of ideas is capable of circumventing any constraint the natural environment may impose on the ability of economies to grow indefinitely. Such models note too that certain forms of investment (e.g., research and development) enjoy cumulative returns because the benefits are durable and can be shared collectively. The models also assume that growth in population leads to an increase in the demand for goods and services. An expansion in the demand and supply of ideas implies that, in the long run, output per capita can be expected to grow at a rate that is itself an increasing function of the rate of growth of population. (It is only when population growth is nil that the long-run rate of growth of output per capita is nil.) According to the models, indefinite population growth is beneficial.[3]

In their pristine form, contemporary growth models assume a positive link between the creation of ideas (technological progress) and population growth in a world where the natural resource base constitutes a fixed, indestructible factor of production (Kremer 1993). The latter assumption is wrong: the natural environment consists of degradable resources (soil, watersheds, fisheries, and sources of fresh water).[4] It may be sensible to make that assumption when studying a period when natural resource constraints did not apply, but not when studying development possibilities open today to people in poor regions.

Contemporary growth theories do not explicitly model the nature of the new products that embody technological progress. But they assume that future innovations will permit indefinite growth in output to make only a finite additional demand on the natural resource base (Jones 2001). I do not know what evidence there is to back that assumption, but I do know of plenty that works against it. If economic growth (growth in GNP) is to be sustainable, capital accumulation and technological progress must counter a declining resource base. But a vanishing resource base would mean a dwindling supply of the multitude of ecosystem services upon which life is dependent.

It may be that the technological progress envisaged in growth models is of such character as to make long-run growth in GNP independent of the natural resource base. However, one problem with that assumption (it has other, more fundamental problems) is that property rights to environmental resources often are either vaguely defined or weakly enforced, meaning that environmental services are frequently underpriced in the market. New technologies can therefore be expected to be rapacious in their use of natural resources: private inventors and innovators have little reason to seek discoveries that economize on their use. It means that new ideas will not offer adequate substitutes for local resources, and that in a world where relocation to escape an intolerable rural situation may be impossible.[5]

In any event, we should be skeptical of a theory that places such reliance on an experience that is not much more than two hundred years old (Maddison 2001). Extrapolation into the past is a sobering exercise. Over the long haul of history (a 5,000-year stretch up to, say, 200 years ago), economic growth even in the currently rich countries was on average not much above zero.[6] The study of possible feedback loops between poverty, demographic behavior, and the character and performance of both human institutions and the natural resource base has not yet entered the research agenda of modern growth theorists.

The Environment and Development Economics

The population–poverty–resource nexus in poor countries is also not a focus of attention among development economists. The subject is largely absent even from studies of the semiarid regions of sub-Saharan Africa and the Indian subcontinent. For example, the authoritative surveys on population growth in poor countries by Birdsall (1988), Kelley (1988), and Schultz (1988) fail to touch on environmental matters. Mainstream demography and agricultural economics have also made light of the environmental constraints facing poor communities in the same regions (e.g., Johnson 2000). Nor does the mainstream literature on poverty in poor countries take population growth and ecological constraints to be significant influences on development (e.g., Stern 1989; Dreze and Sen 1990; Bardhan 1996). Most textbooks on economic development simply point to the Western experience since the industrial revolution and conclude that Malthus got it all wrong.

Scholars should be puzzled by the situation. A prop of the rationale for development economics is the notion that poor countries suffer particularly from institutional failures. But institutional failures in great measure manifest themselves as externalities. To ignore population growth and ecological constraints in the study of poor countries would be to suppose that demographic decisions and resource use there give rise to no externalities of significance. It would also be to suppose that externalities arising from institutional failure have a negligible effect on resource use and demographic behavior. I know of no body of empirical work that justifies these suppositions.

Motivation

Given the neglect of the poverty–population–environment nexus, it should come as no surprise that the measure of economic development is still that old indicator of social well-being—GNP per capita—to which the UN's Human Development Index (HDI) has been added in recent years.[7] The problem is that both GNP and HDI reflect short-run concerns, while the question whether contemporary patterns of development are sustainable requires us to peer into the future.

Consider a question that is often asked: Is our use of the earth's resources reducing the economic possibilities open to our descendants?

In fact there is a remarkable divergence of opinion on the matter, ranging from a straightforward "yes" to a flat "no."[8] One also hears that the question misleads, in that it suggests that resource conflicts are to be found only between "us" and future "thems," whereas, or so it is argued, large pockets of extreme poverty in what is otherwise an increasingly affluent world ensure that there are resource conflicts even among contemporaries (Dasgupta 1993, 2000).

Underlying these intellectual tensions are the conflicting intuitions that have arisen from different empirical perspectives on the sustainability of contemporary economic development, in the poor world and in industrialized countries.[9] On the one hand, if we study historical trends in the price of marketed resources, or the recorded growth in the conventional indices of economic progress in today's rich countries, resource scarcities appear not to have struck yet.[10] On the other hand, if we look at specific resources and services (e.g., freshwater, the atmosphere as a carbon sink), there is evidence that the accelerating pace at which they are utilized is unsustainable.[11]

Recently, a few investigators have developed a theoretical framework that goes some way toward reconciling these conflicting intuitions (Serageldin 1995; Pearce et al. 1996; Dasgupta and Mäler 2000; Dasgupta 2001a, 2001b; Arrow et al. 2003a, 2003b). The theory shows that macroeconomic accounting, when undertaken correctly, is consistent with both microeconomic reasoning and local experiences among the world's poorest. Market prices of depletable resources could be declining (because of falling extraction and refinery costs or discoveries of additional reserves), even while their accounting prices are rising (owing to rising environmental costs associated with their use in production or the disenfranchisement of the poor through usurpation of local resources). Furthermore, at the level of macroeconomic accounting, such indicators of current well-being as GNP per capita and life expectancy at birth could be recording improvements even while an economy's long-run prospects shrink.

In what follows, I study aggregate statistics of economies, for both practical and intellectual reasons. The practical reason is that to do so will enable me to keep the chapter to a readable length. The intellectual reason is that, by making use of social weights for different income categories, I can incorporate both absolute poverty and income inequality in GNP. The real weakness of GNP is not that it is unable to accommodate the phenomena of absolute poverty and economic inequality, but that it is unable to take the future adequately into account. As this chapter is about sustainable development, I focus on the distribution of the standard of living across generations.

Wealth and Well-Being

An economy's long-run prospects are shaped in part by its institutions and by the size and distribution of its capital assets. Together they compose its productive base, the source of its well-being. It is tempting to regard institutions also as capital assets (we often refer to a society's "institutional capital"), but institutions are distinct from capital assets in that they guide the allocation of resources (among which are the capital assets themselves). Improvements in a country's institutions would result in better resource allocation. They offer one route to increases in the standard of living. Another route is the accumulation of capital assets.

There is a long-established name for the value of an economy's capital assets: *wealth*. The notion of wealth I adopt here is a comprehensive one, and the list of assets includes not only those that are manufactured (roads and buildings; machinery and equipment; cables and ports), but also human capital (knowledge and skills) and a wide array of natural capital (oil and natural gas; fisheries and forests; ecosystem services). To say that wealth has increased is to say that, in the aggregate, there has been an accumulation of capital assets. In what follows I shall call the accumulation of capital assets *genuine investment*. This is to contrast recorded investment. As the services of any number of capital assets are missing from national accounts, recorded investment can be positive even while genuine investment is negative.

It can be shown that, given a country's institutions, wealth (or more accurately, a wealth-like index elaborated below) is a measure of a society's well-being, taking both the present and the future of that society into account. By this I mean that, correcting for population change, the well-being of present and future generations, considered together, increases if genuine investment is positive. Changes in wealth over time in a country can be used to identify whether its pattern of development is sustainable. I cannot emphasize strongly enough that these results do *not* rely on specialized assumptions regarding the economy,[12] *nor* do they presume that the government optimizes on behalf of its citizens. The results are therefore applicable in the world we have come to know.[13]

In contrast to wealth, consider GNP. The sum of aggregate consumption and gross investment, GNP is insensitive to the depreciation of capital assets. It is therefore possible for GNP to increase even while the economy's genuine investment is negative and wealth declines. This can happen if, say, increases in GNP are brought about by mining capital assets—for example, degrading ecosystems and depleting oil and mineral deposits—without investing some of the proceeds in substitute forms of capital, such as human capital. So there is little reason to expect movements in GNP to parallel those in wealth. The moral, though banal, is important: GNP cannot be used to identify sustainable development policies. As we will confirm presently, nor can HDI identify them. Adam Smith's classic was an inquiry into the wealth of nations, not the gross national product of nations or the United Nations Development Programme's Human Development Index of nations.

Nature as a Capital Asset

The emphasis I have given to natural capital in the previous paragraph is not accidental. National accounts are highly sophisticated today, but they continue to miss not only the changes brought about by economic activities to the stocks of many natural resources, but also our use of a myriad of Nature's services. The latter include maintaining a genetic library, preserving and regenerating soil, fixing nitrogen and carbon, recycling nutrients, controlling floods, filtering pollutants, assimilating waste, pollinating crops, operating the hydrological cycle, and maintaining the gaseous composition of the atmosphere. The reason such services are frequently missing in national accounts is that they most often do not come with a price tag. The reason for that is that property rights to natural capital are often impossible to establish, let alone to enforce. And the reason for *that* is that natural capital is frequently mobile (birds, insects, river water, and the atmosphere are prototypical). But none of this means that with effort it would not be possible to assign notional prices to Nature's services, prices that would go some way toward reflecting their scarcity values. As matters stand, though, the effects of the interconnectedness of various forms of natural capital often go unrecorded in economic transactions. So it is that those who inflict damage (for example, destroying mangroves in order to create shrimp farms or logging the uplands of watersheds) are not required to compensate those who suffer the damage (fishermen dependent on the mangroves or farmers and fishermen in the downlands of the watersheds).

Rural communities in poor countries recognized this deep underlying problem with Nature's services long ago and developed institutional mechanisms to overcome it in the case of local capital assets. Ponds, tanks, threshing grounds, grazing fields, and woodlands harbor mobile resources, making them unsuitable as private property.[14] In recent years, anthropologists, ecologists, economists, and political scientists have identified a wide variety of nonmarket institutions in rural communities that mediate economic transactions in Nature's services. These institutions are frequently communitarian. Moreover, they correspond to the character of the natural capital under their jurisdiction. For example, communitarian institutions for coastal fisheries are quite different in design from those governing local irrigation systems.[15]

Unhappily, in recent years communitarian institutions have eroded in many of the poorest regions of the world. There are a number of reasons why this has happened; state interference is most prominent, especially in sub-Saharan Africa. Ironically, the growth of marketable goods and services may have contributed as well. When decaying communitarian institutions are neither stayed nor adequately replaced by other institutions, the poorest frequently suffer the most, in particular because their local environmental resource base deteriorates.[16]

When choosing economic policy, decisionmakers need to be sensitive to the interplay of market and nonmarket institutions. Any system, human or otherwise, responds when perturbed. A policy change can create all sorts of effects that ripple through unnoticed by those who are unaffected, because there may be no obvious public signals accompanying them. Tracing the ripples requires an understanding of nonmarket interactions and their interplay with markets. Identifying sustainable development policies involves, among other things, valuing the ripples and, therefore, valuing Nature's services. We can now appreciate in which ways the weaknesses of present-day national accounts mirror the weaknesses in the practice of contemporary policy evaluation. It is reasonable to fear that because Nature's services are typically underpriced, modern economic development has been rapacious in its use of natural capital.

An Application to Poor Countries

We presume then that genuine investment is less than recorded investment. But by how much?

The World Bank has provided estimates of genuine investment in a number of countries by adding net investment in human and natural capital to estimates

Table 14-1. *Genuine Investment and Wealth Accumulation in Selected Regions, 1970–1993*

	I/Y (%)[a]	g(L)[b]	g(W/L)[c]	g(Y/L)[d]	g(HDI)[e]
Bangladesh	−0.3	2.3	−2.40	1.0	+
India	10.7	2.1	−0.50	2.3	+
Nepal	−1.5	2.4	−2.60	1.0	+
Pakistan	8.2	2.9	−1.70	2.7	+
Sub-Saharan Africa	4.7	2.7	−2.00	−0.2	+
China	14.4	1.7	1.09	6.7	−

[a] I/Y = genuine investment as percentage of GNP. (*Source:* Hamilton and Clemens 1999, Tables 3 and 4; and personal communication from Katie Bolt, World Bank). Genuine investment includes total health expenditure (i.e., public plus private), estimated as an average during 1983–1993, from data supplied by the World Health Organization.

[b] g(L) = average annual percentage rate of growth of population, 1965–96. (*Source:* World Bank 1998, Table 1.4).

[c] g(W/L) = average annual percentage rate of change in per capita wealth at constant prices.

[d] g(Y/L) = average annual percentage rate of change in per capita GNP, 1965–1996. (*Source:* World Bank 1998, Table 1.4).

[e] g(HDI) = sign of change in UNDP's Human Development Index, 1987–1997. (*Source:* UNDP 1990–1999). Assumed output–wealth ratio = 0.15 per year.

of investment in manufactured capital (Hamilton and Clemens 1999). A certain awkwardness characterizes the steps the investigators have taken to arrive at their estimates. Their accounts are also incomplete; among the resources that make up natural capital, only commercial forests, oil and minerals, and the atmosphere as a sink for carbon dioxide were included (not water resources, forests as agents of carbon sequestration, fisheries, air and water pollutants, soil, or biodiversity). So they have undercounted, perhaps seriously.[17] Nevertheless, one has to start somewhere. It will prove instructive to use the World Bank figures and assess the character of recent economic development in the poorest regions of the world.

Table 14-1 does that; it covers sub-Saharan Africa, the Indian subcontinent, and China. Taken together, the bulk of the world's 1 billion poorest live there. The first column contains the World Bank's estimates of genuine investment, as a proportion of GNP, during the period 1973–1993. Notice that Bangladesh and Nepal have disinvested: aggregate capital assets declined during the period in question. In contrast, genuine investment has been positive in China, India, Pakistan, and sub-Saharan Africa. So the figures could suggest that the latter countries were wealthier at the end of the period than at the beginning. But when population growth is taken into account, the picture changes.

The second column contains the annual rate of growth of population over the period 1965–1996. All but China experienced rates of growth in excess of 2 percent per year, sub-Saharan Africa and Pakistan having grown at nearly 3 percent per year. On the theory I sketched earlier, we next estimate the average annual change in wealth per capita during 1970–1993. To do this, I have multiplied genuine investment as a percentage of GNP by the average annual output–wealth ratio of an economy to arrive at the investment–wealth ratio, and have then compared changes in the latter ratio to changes in population size.

Since a wide variety of capital assets (for example, human capital and various forms of natural capital) are unaccounted for in national accounts, there is a bias in published estimates of output–wealth ratios, which traditionally have been taken to be something like 0.30 per year. In what follows, I have used 0.15 per year as a check against the bias in traditional estimates for poor countries. Even these figures are almost certainly too high.

The third column contains my estimates of the annual rate of change in the per capita wealth-like index I mentioned earlier. To arrive at these figures, I multiply genuine investment as a proportion of GNP by the output–wealth ratio, and then subtract the population growth rate from that product. This is a crude way to adjust for population change, but more accurate adjustments would involve greater computation.

The striking message of the third column is that in all but China there has been capital *decumulation* during the past 30 years or so. This may not be a surprise in the case of sub-Saharan Africa, which is widely known to have regressed in terms of most socioeconomic indicators, but the figures for Bangladesh, India, Nepal, and Pakistan should cause surprise. Even China, so greatly praised for its progressive economic policies, has just managed to accumulate wealth in advance of population growth. In any event, a more accurate figure for the output–wealth ratio would almost surely be considerably lower than 0.15. Using a lower figure would reduce China's accumulation rate. Moreover, the estimates of genuine investment do not include soil erosion or urban pollution, both of which are thought to be especially problematic in China.

How do changes in wealth per capita compare with changes in conventional measures of the quality of life? The fourth column contains estimates of the rate of change of GNP per capita during 1965–1996; and the fifth column records whether the change in the UN Human Development Index over the period 1987–1997 was positive or negative.

Notice how misleading our assessment of long-term economic development in the Indian subcontinent would be if we were to look at growth rates in GNP per capita. Pakistan's GNP per capita grew at a healthy 2.7 percent per year, implying that the index doubled in value between 1965 and 1993. But the other figures reveal that the real wealth of the average Pakistani declined by some 40 percent during that same period.

Bangladesh too has decumulated capital. The country's GNP per capita grew at a rate of 1 percent per year during 1965–1996. But at the end of the period the average Bangladeshi was about half as wealthy as at the beginning.

The case of sub-Saharan Africa is especially sad. At an annual rate of decline of 2 percent in wealth per capita, wealth of the average person in the region is halved every 35 years. The ills of sub-Saharan Africa are routine reading in today's newspapers and magazines. But the ills are not depicted in terms of a decline in wealth. Table 14-1 reveals that sub-Saharan Africa has experienced an enormous decline in its capital assets over the past three decades.

India avoided a steep decline in wealth per capita. But the country has been at the thin edge of economic development. If the figures are taken literally, the average Indian was slightly poorer in 1993 than in 1970.

What of the Human Development Index? In fact it misleads even more than GNP per capita. As the third and fifth columns show, HDI offers a picture that is the precise opposite of the one we should see when judging the performance of poor countries. The index for sub-Saharan Africa grew during the 1990s and it declined for China. Bangladesh and Nepal have been exemplary in terms of HDI. However, both countries have decumulated their capital assets at a high rate.

The figures in the table are rough and ready, so we should arrive at conclusions tentatively. But they show how accounting for human and natural capital can make substantial differences in our conception of the development process. The implication should be depressing: the Indian subcontinent and sub-Saharan Africa, two of the poorest regions of the world, comprising something like a third of the world's population, have over the past decades become even poorer, some of them a good deal poorer. I do not know whether institutional changes and the importation of knowledge from elsewhere have compensated for the decumulation of wealth in those countries. Data in the next section point to a pessimistic conclusion even on that question.

Sustainable Development and Its Measurement

The International Union for the Conservation of Nature and Natural Resources (1980) and the World Commission on Environment and Development (1987) introduced the concept of sustainable development. The latter defined sustainable development to be "development that meets the needs of the present without compromising the ability of future generations to meet their own needs" (43). Several formulations are consistent with this phrase.[18] But the underlying idea is straightforward enough: we seek a measure that would enable us to judge whether an economy's production possibility set is growing. It can be shown that the requirement that economic development be sustainable implies, and is implied by, the requirement that the economy's productive base be maintained (Dasgupta and Mäler 2000; Dasgupta 2001b; and Arrow et al. 2003a).

Simply requiring that an economy be "sustainable" does not identify a unique economic program. In principle any number of technologically and ecologically feasible economic programs could satisfy the criterion. On the other hand, if substitution possibilities among capital assets are severely limited and technological advances are unlikely to occur, it could be that there is no sustainable economic program open to an economy. Furthermore, even if the government were bent on optimizing social welfare, the chosen program would not correspond to a sustainable path if the future welfare were discounted at too high a rate. It could also be that along an optimum path social welfare declines for a period and then increases thereafter. Optimality and sustainability are thus different notions. The concept of sustainability helps us to better understand the character of economic programs, and is particularly useful for judging the performance of imperfect economies.

What Does Productivity Growth Measure?

Some economists have located signs of economic progress in a statistic very different from either GNP or wealth. In seeking to interpret the expansion of the U.S. economy during the 1990s, *The Economist* (2001) exclaimed:

In judging an economy's prospects, what is the most important measure? Growth in GDP? Inflation? The size of the budget surplus? The level of the stock market? None of the above. Far more important is growth in productivity, which is crucial in itself and which affects all of those things and more.

The idea underlying "productivity," also called "total factor productivity" (TFP), is well known, but it will pay to restate it here.

The aggregate output of an economy is produced by various factors, including labor inputs, the services of manufactured capital, knowledge, and natural resources. We can therefore decompose observed changes in output over time into its sources: how much can be attributed to changes in labor force participation, how much to accumulation of manufactured capital and human capital, how much to the accumulation of knowledge brought about by expenditure in research and development, how much to changes in the use of natural resources, and so on? If a portion of the observed change in output cannot be credited to any of the above factors of production, that portion is called the growth in TFP. Growth in TFP is also known as the "residual," to indicate that it is that bit of growth in output which cannot be explained.

What does the residual measure? The passage quoted above from *The Economist* says that it intimates an economy's prospects. But does it?

Traditionally, labor force participation, manufactured capital, and marketed natural resources have been the recorded factors of production. In recent years, human capital has been added. Attempts have been made also to correct for changes in the quality of manufactured capital resulting from research and development. But national accounts still mostly do not include the use of nonmarketed natural resources—nor, for that matter, nonmarketed labor effort. They do not for the understandable reason that accounting prices of nonmarketed natural resources are extremely hard to estimate. Moreover, how do you estimate unrecorded labor effort? Now imagine that over a period the economy makes increasing use of the natural resource base or of unrecorded labor effort. The residual would be overestimated. In fact, a simple way to increase the residual would be to "mine" the natural resource base at an increasing rate. But this is a perverse thing to do if we seek to measure economic prospects.

What if it is possible to decompose the growth of an economy's aggregate output in a comprehensive manner, by tracing the growth to its sources in all the factors of production? To assume that over the long run the residual could still be positive is to imagine that the country enjoys a "free lunch."

Is the latter a possibility? One way to enjoy a free lunch, for poor countries at least, is to use technological advances made in other countries without paying for them. The residual then reflects increases in freely available knowledge. However, adaptation is not without cost. To meet local conditions, adjustments to product design and production processes are needed; these require appropriate local institutions, frequently nonexistent in poor countries.

For rich countries the residual could reflect serendipitous growth in knowledge. But, as Louis Pasteur observed, serendipity is the good fortune of prepared minds. As preparation involves engagement, time, and resources, we are led back to the

Table 14-2. *Sources of Economic Growth, 1960–1994*

	g(Y/L)[a]	g(K)[b]	g(H)[c]	g(A)[d]
East Asia	4.2	2.5	0.6	1.1
South Asia	2.3	1.1	0.3	0.8
Africa	0.3	0.8	0.2	−0.6
Middle East	1.6	1.5	0.5	−0.3
Latin America	1.5	0.9	0.4	0.2
United States	1.1	0.4	0.4	0.4
Other industrial countries	2.9	1.5	0.4	1.1

[a] g(Y/L) = annual percentage rate of change in GNP per head.

[b] g(K) = share of GNP attributable to manufactured capital multiplied by annual percentage rate of change in manufactured capital.

[c] g(H) = share of GNP attributable to human capital multiplied by annual percentage rate of change in human capital.

[d] g(A) = percentage rate of change in total factor productivity (residual).

Source: Collins and Bosworth (1996).

factors involved in the production of knowledge. Once we take into account the contribution of those factors of production, the residual could well be negligible.

Growth economists increasingly offer intercountry differences in TFP and the "residual" as (proximate) explanations for differences in their economic performance over the past few decades.[19] But there appears to be no consensus on what TFP reflects. Easterly and Levine (2001, *178*) write, "Different theories offer very different conceptions of TFP. These range from changes in technology (the instructions for producing goods and services) to the role of externalities, changes in the sector composition of production, and the adoption of lower-cost production methods." The article does not explicitly mention the variety of civic institutions that could decide the extent to which people trust one another for purposes of economic transactions. Microeconomic explorations of the idea of "social capital" suggest that TFP reflects not only technology, government policies, and public institutions, but also civic institutions.[20]

In this chapter I have been concerned not with differences in economic performance between countries, but with the changing economic prospects facing each of a number of poor regions over time. TFP can exhibit short bursts of growth in imperfect economies. Imagine, for example, that a government reduces economic inefficiencies by improving the enforcement of various types of public, private, and communal property rights (improving public institutions) or by reducing centralized regulations, such as import quotas, price controls, and so forth (improving macroeconomic policies). We would expect the factors of production to find a better use. As factors realign in a more productive fashion, TFP would increase.

Instead, the TFP could decline. Increased government corruption could be a cause, or civil strife, which not only destroys capital assets, but also damages a country's institutions. When institutions deteriorate, assets are used even more inefficiently than previously: TFP declines. This would appear to have happened in sub-Saharan Africa over the past 40 years (see Table 14-2).

As the name suggests, the residual reflects our ignorance of the sources of change in aggregate output, be that change positive or negative. The originators of the concept did not interpret it as a measure of an economy's prospects, and they certainly did not mean it as a measure of prospects over the long run (Abramovitz 1956; Solow 1957). They were right not to have done so.

Table 14-2, taken from Collins and Bosworth (1996), gives estimates of the annual rate of growth of GNP per capita and its breakdown between two factors of production (manufactured and human capital) in various regions of the world. The estimates are given in the first three columns. The period was 1960–1994. The fourth column represents the residual in each region. This is simply the difference between figures in the first column and the sum of the figures in the second and third columns.[21] Collins and Bosworth did not include Nature's services as factors of production. If the use of those services has grown during the period in question (a most likely possibility), we should conclude that the residual is an overestimate. Even so, the residual in Africa was negative (−0.6 percent annually). The true residual was in all probability even lower. The residual in South Asia, the other really poor region of the world, was 0.8 percent annually, but as this is undoubtedly an overestimate, I am uncertain whether there has been any growth in total factor productivity in that part of the world. One can but conclude that two of the poorest regions of the world (the Indian subcontinent and sub-Saharan Africa) have not improved their institutional capabilities over four decades, nor have they been able to improve productivity by making free use of knowledge acquired in advanced industrial nations.

Acknowledgments

Over the past several years Kenneth Arrow and Karl-Göran Mäler have greatly influenced my thinking on the problems discussed here. The theoretical framework underlying this chapter was developed jointly with them. For their encouragement and advice, I am very grateful.

Notes

1. The view's origin can be traced to World Bank (1992), which reported that in cross-country studies the emission of sulfur oxides has been found to be related to GNP per capita in the form of an inverse-U (see also Cropper and Griffiths 1994; Grossman and Krueger 1995). It is tempting, but wrong, to extrapolate from this an inverse-U relationship between the use of all natural resources and GNP per capita. See the comments in Arrow et al. (1995) on the inverse-U relationship and the responses it elicited in symposia built round the article in *Ecological Economics,* 1995, 15(2); *Ecological Applications,* 1996, 6(1); and *Environment and Development Economics,* 1996, 1(1). See also the special issue of *Environment and Development Economics,* 1997, 2(4).

2. I have explored such pathways elsewhere. See, for example, Dasgupta (1993, 1997, 1998, 2000, 2001a, 2001b). By "harmful spillovers" I mean "negative externalities," and by "externalities" I mean the effects of human activities that occur without the agreement of the people who are affected. Logging in the upland forests of watersheds can cause water runoff and inflict damage on farmers and fishermen in the downlands. The damage is an externality if those who suffer damage are not compensated by mutual agreement. Free riding on common property resources is another example of an activity that gives rise to externalities. The former example

involves a "unidirectional" externality, while the latter reflects "reciprocal" externalities (Dasgupta 1982).

3. The intellectual precursor of modern theories of growth is Simon (1981). See Jones (2001) for an account of contemporary growth models.

4. Daily (1997) is a collection of essays on the character of ecosystem services. See also Levin (2001) for an exhaustive collection of summaries of what is currently known about biodiversity and the role it plays as a productive asset.

5. See Agarwal (1986), Kalipeni (1994), and Chopra and Gulati (2001) for fine empirical studies on rural poverty and resource depletion in semiarid regions.

6. See Landes (1969, 1998), Fogel (1994, 1999), Johnson (2000), and especially Maddison (2001). The claim holds even if the past two hundred years are included. The rough calculation is simple enough: world per capita output today is about 5,000 U.S. dollars. The World Bank regards one dollar a day to be about as bad as it can be. People would not be able to survive on anything substantially less than that. It would then be reasonable to suppose that 2,000 years ago per capita income was not less than a dollar a day. So, let us assume that it was a dollar a day. This would mean that per capita income 2,000 years ago was about 350 dollars a year. Rounding off numbers, this means, very roughly speaking, that per capita income has risen about 16 times since then. This in turn means that world income per capita has doubled every 500 years, which in its turn means that the average annual rate of growth has been about 0.14 percent per year, a figure not much in excess of zero.

7. HDI is a combined index of GNP per capita, life expectancy at birth, and literacy. Country estimates of HDI are offered annually in the *Human Development Report* of the United Nations Development Programme. Because the weaknesses that I identify below in GNP as a measure of social well-being are shared by HDI, I shall not comment on the latter here. For an account of HDI's particular weaknesses, see Dasgupta (2001b).

8. Gore (1992) and Lomborg (2001) are prominent statements of the two viewpoints, respectively, by authors whose specializations lie outside the ecological and economic sciences.

9. For a good illustration of the conflicting intuitions, see the debate between Norman Myers and the late Julian Simon in Myers and Simon (1994).

10. See, for example, Barnett and Morse (1963), Simon (1981), and Johnson (2000).

11. See, for example, Vitousek et al. (1986), Postel et al. (1996), and Vitousek et al. (1997) for global estimates; and Agarwal (1986), Kalipeni (1994), Chopra and Gulati (2001), and Jodha (2001) for spatially localized studies.

12. In technical terms, the results do not rely on the assumption that the economy is "convex," meaning that diminishing returns characterize all production possibilities.

13. See Dasgupta and Mäler (2000), Dasgupta (2001b), and Arrow et al. (2003a, 2003b).

14. There are other reasons why they were found to be unsuitable as private property. In the text I am focusing on mobility.

15. The literature on this is now huge. See Dasgupta (2001b) for references.

16. For why and how, see Dasgupta (1993, 2001b).

17. Arrow et al. (2003a) have developed a catalog of results concerning ways to estimate the accounting prices of particular capital assets in a variety of institutional settings.

18. Pezzey (1992) contains an early, but thorough, account.

19. See, for example, the symposium "What Have We Learned from a Decade of Empirical Research on Growth?" in *World Bank Economic Review* (2001).

20. These ideas are developed at much greater length in Dasgupta (1999, 2003). The empirical classic on the subject of social capital is Putnam et al. (1993). However, in trying to explain differences in the economic performance of Italy's 20 provinces over the past few decades in terms of differences in their civic institutions, Putnam did not try to connect the workings of civic institutions to realized TFP.

21. Subject to rounding errors.

References

Abramovitz, M. 1956. Resource and Output Trends in the United States since 1870. *American Economic Review* 46 (Papers & Proceedings): 5–23.

Agarwal, B. 1986. *Cold Hearths and Barren Slopes: The Woodfuel Crisis in the Third World.* New Delhi: Allied Publishers.

Arrow, K.J., B. Bolin, R. Costanza, P. Dasgupta, C. Folke, C.S. Holling, B.-O. Jansson, S. Levin, K.-G. Mäler, C. Perrings, and D. Pimentel. 1995. Economic Growth, Carrying Capacity, and the Environment. *Science* 268(5210): 520–521.

Arrow, K.J., P. Dasgupta, and K.-G. Maler. 2003a. Evaluating Projects and Assessing Sustainable Development in Imperfect Economies. *Environmental and Resource Economics* 26(4): 647–685.

———. 2003b. The Genuine Saving Criterion and the Value of Population. *Economic Theory* 21(2): 217–225.

Bardhan, P. 1996. Research on Poverty and Development Twenty Years after *Redistribution with Growth. Proceedings of the Annual World Bank Conference on Development Economics, 1995.* Supplement to *World Bank Economic Review* and *World Bank Research Observer*, 59–72.

Barnett, H., and C. Morse. 1963. *Scarcity and Growth: The Economics of Natural Resource Availability.* Baltimore: Johns Hopkins University Press for Resources for the Future.

Birdsall, N. 1988. Economic Approaches to Population Growth. In *Handbook of Development Economics,* vol. 1, edited by H. Chenery and T.N. Srinivasan. Amsterdam: North Holland.

Chopra, K., and S.C. Gulati. 2001. *Migration, Common Property Resources and Environmental Degradation.* New Delhi: Sage.

Collins, S., and B. Bosworth. 1996. Economic Growth in East Asia: Accumulation versus Assimilation. *Brookings Papers on Economic Activity* 2: 135–191.

Cropper, M.L., and C. Griffiths. 1994. The Interaction of Population Growth and Environmental Quality. *American Economic Review* 84 (Papers & Proceedings): 250–254.

Cropper, M.L., and W. Oates. 1992. Environmental Economics: A Survey. *Journal of Economic Literature* 30(2): 675–740.

Daily, G. (ed.). 1997. *Nature's Services: Societal Dependence on Natural Ecosystems.* Washington, DC: Island Press.

Dasgupta, P. 1982. *The Control of Resources.* Cambridge, MA: Harvard University Press.

———. 1993. *An Inquiry into Well-Being and Destitution.* Oxford: Clarendon Press.

———. 1997. Environmental and Resource Economics in the World of the Poor. 45th Anniversary Lecture. Washington, DC: Resources for the Future.

———. 1998. The Economics of Poverty in Poor Countries. *Scandinavian Journal of Economics* 100(1): 41–68.

———. 1999. Economic Progress and the Idea of Social Capital. In *Social Capital: A Multifaceted Perspective,* edited by P. Dasgupta and I. Serageldin. Washington, DC: World Bank.

———. 2000. Population and Resources: An Exploration of Reproductive and Environmental Externalities. *Population and Development Review* 26(4): 643–689.

———. 2001a. Valuing Objects and Evaluating Policies in Imperfect Economies. *Economic Journal* 111(Conference Issue): 1–29.

———. 2001b. *Human Well-Being and the Natural Environment.* Revised edition 2004. Oxford: Oxford University Press.

———. 2003. Social Capital and Economic Performance: Analytics. In *Foundations of Social Capital,* edited by E. Ostrom and T.-K. Ahn. Cheltenham, UK: Edward Elgar.

Dasgupta, P., and K.-G. Mäler. 2000. Net National Product, Wealth, and Social Well-Being. *Environment and Development Economics* 5(2): 69–93.

Dreze, J., and A. Sen. 1990. *Hunger and Public Action.* Oxford: Clarendon Press.

Easterly, W., and R. Levine. 2001. It's Not Factor Accumulation: Stylized Facts and Growth Models. *World Bank Economic Review* 15(2): 177–219.

The Economist. 2001. Revisiting the "New Economy." February 10–16, 22.

Fogel, R.W. 1994. Economic Growth, Population Theory, and Physiology: The Bearing of Long-Term Processes on the Making of Economic Policy. *American Economic Review* 84(3): 369–395.

————. 1999. Catching Up with the Economy. *American Economic Review* 89(1): 1–19.

Gore, A. 1992. *Earth in the Balance: Forging a New Common Purpose.* London: Earthscan.

Grossman, G.M., and A.B. Krueger. 1995. Economic Growth and the Environment. *Quarterly Journal of Economics* 110(2): 353–377.

Hamilton, K., and M. Clemens. 1999. Genuine Savings Rates in Developing Countries. *World Bank Economic Review* 13(2): 333–356.

International Union for the Conservation of Nature and Natural Resources. 1980. *The World Conservation Strategy: Living Resource Conservation for Sustainable Development.* Geneva: IUCN.

Jodha, N.S. 2001. *Living on the Edge: Sustaining Agriculture and Community Resources in Fragile Environments.* New Delhi: Oxford University Press.

Johnson, D.G. 2000. Population, Food, and Knowledge. *American Economic Review* 90(1): 1–14.

Jones, C.I. 2001. *Introduction to Economic Growth.* 2nd edition. New York: W.W. Norton.

Kalipeni, E. (ed.). 1994. *Population Growth and Environmental Degradation in Southern Africa.* Boulder, CO: Lynne Rienner.

Kelley, A.C. 1988. Economic Consequences of Population Change in the Third World. *Journal of Economic Literature* 26(4): 1685–1728.

Kneese, A., and J. Sweeney. 1985–1993. *Handbook of Natural Resource and Energy Economics,* Vols. 1–3. Amsterdam: North Holland.

Kremer, M. 1993. Population Growth and Technological Change: One Million B.C. to 1990. *Quarterly Journal of Economics* 108(3): 681–716.

Landes, D. 1969. *The Unbound Prometheus.* Cambridge: Cambridge University Press.

————. 1998. *The Wealth and Poverty of Nations: Why Some Are So Rich and Some So Poor.* New York: W.W. Norton.

Levin, S.A. (ed.). 2001. *Encyclopedia of Biodiversity.* New York: Academic Press.

Lomborg, B. 2001. *The Skeptical Environmentalist.* Cambridge: Cambridge University Press.

Maddison, A. 2001. *The World Economy: A Millennial Perspective.* Paris: Organisation for Economic Co-operation and Development.

Myers, N., and J.L. Simon. 1994. *Scarcity or Abundance? A Debate on the Environment.* New York: W.W. Norton.

Oates, W. (ed.). 1992. *The International Library of Critical Writings in Economics: The Economics of the Environment.* Cheltenham:, UK Edward Elgar.

Pearce, D., K. Hamilton, and G. Atkinson. 1996. Measuring Sustainable Development: Progress on Indicators. *Environment and Development Economics* 1: 85–101.

Pezzey, J. 1992. Sustainable Development Concepts: An Economic Analysis. World Bank Environment Paper No. 2.

Postel, S.L., G. Daily, and P.R. Ehrlich. 1996. Human Appropriation of Renewable Fresh Water. *Science* 271: 785–788.

Putnam, R.D., R. Leonardi, and R.Y. Nanetti. 1993. *Making Democracy Work: Civic Traditions in Modern Italy.* Princeton: Princeton University Press.

Schultz, T.P. 1988. Economic Demography and Development. In *The State of Development Economics,* edited by G. Ranis and T.P. Schultz. Oxford: Basil Blackwell.

Serageldin, I. 1995. Are We Saving Enough for the Future? In *Monitoring Environmental Progress.* Report on Work in Progress, Environmentally Sustainable Development. Washington, DC: World Bank.

Simon, J. 1981. *The Ultimate Resource.* Oxford: Martin Robinson.

Solow, R.M. 1957. Technical Change and the Aggregate Production Function. *Review of Economics and Statistics* 39(3): 312–320.

Stern, N. 1989. The Economics of Development: A Survey. *Economic Journal* 99(2): 597–685.

UNDP (United Nations Development Programme). 1990–1999. *Human Development Report.* New York: Oxford University Press.

Vitousek, P., P.R. Ehrlich, A.H. Ehrlich, and P. Matson. 1986. Human Appropriation of the Product of Photosynthesis. *BioScience* 36: 368–373.

Vitousek, P.M., H.A. Mooney, J. Lubchenco, and J.M. Melillo. 1997. Human Domination of Earth's Ecosystem. *Science* 277: 494–499.

World Bank. 1992. *World Development Report.* Washington, DC: World Bank.

World Bank. 1998. *World Development Indicators.* Washington, DC: World Bank.

World Bank Economic Review. 2001. What Have We Learned from a Decade of Empirical Research on Growth? Symposium. 15(2).

World Commission on Environment and Development. 1987. *Our Common Future.* New York: Oxford University Press.

Index

Note: Page numbers in italics indicate figures and tables. Page numbers followed by an 'n' indicate notes.